Advances in Numerical Computation and Mathematical Modelling for Geotechnical Engineering

Advances in Numerical Computation and Mathematical Modelling for Geotechnical Engineering

Guest Editors

Jiadong Qiu
Changtai Zhou
Yichao Rui
Fan Feng

MDPI

Basel • Beijing • Wuhan • Barcelona • Belgrade • Novi Sad • Cluj • Manchester

Guest Editors

Jiadong Qiu
School of Resource Environment and Safety Engineering
University of South China
Hengyang
China

Changtai Zhou
Department of Geotechnical Engineering
Tongji University
Shanghai
China

Yichao Rui
State Key Laboratory of Coal Mine Disaster Dynamics and Control
Chongqing University
Chongqing
China

Fan Feng
College of Energy and Mining Engineering
Shandong University of Science and Technology
Qingdao
China

Editorial Office
MDPI AG
Grosspeteranlage 5
4052 Basel, Switzerland

This is a reprint of the Special Issue, published open access by the journal *Applied Sciences* (ISSN 2076-3417), freely accessible at: https://www.mdpi.com/si/applsci/3W95M28F51.

For citation purposes, cite each article independently as indicated on the article page online and as indicated below:

Lastname, A.A.; Lastname, B.B. Article Title. *Journal Name* **Year**, *Volume Number*, Page Range.

ISBN 978-3-7258-6015-9 (Hbk)
ISBN 978-3-7258-6016-6 (PDF)
https://doi.org/10.3390/books978-3-7258-6016-6

Contents

About the Editors

Jiadong Qiu

Jiadong Qiu is an Associate Professor and Master's supervisor at the University of South China and serves as a Council Member of the Hunan Rock Mechanics Association. He earned both his M.Eng. and Ph.D. in Mining Engineering from Central South University and completed post-doctoral research at Shenzhen University. His research interests are the rock mass mechanics in deep mining, rock dynamics under engineering disturbances, and multi-field coupled rock fracturing. He has published over 20 high-quality academic papers as the first or corresponding author, and has led multiple projects funded by national and provincial agencies. His scholarly contributions have been recognized with the Second Prize of the China Green Mine Science & Technology Award and the Second Prize in the National Young Teachers' Lecture Competition for Mining Majors. In addition, he serves as a Young Editor or Guest Editor for several academic journals.

Changtai Zhou

Changtai Zhou is currently a research fellow at the Hong Kong University of Science and Technology, where he received his PhD in Geotechnical Engineering from the University of Adelaide, Australia. His research interests focus on rock mechanics, discrete element modeling, discrete fracture networks, machine learning, and rock dynamics. Dr. Zhou has published over 40 academic papers in leading industry journals including the *International Journal of Rock Mechanics and Mining Sciences, Rock Mechanics and Rock Engineering, Tunnelling and Underground Space Technology, International Journal of Heat and Mass Transfer*, and *International Journal for Numerical and Analytical Methods in Geomechanics*, with 24 papers as first or corresponding author. He has applied for 14 invention patents with 6 granted, and serves as the principal investigator for the National Natural Science Foundation of China Youth Project and two "Pearl River Talent Program" youth talent projects. He participates in multiple National Key Research and Development Programs and National Natural Science Foundation projects. Dr. Zhou serves as an editorial board member for the *International Journal Discover Civil Engineering*, and as a youth editorial board member for *Deep Underground Science and Engineering and Rock Mechanics Bulletin*. He has also served as a reviewer for over ten authoritative SCI journals in the field of geotechnical mechanics, including the *International Journal of Rock Mechanics* and *Mining Science and Rock Mechanics and Rock Engineering*.

Yichao Rui

Yichao Rui is currently an assistant researcher and master's supervisor at the College of Resources and Safety, Chongqing University. His main research areas include disaster source monitoring and intelligent early warning technology, as well as intelligent detection technology for geological information while drilling. He has presided over 6 national-level and 1 provincial and ministerial-level key projects. He has presided over 6 other horizontal and vertical scientific research projects; In addition, as the main person in charge, he has participated in the National 973 Program, the National Key Research and Development Program (Young Scientist) Project, etc. A total of 42 SCI papers have been published in internationally renowned journals, among which 30 are SCI first-tier papers as the main author (22 are in TOP journals, 4 are ESI highly cited, 6 have an IF $>$ 10, and 8 have an IF $>$ 8). A total of 33 national invention patents have been publicly disclosed (22 are first-half inventor patents, 16 of which have been authorized), and 1 book has been published as the associate editor. Two software Copyrights have been authorized. He is a part-time young editorial board member of international journals *"Int J Min Sci Techno"*, *"Geohazard Mech"*, and *"Earth Energy Science"*. He has served as a reviewer for 14 internationally renowned Sci journals such as *"Int J Min Sci Techno"*, *"Ocean Eng"*, *"Measurement"*, *"Mech Syst Signal Pr"*, and *"Int J Elec Power"*. He was selected for the China Association for Science and Technology's Youth Talent Support Program and awarded the First Prize of the Science and Technology Award of the China Communications and Transportation Association, the First Prize of the Science and Technology Progress Award of the China Association for Work Safety, and other honors.

Fan Feng

Fan Feng obtained his PhD degree in Mining Engineering from Central South University in 2018. He worked as a Research Assistant at the Earth Mechanics Institute, Colorado School of Mines, from 2017 to 2018. He is currently working as a Professor at Shandong University of Science and Technology. His research interests include (1) deep hard rock mechanics and engineering; (2) prevention and control of mine ground pressure disasters; and (3) underground metal mining methods and crafts. He hosted a total of 21 scientific projects, including National Natural Science Foundation of China (NSFC) and the Outstanding Young Science Foundation of Shandong Province. As the first author/corresponding author, he has published 36 papers indexed by SCI/EI. He has won the Green Mine Innovation Contribution Award, Youth Science and Technology Award of Qingdao City, Leader of Youth Entrepreneurship Teams in Shandong Province's Universities, and 17 provincial and ministerial science and technology awards.

applied sciences

Editorial

Advances in Numerical Computation and Mathematical Modeling for Geotechnical Engineering

Jiadong Qiu [1], Changtai Zhou [2,*], Yichao Rui [3] and Fan Feng [4]

1 School of Resources, Environment and Safety Engineering, University of South China, Hengyang 421001, China
2 Department of Architecture and Civil Engineering, City University of Hong Kong, Tat Chee Avenue, Kowloon, Hong Kong, China
3 School of Resources and Safety Engineering, Chongqing University, Chongqing 400044, China
4 College of Energy and Mining Engineering, Shandong University of Science and Technology, Qingdao 266590, China
* Correspondence: chouchangtai@126.com

1. Introduction

The field of geotechnical engineering is experiencing a transformative period driven by the increasing complexity of in situ projects and the need for operations in challenging geological environments [1–4]. As engineering activities extend into deeper strata and more complex geological conditions, practitioners face unprecedented challenges, including groundwater seepage control, dynamic rock mass failure analysis, geothermal effects on rock stability, and microseismic monitoring [5–9]. These multifaceted problems demand sophisticated numerical computation, mathematical modeling, and engineering monitoring methodologies that can provide reliable solutions for analysis, evaluation, and design processes [10,11].

This Special Issue, "Advances in Numerical Computation and Mathematical Modeling for Geotechnical Engineering", addresses the critical need for advanced computational tools in geotechnical engineering practice. The collection focuses on the development and application of cutting-edge numerical computation and modeling methods for geotechnical monitoring, analysis, and theoretical prediction. By assembling high-quality original research papers, comprehensive case studies, and innovative methodological contributions, this Special Issue aims to bridge the gap between fundamental mathematical theories and practical computer-based applications in geotechnical engineering.

The rapid pace of global urbanization and infrastructure development has generated an unprecedented demand for reliable geotechnical solutions [12]. Modern projects such as deep excavations for underground transportation systems, high-rise building foundations in challenging soil conditions, and the exploitation of deep mineral resources all require accurate prediction of ground behavior under complex, multi-axial loading conditions. While traditional empirical methods remain valuable, they often prove inadequate when addressing the multi-physics nature of contemporary geotechnical problems, which involve coupled processes, material heterogeneity, and time-dependent behavior [13,14].

Recent advances in computational power and numerical algorithms have revolutionized our understanding and predictive capabilities in geotechnical behavior analysis. The development of sophisticated constitutive models capable of capturing complex stress–strain relationships in soils and rocks, combined with advanced numerical methods including the Finite Element Method (FEM), Discrete Element Method (DEM), and innovative hybrid approaches, has fundamentally transformed our analytical capabilities [10,11,15–18]. These computational tools enable engineers to comprehensively consider previously intractable factors, such as material heterogeneity, non-linear behavior, time-dependent effects, and complex coupled processes.

2. An Overview of Published Articles

This Special Issue encompasses a comprehensive range of research topics that reflect the current frontiers and emerging trends in computational geotechnics. These contributions can be categorized into several key thematic areas that collectively advance the state-of-the-art in numerical modeling and mathematical analysis.

Zhang et al. [19] utilized laboratory experiments and the Universal Distinct Element Code (UDEC) to explore the influence of the compressive strength of joint walls on the shear deformation behavior of rock discontinuities. The numerical results demonstrated good consistency with the experimental findings. At the microscopic level, the study revealed that roughness and normal stress are the main factors influencing the shear dilation process of discontinuities, and parameters such as dilation magnitude and shear strength are constrained by joint wall strength.

The application of advanced mathematical models in geotechnical engineering represents a mature yet continuously evolving field that successfully bridges theoretical analysis and practical engineering applications [20]. Liao et al. [21] proposed an innovative comprehensive evaluation methodology that integrates an improved grey correlation–Delphi model with validity and reliability coefficients. This sophisticated approach is specifically designed to address the inherent challenges in both quantitative and qualitative evaluation of loess slope stability—a critical issue in many regions worldwide where loess deposits are prevalent. The proposed method effectively overcomes the subjective dependence and index selection redundancy that characterize traditional evaluation approaches, demonstrating high reliability and providing a valuable mathematical framework for comprehensive geological slope assessment. This contribution is particularly significant given the increasing frequency of slope failures in loess regions and the need for more objective, scientifically based evaluation tools.

The integration of artificial intelligence and machine learning techniques with traditional numerical methods represents one of the most exciting and rapidly developing areas in computational geotechnics [22–25]. Sui, et al. [26] introduced a groundbreaking artificial intelligence algorithm based on a stacking integration strategy for predicting fragmentation size in open-pit bench blasting operations. This innovative AI algorithm effectively integrates Random Forest and XGBoost models, achieving significantly higher predictive accuracy compared to traditional single machine learning approaches. The research demonstrates the potential of ensemble learning methods in addressing complex geotechnical prediction problems where multiple variables and non-linear relationships are involved. This work has direct applications in mining engineering, where accurate prediction of blast fragmentation is crucial for optimizing extraction processes and minimizing environmental impact.

3. Conclusions

This Special Issue represents a significant contribution to the field of computational geotechnics, showcasing innovative research that addresses both fundamental scientific questions and practical engineering challenges. These papers presented here demonstrate the power of numerical computation and mathematical modeling in solving complex geotechnical problems, from the microscale behavior of soil particles to the large-scale response of geological systems. As we face increasingly complex engineering challenges in an era of climate change and sustainable development, the role of advanced computational methods in geotechnical engineering will only grow in importance. The research presented in this Special Issue provides the foundation for future developments, offering novel tools and insights that will enable engineers to design safer, more efficient, and more sustainable geotechnical solutions.

Author Contributions: Conceptualization, J.Q.; writing—original draft preparation, J.Q. and C.Z.; writing—review and editing, Y.R. and F.F. All authors have read and agreed to the published version of the manuscript.

Funding: The work was financially supported by the National Natural Science Foundation of China (52304091, 52374095) and the Shandong Province Youth Innovation Team Science and Technology Support Program for Higher Education Institutions (2024KJH068).

Conflicts of Interest: The authors declare no conflicts of interest.

References

1. Im, K.; Saffer, D.; Marone, C.; Avouac, J.P. Slip-rate-dependent friction as a universal mechanism for slow slip events. *Nat. Geosci.* **2020**, *13*, 705–710. [CrossRef]
2. Juang, C.H.; Gong, W.; Martin, J.R.; Chen, Q. Model selection in geological and geotechnical engineering in the face of uncertainty—Does a complex model always outperform a simple model? *Eng. Geol.* **2018**, *242*, 184–196. [CrossRef]
3. Meng, W.; Wu, W. Machine Learning-Aided Prediction of the Mechanical Properties of Frozen Fractured Rocks. *Rock Mech. Rock Eng.* **2023**, *56*, 261–273. [CrossRef]
4. Zhou, C.; Rui, Y.; Qiu, J.; Wang, Z.; Zhou, T.; Long, X.; Shan, K. The role of fracture in dynamic tensile responses of fractured rock mass: Insight from a particle-based model. *Int. J. Coal Sci. Technol.* **2025**, *12*, 39. [CrossRef]
5. Li, Y.; Dai, X.; Yang, R.; Zhu, Y.; Li, R.; Zhou, C.; Xie, L.; Li, W. Failure Mechanism and Support Technology of Weakly Cemented Soft Rock Roadway in Water-Rich Environment. *Geotech. Geol. Eng.* **2025**, *43*, 213. [CrossRef]
6. Hu, J.; Xie, H.; Li, C.; Liu, G. Evolution mechanism of permeability of hot dry rock under coupled effect of thermal fatigue and seawater interaction during coastal geothermal development. *Renew. Sustain. Energy Rev.* **2024**, *189*, 114061. [CrossRef]
7. Hu, J.; Xie, H.; Gao, M.; Li, C.; Sun, Q. Damage mechanism and heat transfer characteristics of limestone after thermal shock cycle treatments based on geothermal development. *Int. J. Rock Mech. Min. Sci.* **2022**, *160*, 105269. [CrossRef]
8. Rutqvist, J. The geomechanics of CO_2 storage in deep sedimentary formations. *Geotech. Geol. Eng.* **2012**, *30*, 525–551. [CrossRef]
9. Griffiths, L.; Lengliné, O.; Heap, M.J.; Baud, P.; Schmittbuhl, J. Thermal Cracking in Westerly Granite Monitored Using Direct Wave Velocity, Coda Wave Interferometry, and Acoustic Emissions. *J. Geophys. Res. Solid Earth* **2018**, *123*, 2246–2261. [CrossRef]
10. Jing, L. A review of techniques, advances and outstanding issues in numerical modelling for rock mechanics and rock engineering. *Int. J. Rock Mech. Min. Sci.* **2003**, *40*, 283–353. [CrossRef]
11. Lisjak, A.; Grasselli, G. A review of discrete modeling techniques for fracturing processes in discontinuous rock masses. *J. Rock Mech. Geotech. Eng.* **2014**, *6*, 301–314. [CrossRef]
12. Ranjith, P.G.; Zhao, J.; Ju, M.; De Silva, R.V.S.; Rathnaweera, T.D.; Bandara, A.K.M.S. Opportunities and Challenges in Deep Mining: A Brief Review. *Engineering* **2017**, *3*, 546–551. [CrossRef]
13. Mas Ivars, D.; Pierce, M.; Darcel, C.; Reyes-Montes, J.; Potyondy, D.; Young, R.; Cundall, P. The synthetic rock mass approach for jointed rock mass modelling. *Int. J. Rock Mech. Min. Sci.* **2011**, *48*, 219–244. [CrossRef]
14. Liu, J.; Feng, Y.; Jiao, Y. The Phlogiston Theory of Rock Mass Classification: Philosophical and Mathematical Critique of Ordinal Data Usage. *Rock Mech. Bull.* **2025**, *4*, 100205. [CrossRef]
15. Qiu, J.; Huang, R.; Wang, H.; Wang, F.; Zhou, C. Rate-dependent tensile behaviors of jointed rock masses considering geological conditions using a combined BPM-DFN model: Strength, fragmentation and failure modes. *Soil Dyn. Earthq. Eng.* **2025**, *195*, 109393. [CrossRef]
16. Saadat, M.; Taheri, A. A numerical approach to investigate the effects of rock texture on the damage and crack propagation of a pre-cracked granite. *Comput. Geotech.* **2019**, *111*, 89–111. [CrossRef]
17. Shen, Z.; Jiang, M.; Fang, N.; Zhou, F. A damage evolution law enriched by microscopic mechanisms for structured sand in mechanical loading. *Acta Geotech.* **2019**, *14*, 1905–1924. [CrossRef]
18. Yin, Y.; Song, M.; Feng, Y.; Liu, Z.; Chen, X.; Sun, Q. Theoretical and numerical investigation of the effects of in-situ stresses and dual-borehole combinations in eccentric decoupled charge blasting. *Rock Mech. Bull.* **2025**, *4*, 100191. [CrossRef]

19. Zhang, X.; Ouyang, S.; Fan, C.; Yi, L.; Liu, D. Shear Properties and Failure Mechanism of Matched Discontinuities Between Two Different Rock Types Under Direct Shear. *Appl. Sci.* **2025**, *15*, 5289. [CrossRef]
20. Langenfeld, K.; Mosler, J. A micromorphic approach for gradient-enhanced anisotropic ductile damage. *Comput. Methods Appl. Mech. Eng.* **2019**, *357*, 112717. [CrossRef]
21. Liao, J.; Sun, H.; An, J. Research on Screening Method of Loess Slope Stability Evaluation Indexes Based on Validity and Reliability Coefficient. *Appl. Sci.* **2025**, *15*, 6216. [CrossRef]
22. Wessels, H.; Weißenfels, C.; Wriggers, P. The neural particle method—An updated Lagrangian physics informed neural network for computational fluid dynamics. *Comput. Methods Appl. Mech. Eng.* **2020**, *368*, 113127. [CrossRef]
23. Singh, J.; Pradhan, S.P.; Vishal, V.; Singh, M. Characterization of a fractured rock mass using geological strength index: A discrete fracture network approach. *Transp. Geotech.* **2023**, *40*, 100984. [CrossRef]
24. Liu, L.; Song, Z.; Li, X. Artificial intelligence in tunnel construction: A comprehensive review of hotspots and frontier topics. *Geohazard Mech.* **2024**, *2*, 1–12. [CrossRef]
25. Konurin, A.; Sergey, N.; Alexandr, N.; Dmitry, O.; Ivan, Z.; Maria, K. Application of artificial neural networks for stress state analysis based on the photoelastic method. *Geohazard Mech.* **2023**, *1*, 128–139. [CrossRef]
26. Sui, Y.; Zhou, Z.; Zhao, R.; Yang, Z.; Zou, Y. Open-Pit Bench Blasting Fragmentation Prediction Based on Stacking Integrated Strategy. *Appl. Sci.* **2025**, *15*, 1254. [CrossRef]

applied sciences

Article

Research on Screening Method of Loess Slope Stability Evaluation Indexes Based on Validity and Reliability Coefficient

Jianlong Liao [1,*], Hongjun Sun [1] and Jianchao An [2]

1 School of Civil and Architectural Engineering, Liaoning University of Technology, Jinzhou 121001, China; sunhongjun_2006@163.com
2 Jinzhou Water (Group) Co., Ltd., Jinzhou 121002, China; 15604063205@163.com
* Correspondence: jianlongliao@163.com

Abstract: Aiming at the problems of intense subjectivity and high redundancy in the screening of indicators in the stability evaluation of loess slopes, this study proposes an evaluation method integrating the validity and reliability coefficients. Following an initial screening of the indexes based on the engineering geological characteristics of loess slopes and literature research, 22 significant indicators were kept following a qualitative screening process (the principles of uniqueness, purpose, etc.); combined with the improved grey correlation-Delphi model for quantitative screening, the validity coefficient ($\beta = 0.0816$) and reliability coefficient ($\rho = 0.9609$) were introduced to validate the scientificity and consistency of the indicator system. The results showed that the 10 core indicators, including Cohesion, Internal Friction Angle, Maximum Monthly Rainfall, Rock Mass Structure, and Anthropogenic Engineering Activities, had a significant influence on loess slope stability, and the screening process effectively reduced the subjective bias and information redundancy. The method provides a data-driven theoretical framework for eolian slope risk assessment, which can improve the accuracy of landslide warning and the reliability of engineering protection design, and the engineering applicability of the model can be further optimized by combining the dynamic environmental parameters and multi-source monitoring data.

Keywords: slope stability; grey correlation; validity coefficient; reliability coefficient; loess slope

1. Introduction

One of the most damaging geological risks in the region where loess is distributed worldwide is loess slope instability. Under the combined influence of rainfall infiltration, freeze–thaw cycles, and human engineering activities (such as irrigation and excavation), loess is extremely vulnerable to landslide hazards due to its porous structure, susceptibility to wetting, and high sensitivity to hydrological conditions. These factors pose a serious threat to residential areas, farmland infrastructures, and traffic arterials [1–4]. At the same time, the suddenness and multi-scale damage characteristics of such disasters (e.g., cave collapse, slope shoulder crack extension) further highlight the complexity of stability evaluation [5,6]. About one-third of the annual geological disasters in China occur in the Loess Plateau, which not only causes casualties and infrastructure damage but also leads to huge direct economic losses [7,8]. Moreover, with the fluctuation of groundwater levels and the frequent occurrence of extreme weather events, the risk of slope landslides and other disasters in the Loess Plateau region is also increasing [4,9,10]. However, traditional analysis methods, such as the limit equilibrium method and finite element numerical simulation, make it difficult to quantify the nonlinear interaction effects among indicators due to the over-reliance on idealized assumptions (e.g., homogeneous soils, fixed slip fracture surfaces) [7,11,12]. This makes the theory often inconsistent with the actual situation in practical applications and, thus, there is an urgent need to construct a novel analytical framework that can integrate multiple data sources.

As the core of novel evaluation methods, the indicator screening mechanism is a widespread challenge. Among the existing methods, the subjective assignment method is susceptible to expert cognitive bias, leading to indicator redundancy or omission of key factors [13]. Objective methods such as principal component analysis can extract data features, but may screen out statistically significant indicators but lack engineering interpretability [14]; traditional methods have obvious human intervention in the screening process, such as the resolution coefficient in the grey correlation model, which needs to rely on empirical settings. In this example, the resolution coefficient in the grey correlation model relies on the empirical setting, and when the coefficient varies within a certain range, the variability of the correlation order of indicators is high [15]. However, the selection of indexes for slope stability evaluation typically depends too much on the knowledge and judgment of experts and is not supported by systematic verification procedures, raising questions about the repeatability and scientificity of index screening results [16,17]. When redundant parameters are added, computational complexity is needlessly increased, and the reliability of the evaluation model output is directly impacted by the significant decline in the identification accuracy of important driving factors under the coupling effect of multiple parameters [18,19]. Therefore, to minimize the subjectivity of index screening, guarantee the objectivity of the screening findings, and assure the correctness of slope stability evaluation, it is imperative to develop an objective index screening algorithm framework and a corresponding result verification method.

This study developed a quantitative screening mechanism for the joint control of validity and reliability and creatively suggested a quantitative screening mechanism based on the Delphi method and the improved grey correlation model to address the main issues in the stability evaluation of loess slope.The two-stage framework used to optimize the index system eliminated redundancy through qualitative screening, fused subjective and objective data through quantitative screening, and quantified the quality of the fusion using validity and reliability coefficients. This helped to validate expert consensus and statistical effectiveness, achieve multi-dimensional optimization of the index system, and effectively improve its applicability in loess slope scenarios with complex geological parameters and limited historical samples. Simultaneously, the two-stage screening framework may be adjusted to meet the requirements of regionally diverse evaluation by modifying the validity–reliability threshold parameters. A new methodological framework for the optimization of similar complex system index systems is provided by the collaborative control strategy based on subjective and objective data fusion, which transcends the boundaries of traditional analysis. Its core algorithm is not specific to any particular slope, so that all different types of slopes can be screened by the model, and this method can be applied tomulti-source heterogeneous data processing and complex system index optimization in geotechnical engineering.

2. Regional Geological Setting

2.1. Topography and Geomorphology

Situated in the southern region of the Loess Plateau of Longdong, in Lingtai County, Pingliang City, Gansu Province (Figure 1), the study area is a typical loess hilly-gully landform with a total area of 2039 km^2 and geographic coordinates ranging from 107°00′ to 107°57′ east longitude and 34°02′ to 34°23′ north latitude. The topography of the region is dominated by loess beam-pebble hills and erosion-accumulation valley plains. It has loess beams and a pebble hills base for the Cretaceous sandstone and gravel, the surface layer is covered with Malan loess, the formation of beams and pebbles interlace terrain, elevation is 910 to 1020 m, the gully is a V-shaped cut, depth is 20 to 150 m, a valley slope gradient of 15 to 40 degrees, and development of the gully and the intensive local collapse landslide. Erosion-accumulation of the river valley plains is distributed in the valley of the Daxi River and its tributaries, the development of river beds, river rambling beach, and multi-level terrace. The first and second terraces are flat and broad, while the third and fourth terraces are sporadically distributed and are base terraces, with a difference of 20 to 40 m in height between the terrace surfaces. Micro-geomorphology in the area is dominated by waterfall holes and gullies, with waterfall holes of 0.2–1.5 m in diameter, which are concentrated in the upper and middle slopes due to precipitation or irrigation seepage, and gullies have proliferated due to the frequency of torrential rains in recent years, aggravating the erosion and damage of slopes.

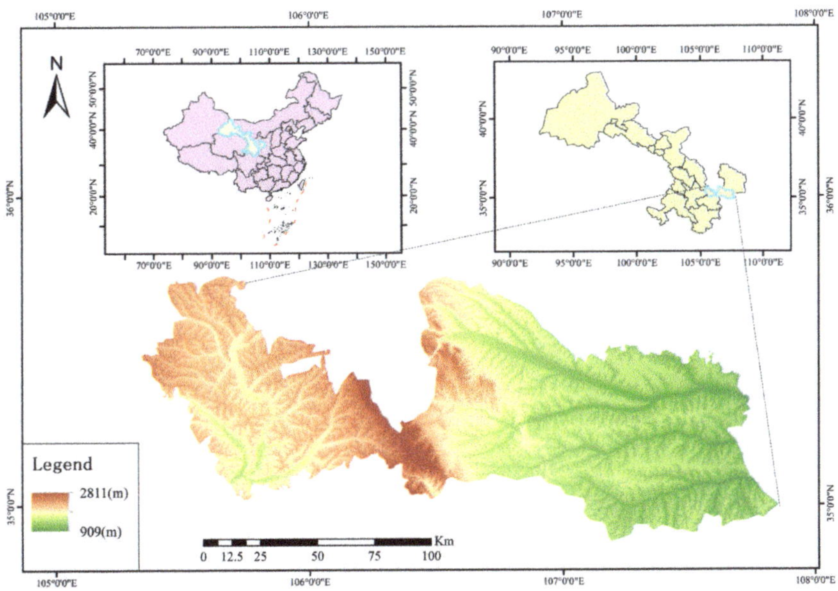

Figure 1. Location map of the study area.

2.2. Lithological Characteristics

Cretaceous bedrock and Quaternary loose sedimentary layers dominate the research area's geological structure. As the Quaternary basement, the Cretaceous strata are primarily found on both banks of the Daxi River and its tributaries, including the Luohandong Formation (K_1lh) and the Zhidan Group Huanhe Formation (K_1h). The brownish-red sandy mudstone of the Cretaceous Huanhe Formation (K_1h) is interbedded with sandstone and has an argillaceous cementation that is easily dissolved and softened by water. With small fissures, purple-red sandstone, mudstone, and conglomerate make up the majority of the Luohan Cave Formation (K_1lh). The Quaternary strata cover a wide area. Among them, the Lower Pleistocene ($Q_p{}^{1al}$) is a gravel and clay layer, which is distributed in tributary river valleys. The Middle Pleistocene (Q_{p2}) is dominated by lithic loess (Q_{p2eol}), with a thickness of 60∼150 m, developed vertical joints, and multi-layered paleosoil, which is prone to collapse and landslide. The Upper Pleistocene (Q_{p3}) includes the Malan loess ($Q_p{}^{3eol}$, collapsible amount 97.4∼216.2 mm) with moderate collapsibility and loose gravelly loess silt ($Q_p{}^{3eol+pl}$). The Holocene (Q_{p4}) is dominated by sand, gravel, and plain fill (Q_{p4ml}), and its mechanical properties are extremely unstable. Details are shown in Figure 2.

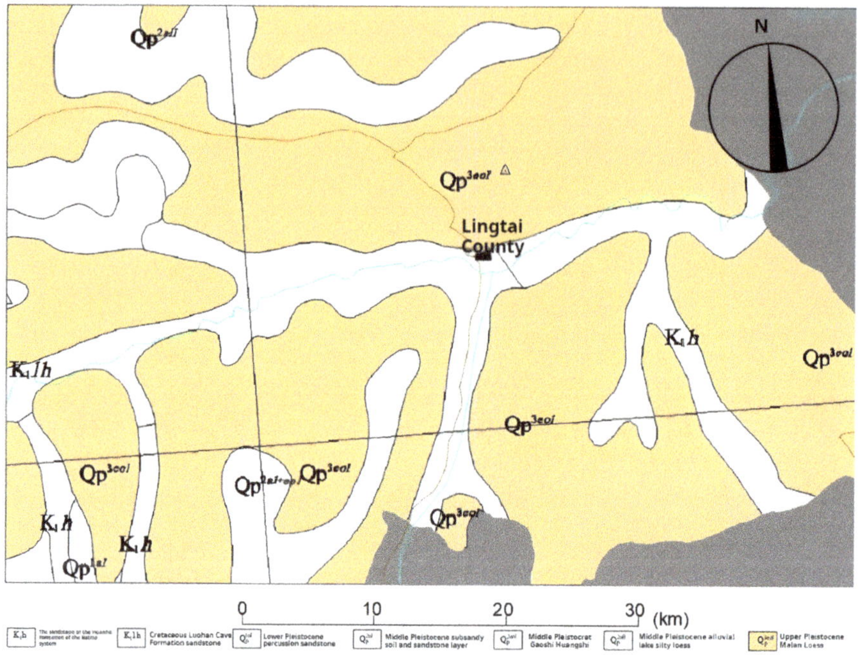

Figure 2. Geological map of Lingtai County and surrounding areas [20].

2.3. Tectonic Features

According to the investigations of relevant literature [20], the study area is located in the stable tectonic unit of the southwestern edge of the Ordos Basin, with weak tectonic activities, no regional rupture, and the development of five gentle folds (four oblique and one dorsal). The new tectonic movement is dominated by intermittent uplift, forming the four-level terraces and bedrock outcrops in the river valley. Earthquakes are controlled by strong peripheral earthquakes, and the 1920 Haiyuan 8.5-magnitude earthquake triggered secondary disasters. The regional seismic defense intensity is VII degrees (peak acceleration of ground shaking 0.10 g), and the Earth's crust is stable as a whole, but the loess slopes are weak in seismic performance, so it is necessary to carry out the analysis of anti-slip stability and reinforcement design under seismic conditions.

2.4. Meteorology and Hydrology

With an average annual precipitation of 654.4 mm, a historical maximum daily precipitation of 150.1 mm, and an average annual evaporation of 1145 mm, the study region has warm-temperate semi-humid continental monsoon weather. Details are shown in Figure 3. With an average yearly runoff of 53.28 million cubic meters for the Daxi River and 33 million cubic meters for the Heihe River, the two main rivers of the Jinghe River System are known to have a mineralization of 0.3 to 0.46 g/L and a water chemistry type that is dominated by HCO_3^-, Ca^{2+}, and Mg^{2+}. The flood peak flow of the east ditch and west ditch in the area reaches 65 m^3/s and 55 m^3/s, respectively, during heavy rainfall, which makes it easy to trigger a mudflow disaster. Hydrogeological conditions are significantly affected by irrigation seepage and heavy rainfall infiltration, and the water table elevation and loess engineering characteristics jointly exacerbate the hydraulic erosion of slopes.

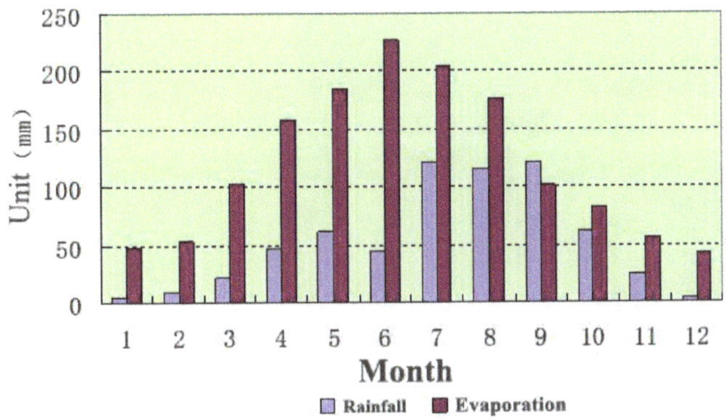

Figure 3. The average rainfall and evaporation in Lingtai County for many years [20].

2.5. Human Activities

In the study area on the south side of Jingshan Mountain in Lingtai County, the risk of geological disasters has been significantly exacerbated by human engineering activities, which are mainly reflected in slope cutting, green irrigation, and air-raid shelters. Due to the large number of outsiders in the densely populated area, a large number of slopes were cut to build houses, which destroyed the original slope, increased the slope and slope weight, and directly destroyed the stability of the original slope. This is shown in Figure 4. The greening irrigation on the top of the slope causes water to infiltrate along the vertical joints and sinkholes, which raises the groundwater level weakens the shear strength of rock and soil, and weakens the stability of the slope. In addition, although the remaining air-raid shelters have been blocked, there are still hidden seepage channels, and the mechanical properties of the rock and soil around the caves are accelerated by water intrusion during rainfall. These activities are the core anthropogenic drivers of slope instability by altering the topographic structure, permeability conditions, and geophysical properties. This is shown in Figure 5.

Figure 4. The phenomenon of cutting slopes to build houses around Jingshan [20].

Figure 5. An air-raid shelter at the foot of the slope of the study area [20].

3. Research Method

This study focuses on the stability of loess slopes and proposes the four-step method of "preliminary selection, qualitative screening, quantitative screening, and verification" to construct the slope stability evaluation index system and combines the improved grey correlation model with the statistical validation method to systematically analyze the validity and reliability of the indexes. The specific operation process is shown in Figure 6.

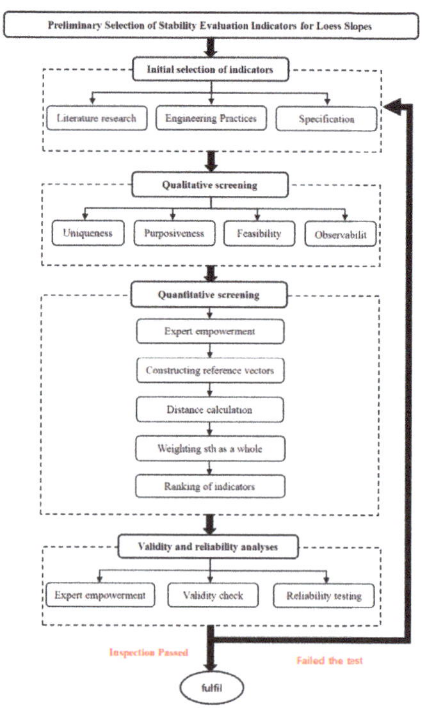

Figure 6. Flowchart of the screening of indicators for loess slopes.

3.1. Preliminary Selection and Qualitative Screening of Evaluation Indicators

The stability state and evaluation criteria of loess slopes under various geological conditions differ greatly, and the influencing factors of slope stability are extremely varied and complex. These factors must be combined with the unique mechanical and physical characteristics of loess slopes for focused analysis. To systematically identify the core influencing factors of loess slope stability, the study is based on the research results of loess engineering geology and slope stability at home and abroad [21–26] and comprehensively refers to China's "Code for Construction in Wet Trapped Loess Areas" (GB 50025-2018) [27] and "Highway Subgrade Design Code" (JTG D30-2015) [28]. Through theoretical analysis and engineering case verification, a preliminary loess slope stability evaluation index system containing several key parameters was constructed.

However, a large number of correlation indicators not only weaken the importance of the core parameters but also reduce the calculation efficiency due to the repetition of information, which is a result of this indicator system's problem with an excessive number and redundancy of indicators. In this study, based on the analysis of the main control factors of loess slope stability in the study area, the four basic principles of uniqueness, purpose, feasibility, and observability are used as the screening framework to qualitatively screen several preliminary indicators. The principle of uniqueness requires the elimination of redundant and invalid indicators; the purpose emphasizes that the indicators should directly reflect the characteristics of loess slope stability, the feasibility ensures that the indicator data are accessible and operable, and the observability requires that the meaning of the indicators is clear and the parameters can be quantified.

In the stability assessment of loess slope, all four of these principles can efficiently eliminate redundant indicators, highlight key factors, increase computation efficiency, and guarantee that the indicators accurately reflect the stability characteristics, all of which can improve evaluation accuracy. This approach, however, could result in the loss of information from the unintentional deletion of duplicate signs with minor variations or the exclusion of important indicators owing to oversimplification. Furthermore, the choice of indicators might be influenced subjectively by the assessors. Thus, the initial screening findings for the core indicators were obtained using this method. After removing indications with poor relevance, complicated data collecting, and unclear representation, quantitative analyses are required to guarantee the accuracy of the evaluation.

3.2. Quantitative Screening of Evaluation Indicators

The quantitative screening stage is based on the improved grey correlation-Delphi model, aiming at mining the core influencing factors from the qualitatively screened indicators through the combination of mathematical modeling and expert empowerment. The specific process is as follows:

Improved Grey Correlation-Delphi Model

Grey correlation analysis is a research method in grey system theory, which assesses the strength of correlation by quantifying the similarity of the dynamically changing characteristics of data series [29]. The core step is to divide the target indicator into reference and comparison sequences and analyze the degree of coincidence between the two change trends; the higher the correlation, the greater the contribution of the indicator to the system's behavioral pattern. With the core function of identifying the main controlling factors of the system, this method has been widely used in engineering practice, ecological research, economic analysis, and other fields [30–32]. However, in the traditional grey correlation model, the value of the discrimination coefficient λ has a strong subjective dependence due to the lack of objective criteria. In slope stability assessment, the same index may be classified as a core or secondary factor due to the different values of λ, which leads to bias in the development of engineering solutions [33]. Moreover, a slight adjustment of the λ value can trigger a large fluctuation of the correlation coefficient, resulting in a significant shift in the weighting of indicators and a non-stationary fluctuation of decision-making results. Aiming at this limitation of the traditional model, this study constructed an improved grey correlation analysis framework [34], and its improvement process is described as follows:

1. The Delphi method, also known as the expert survey method, is a process for making decisions that involves anonymous rounds of expert consultation. It involves screening and optimizing the evaluation index system using principal component analysis, hierarchical analysis, and other techniques [35–37]. The implementation process includes designing a structured questionnaire, inviting experts to independently rate or suggest, and adjusting opinions after multiple rounds of data statistics and anonymous feedback until the expert group reaches a consensus. The method reduces subjective bias through anonymous interaction and guarantees objective and reliable conclusions. In the application, m experts can anonymously rate the weights of n indicators and construct the decision matrix D characterizing the weights of the indicators. This is shown in Equation (1):

$$D = \begin{bmatrix} d_{11} & d_{12} & \cdots & d_{1m} \\ d_{21} & d_{22} & \cdots & d_{2m} \\ \vdots & \vdots & \ddots & \vdots \\ d_{n1} & d_{n2} & \cdots & d_{nm} \end{bmatrix} \tag{1}$$

   where D is the matrix of indicator weights and d_{nm} is the value of the mth expert's assessment of the weight of the nth indicator to be screened.

2. Construct the reference vector D_0 by extracting the maximum value of each column of the matrix D, as shown in Equation (2):

$$D_0 = \begin{pmatrix} d_{01} & d_{02} & \cdots & d_{0m} \end{pmatrix} \tag{2}$$

3. According to Equation (3), the spatial difference value between each evaluation index and the reference series can be calculated, which intuitively characterizes the degree of dynamic shift of the index, and provides a quantitative basis for weight allocation and correlation analysis.

$$D_{0i} = \sum_{k=1}^{m} (d_{0k} - d_{ik})^2 \tag{3}$$

   where d_{ik} is the weight that the Kth expert awarded to the ith indicator, and D_{0i} is the distance between the ith indicator and the reference indicator.

4. As shown in Equations (4) and (5), we calculate the weight coefficients of each assessment index using the standardized processed data matrix.

$$w_i = \frac{1}{1 + D_{0i}} \tag{4}$$

$$\overline{w}_i = \frac{w_i}{\sum_{l=1}^{n} w_l} \tag{5}$$

3.3. Validity and Reliability Verification

This study uses statistical analysis techniques to build a quantitative assessment model and uses a two-parameter validation mechanism of the validity coefficient β and the reliability coefficient ρ to systematically test the level of confidence, structural validity, and data stability of the screening indexes to guarantee the evaluation index system's scientific and engineering applicability [38].

3.3.1. Validity Test of Evaluation Indicator System

Assessing whether the index system can accurately reflect the objective reality requires the use of the validity test. Based on statistical theory, this study addressed the issues of subjective evaluation uncertainty and redundancy or lack of indicators by introducing the validity coefficient β as a tool for quantifying deviations and measuring the fit between the index system and the real scenario using expert scoring data. Compared to costly field measurements and machine learning approaches that rely heavily on data quality, validity tests achieve a balance between scientific rigor and engineering feasibility, demonstrating distinct methodological advantages. For the 10 key indicators established in the pre-screening, the validity coefficient test is used to conduct empirical research, and the specific modeling process is as follows:

1. Data collection and standardization: Let the evaluation index system be $Z = \{z_1, z_2, z_3, \ldots, z_n\}$ containing a total of n indicators. Invite M experts to participate in independent scoring. According to the relative importance of the indicators, the experts need to assign scores to the indicators within the $[0, 1]$ standardization interval, forming the scoring matrix $X = \{x_{1j}, x_{2j}, x_{3j}, \ldots, x_{nj}\}$,, where x_{ij} indicates the scoring value of the jth expert for the indicator z_i.
2. Calculation of the validity coefficient β_i of the single indicator: Define the validity coefficient for indicator z_i as:

$$\beta_i = \sum_{j=1}^{s} \left| \frac{\overline{x_i} - x_{ij}}{MV} \right| \tag{6}$$

where V is the theoretical optimal value and x_i is the average score value of the evaluation indicator z_i, which can be calculated using Equation (7):

$$\overline{x_i} = \sum_{j=1}^{s} x_{ij}/M \tag{7}$$

3. Global validity coefficient synthesis: Integrate the validity coefficients of each index through the arithmetic average method to get the overall validity coefficient β of the evaluation system:

$$\beta = \sum_{j=1}^{n} \beta_j/n \tag{8}$$

4. Validity determination criteria: The validity coefficient β is often used as a core parameter to measure the scientificity of the evaluation system, and when the β value is lower than the empirical threshold (routinely set at 0.15), the consensus of the expert group on the evaluation dimensions is significantly enhanced, and the validity of the indicator system in reflecting the objective reality is also enhanced.

3.3.2. Reliability Test of Evaluation Index System

In the process of constructing the evaluation index system, the subject of evaluation is affected by the heterogeneity of professionalism, cognitive framework, and knowledge reserve, and its scoring behavior is easily interfered with by subjective preferences, which leads to systematic bias in the assessment results. Such subjective cognitive differences will weaken the anti-interference and data reproduction ability of the indicator system, which in turn triggers controversy over the validity of the indicator screening. To solve this problem, this study proposes a reliability coefficient model (ρ) based on the strength of statistical association, which quantifies the intrinsic correlation between indicators to achieve validity verification, and the specific modeling process is as follows:

1. Calculation of the mean value of a single indicator: For expert rating data processing of the jth indicator z_i in the evaluation system, its arithmetic mean can be determined using Equation (9). This calculation process integrates the results of the multidimensional assessment of the members of the expert group for a particular indicator.

$$\overline{x_j} = \sum_{i=1}^{n} x_{ij}/n \tag{9}$$

2. Calculation of the global mean for the data set: Equation (10) illustrates how the baseline scoring standards of the expert panel for each indicator are first established to create the comprehensive evaluation model. $Y = \{y_1, y_2, \ldots, y_n\}$ represents the expert scoring dataset corresponding to an evaluation system with n indicators. Based on this, the comprehensive mean score of this dataset must be determined using Equation (11), a formula system that integrates expert assessment data from multiple dimensions to achieve the quantitative characterization of the overall level of the evaluation system.

$$\mathbf{y}_i = \frac{1}{M}\sum_{j=1}^{M} x_{ij} \tag{10}$$

$$\overline{\mathbf{y}} = \frac{1}{n}\sum_{i=1}^{n} \mathbf{y}_i \tag{11}$$

3. Calculating the single-indicator reliability coefficient ρ_j.$\rho_j$can be done using Equation (12):

$$\rho_j = \frac{\sum_{i=1}^{n}(x_{ij}-\overline{x}_j)(y_j-\overline{y})}{\sqrt{\sum_{i=1}^{n}(x_{ij}-\overline{x}_j)^2\sum_{i=1}^{n}(y_j-\overline{y})^2}} \tag{12}$$

   Among these, when $\rho_j > 0$, it means that z_j has a positive correlation with the whole evaluation and that the score is consistent; the more closely ρ_j approaches one, the more reliable the indicator z_j is and the more powerful it is at explaining the total results.

4. The synthetic total reliability coefficient can be calculated using Equation (13). This value synthetically reflects the stability and internal consistency of the whole indicator system. When is high, it indicates that the indicators within the indicator system are highly correlated and the degree of consensus of expert scores is high; when ρ is low, it means that the logic between the indicators is loose or there is redundancy, and the structure needs to be optimized.

$$\rho = \frac{1}{M}\sum_{j=1}^{M} \rho_j \tag{13}$$

5. Reliability level determination: Generally speaking, when $0.90 \le \rho < 0.95$, it indicates that the reliability of the evaluation index system is high, when $0.80 \le \rho < 0.90$, its reliability is at a general level, and when $0 \le \rho < 0.80$, its reliability is relatively low.

4. Engineering Application

In this paper, the loess slope on the south side of Jing Mountain in Lintai County, Gansu Province [20] is taken as the study area, and the stability evaluation indexes are selected for this slope. The area belongs to the Longdong loess hilly geomorphological area, dominated by natural loess gullies and artificially modified steep slopes. Regional tectonics are relatively stable, but it is strongly influenced by external dynamic geological effects such as weathering, gravity, surface water scouring, and groundwater activities, superimposed on the developed surface water system, high annual rainfall, and frequent human engineering activities, resulting in a significant risk of geological hazards. Several unstable loess slopes in the study area are near residential areas and public facilities, posing a high potential threat. Details are shown in Figures 7 and 8.

Figure 7. Studying the location of the area and the building [20].

Figure 8. The landslide's location in relation to the building [20].

A preliminary construction of a loess slope stability evaluation index system that multi-dimensional indexes of loess stratification and structure, hydrogeological conditions, soil mechanical parameters, and human interference, was conducted in accordance with the principle of primary selection of evaluation indicators. Based on the four basic principles of uniqueness, purposefulness, feasibility, and observability, the qualitative screening of a number of indicators in the primary selection was carried out. Ultimately, as Table 1 illustrates, a simplified assessment framework comprising 22 fundamental indicators was developed.

In order to extract the primary influencing elements from the subjectively screened indicators, the assessment indicators were quantitatively screened using the enhanced grey correlation model. Five specialists who are knowledgeable about the geological state of loess slopes were gathered using the Delphi technique to weigh each assessment index in accordance with Table 2 and create the index weight matrix D of Equation (14). The primary elements impacting the stability of loess slopes were determined by calculating and ranking the weights of the assessment indicators.

$$D = \begin{bmatrix} 0.08 & 0.08 & 0.06 & 0.05 & 0.05 \\ 0.03 & 0.03 & 0.025 & 0.02 & 0.035 \\ 0.01 & 0.015 & 0.015 & 0.01 & 0.02 \\ 0.12 & 0.12 & 0.10 & 0.11 & 0.10 \\ 0.02 & 0.025 & 0.025 & 0.02 & 0.03 \\ 0.09 & 0.09 & 0.08 & 0.07 & 0.07 \\ 0.025 & 0.03 & 0.03 & 0.02 & 0.025 \\ 0.14 & 0.14 & 0.12 & 0.15 & 0.13 \\ 0.13 & 0.13 & 0.12 & 0.11 & 0.12 \\ 0.25 & 0.20 & 0.25 & 0.18 & 0.15 \\ 0.10 & 0.10 & 0.09 & 0.12 & 0.11 \\ 0.01 & 0.015 & 0.015 & 0.01 & 0.015 \\ 0.015 & 0.015 & 0.02 & 0.015 & 0.02 \\ 0.02 & 0.02 & 0.01 & 0.02 & 0.015 \\ 0.03 & 0.035 & 0.035 & 0.03 & 0.04 \\ 0.02 & 0.02 & 0.015 & 0.02 & 0.025 \\ 0.11 & 0.11 & 0.10 & 0.12 & 0.10 \\ 0.025 & 0.025 & 0.03 & 0.02 & 0.03 \\ 0.03 & 0.03 & 0.02 & 0.025 & 0.02 \\ 0.14 & 0.14 & 0.12 & 0.15 & 0.13 \\ 0.015 & 0.02 & 0.02 & 0.015 & 0.015 \\ 0.20 & 0.18 & 0.20 & 0.24 & 0.18 \end{bmatrix} \tag{14}$$

The reference vector D_0 is constructed by extracting the maximum value of each column of the matrix D, as shown in Equation (15).

$$D_0 = \left(0.25, \quad 0.20, \quad 0.25, \quad 0.24, \quad 0.18\right) \tag{15}$$

Table 3 displays the weights of the assessment indicators following qualitative screening, which were calculated by combining Equations (3)–(5).

Analysis of the weight distribution in Table 2 reveals that the key influencing factors of loess slope stability (screening threshold of 0.0450) can be summarized into the following 10 core parameters: Rock Mass Structure (Indicator 1), Slope Height (Indicator 4), Underground Water Body (Indicator 6), Internal Friction Angle (Indicator 8), Cohesion (Indicator 9), Maximum Monthly Rainfall (Indicator 10), Slope Gradient (Indicator 11), Seismic Intensity (Indicator 17), Hydrological Conditions (Indicator 20), and Anthropogenic Engineering Activities (Indicator 22). Details are shown in Table 3. The Maximum Monthly Rainfall (indicator 10) and Anthropogenic Engineering Activities (indicator 22) have the highest weight among them. This is in line with the region's actual conditions of high precipitation and robust human engineering activity, which confirms the accuracy and rationality of the weight calculation results.

Table 1. Evaluation indicators for qualitative screening.

Indicator Number	Indicator Name	Indicator Number	Indicator Name
Indicator 1	Rock Mass Structure	Indicator 12	Permeability Coefficient of Loess
Indicator 2	Thickness of Loess Stratum	Indicator 13	Moisture Sensitivity of Loess
Indicator 3	Geological Formation	Indicator 14	Degree of vertical fracture development
Indicator 4	Slope Height	Indicator 15	Vertical fracture connectivity
Indicator 5	Irrigation-Induced Seepage Intensity	Indicator 16	Surface Runoff Intensity
Indicator 6	Underground Water Body	Indicator 17	Seismic Intensity
Indicator 7	Surface Water Infiltration Capacity	Indicator 18	Cyclic Freeze–Thaw Degradation Effects
Indicator 8	Internal Friction Angle	Indicators 19	Gully Erosion Depth
Indicator 9	Cohesion	Indicators 20	Hydrological Conditions
Indicator 10	Maximum Monthly Rainfall	Indicator 21	Rainfall-Induced Erosion
Indicator 11	Slope Gradient	Indicator 22	Anthropogenic Engineering Activities

Table 2. Evaluation indicator weights based on improved grey correlation.

Marker	Value	Symbol	Value	Symbol	Value
D_1	0.1324	W_1	0.8831	$\bar{W}_1$	0.0456
D_2	0.1973	W_2	0.8352	$\bar{W}_2$	0.0431
D_3	0.2255	W_3	0.8160	$\bar{W}_3$	0.0421
D_4	0.0691	W_4	0.9354	$\bar{W}_4$	0.0483
D_5	0.2051	W_5	0.8298	$\bar{W}_5$	0.0429
D_6	0.1076	W_6	0.9029	$\bar{W}_6$	0.0466
D_7	0.2004	W_7	0.8331	$\bar{W}_7$	0.0430
D_8	0.0432	W_8	0.9586	$\bar{W}_8$	0.0495
D_9	0.0567	W_9	0.9463	$\bar{W}_9$	0.0489
D_{10}	0.0045	W_{10}	0.9955	$\bar{W}_{10}$	0.0514
D_{11}	0.0774	W_{11}	0.9282	$\bar{W}_{11}$	0.0479
D_{12}	0.2272	W_{12}	0.8149	$\bar{W}_{12}$	0.0421
D_{13}	0.2186	W_{13}	0.8206	$\bar{W}_{13}$	0.0424
D_{14}	0.2185	W_{14}	0.8207	$\bar{W}_{14}$	0.0424
D_{15}	0.1855	W_{15}	0.8445	$\bar{W}_{15}$	0.0436
D_{16}	0.2130	W_{16}	0.8244	$\bar{W}_{16}$	0.0426
D_{17}	0.0710	W_{17}	0.9337	$\bar{W}_{17}$	0.0482
D_{18}	0.2006	W_{18}	0.8330	$\bar{W}_{18}$	0.0430
D_{19}	0.2020	W_{19}	0.8319	$\bar{W}_{19}$	0.0430
D_{20}	0.0432	W_{20}	0.9586	$\bar{W}_{20}$	0.0495
D_{21}	0.2184	W_{21}	0.8208	$\bar{W}_{21}$	0.0424
D_{22}	0.0054	W_{22}	0.9946	$\bar{W}_{22}$	0.0514

Table 3. Evaluation indicators for quantitative screening.

Indicator Number	Indicator Name	Indicator Number	Indicator Name
Indicator 1	Rock Mass Structure	Indicator 10	Maximum Monthly Rainfall
Indicator 4	Slope Height	Indicator 11	Slope Gradient
Indicator 6	Underground Water Body	Indicator 17	Seismic Intensity
Indicator 8	Internal Friction Angle	Indicator 20	Hydrological Conditions
Indicator 9	Cohesion	Indicator 22	Anthropogenic Engineering Activities

5. Conclusions

To address the problem of stability evaluation of loess slopes on the south side of Jingshan Mountain in Lintai County, Gansu Province, this study proposes an evaluation index screening method that combines qualitative and quantitative analyses and integrates the improved grey correlation model through the introduction of the coefficient of validity (β) and the coefficient of reliability (ρ), which significantly improves the redundancy of the indexes and subjective dependence that exists in the traditional method. The main research results are as follows:

1. Multi-dimensional indicator screening framework: A qualitative analysis framework was constructed based on engineering geology, hydrogeology, and anthropogenic factors, and the preliminary selection of 60 indicators was quantitatively screened by combining it with the improved grey correlation model, and 10 core indicators were finally identified, including rock structure, slope height, groundwater body, internal friction angle, cohesion, maximum monthly rainfall, and geological structure. The framework achieves the precision and efficiency of the indicator system through a hierarchical screening mechanism, providing a scientific basis for the evaluation process.
2. Evaluation system validity and reliability test: By calculating the validity coefficient ($\beta = 0.0816$) and reliability coefficient ($\rho = 0.9609$), the logical rationality and data consistency of the screening indicator set were verified. The results show that the method takes into account expert experience and objective data analysis, significantly reduces human bias, and can provide a highly credible indicator basis for slope stability evaluation.
3. Engineering practice value: The research results provide a standardized analysis tool for loess slope stability assessment, which can systematically identify potential risk factors, guide the deployment of the monitoring network, optimize the reinforcement scheme, and help improve the refinement of mine safety management. The method also has cross-regional adaptability and can be extended to the evaluation of loess slopes with different geological backgrounds.

Loess slopes in the southern Jingshan region of Lingtai County, Gansu Province, are the subject of this study. The model's broad applicability needs more validation due to the limited expert samples and data coverage, especially across varied geological units including steep high slopes and composite stratigraphy. The proposed indicator screening model could be expanded in future research through the integration of critical parameters including extreme rainfall measurements, seismic indices, and freeze–thaw cycles into existing evaluation systems while adopting machine learning techniques to quantify coupling effects among these factors. Additionally, comprehensive model validation and numerical simulations across various geomorphological units of the Loess Plateau could elucidate the evolution mechanisms of slope collapse under extreme loading conditions. These enhancements would substantially improve the model's adaptability to complex scenarios involving intense rainfall and seismic activities in loess terrains while strengthening its threshold identification capability.

Author Contributions: Conceptualization, H.S. and J.A.; Writing—original draft, J.L. All authors have read and agreed to the published version of the manuscript.

Funding: 2024 Liaoning Provincial Science and Technology Plan Joint Programme Project [2024011314-JH2/1026].

Institutional Review Board Statement: Not applicable.

Informed Consent Statement: Not applicable.

Data Availability Statement: The original contributions presented in this study are included in the article. Further inquiries can be directed to the corresponding author.

Conflicts of Interest: Author Jianchao An was employed by the company Jinzhou Water (Group) Co., Ltd. The remaining authors declare that the research was conducted in the absence of any commercial or financial relationships that could be construed as a potential conflict of interest.

References

1. Zhuang, J.; Peng, J.; Zhu, Y. Study of the effects of clay content on loess slope failure mode and loess strength. *Bull. Eng. Geol. Environ.* **2021**, *80*, 1999–2009. [CrossRef]
2. Xu, J.; Wang, Z.Q.; Ren, J.W.; Wang, S.H.; Jin, L. Mechanism of slope failure in loess terrains during spring thawing. *J. Mt. Sci.* **2018**, *15*, 845–858. [CrossRef]
3. Li, X.; Lu, Y.; Zhang, X.; Fan, W.; Lu, Y.; Pan, W. Quantification of macropores of Malan loess and the hydraulic significance on slope stability by X-ray computed tomography. *Environ. Earth Sci.* **2019**, *78*, 522. [CrossRef]
4. Zhuang, J.; Peng, J.; Wang, G.; Iqbal, J.; Wang, Y.; Li, W.; Xu, Q.; Zhu, X. Prediction of rainfall-induced shallow landslides in the Loess Plateau, Yan'an, China, using the TRIGRS model. *Earth Surf. Process. Landforms* **2017**, *42*, 915–927. [CrossRef]
5. Wu, Z.J.; Zhao, D.Y.; Che, A.L.; Chen, D.W.; Liang, C. Dynamic response characteristics and failure mode of slopes on the loess tableland using a shaking-table model test. *Landslides* **2020**, *17*, 1561–1575. [CrossRef]
6. Peng, J.; Sun, P.; Igwe, O.; Li, X. Loess caves, a special kind of geo-hazard on loess plateau, northwestern China. *Eng. Geol.* **2018**, *236*, 79–88. [CrossRef]
7. Zhuang, J.; Peng, J.; Wang, G.; Javed, I.; Wang, Y.; Li, W. Distribution and characteristics of landslide in Loess Plateau: A case study in Shaanxi province. *Eng. Geol.* **2018**, *236*, 89–96. [CrossRef]
8. Peng, J.; Fan, Z.; Wu, D.; Huang, Q.; Wang, Q.; Zhuang, J.; Che, W. Landslides triggered by excavation in the loess plateau of China: A case study of Middle Pleistocene loess slopes. *J. Asian Earth Sci.* **2019**, *171*, 246–258. [CrossRef]
9. Gao, H.; Zhang, X. Landslide susceptibility assessment considering landslide volume: A case study of Yangou watershed on the Loess Plateau (China). *Appl. Sci.* **2022**, *12*, 4381. [CrossRef]
10. Ma, J.; Yao, Y.; Wei, Z.; Meng, X.; Zhang, Z.; Yin, H.; Zeng, R. Stability analysis of a loess landslide considering rainfall patterns and spatial variability of soil. *Comput. Geotech.* **2024**, *167*, 106059. [CrossRef]
11. Sengani, F.; Allopi, D. Accuracy of two-dimensional limit equilibrium methods in predicting stability of homogenous road-cut slopes. *Sustainability* **2022**, *14*, 3872. [CrossRef]
12. Wang, G.; Zhao, B.; Wu, B.; Zhang, C.; Liu, W. Intelligent prediction of slope stability based on visual exploratory data analysis of 77 in situ cases. *Int. J. Min. Sci. Technol.* **2023**, *33*, 47–59. [CrossRef]
13. Wang, G.; Liu, Y.; Hu, Z.; Lyu, Y.; Zhang, G.; Liu, J.; Liu, Y.; Gu, Y.; Huang, X.; Zheng, H.; et al. Flood risk assessment based on fuzzy synthetic evaluation method in the Beijing-Tianjin-Hebei metropolitan area, China. *Sustainability* **2020**, *12*, 1451. [CrossRef]
14. Li, L.M.; Cheng, S.K.; Wen, Z.Z. Landslide prediction based on improved principal component analysis and mixed kernel function least squares support vector regression model. *J. Mt. Sci.* **2021**, *18*, 2130–2142. [CrossRef]
15. Dong, Q.; Yi, P.; Li, W.; Wang, L. Evaluation of city sustainability using the HGRW method: A case study of urban agglomeration on the West Side of the Straits, China. *J. Clean. Prod.* **2022**, *358*, 132008. [CrossRef]
16. Chen, L.; Guo, H.; Gong, P.; Yang, Y.; Zuo, Z.; Gu, M. Landslide susceptibility assessment using weights-of-evidence model and cluster analysis along the highways in the Hubei section of the Three Gorges Reservoir Area. *Comput. Geosci.* **2021**, *156*, 104899. [CrossRef]
17. Gong, B. Study of PLSR-BP model for stability assessment of loess slope based on particle swarm optimization. *Sci. Rep.* **2021**, *11*, 17888. [CrossRef]
18. Mahmoodzadeh, A.; Mohammadi, M.; Farid Hama Ali, H.; Hashim Ibrahim, H.; Nariman Abdulhamid, S.; Nejati, H.R. Prediction of safety factors for slope stability: Comparison of machine learning techniques. *Nat. Hazards* **2022**, *111*, 1771–1799. [CrossRef]
19. Li, X.; Huang, F.; Yang, Z. Multisource monitoring data-driven slope stability prediction using ensemble learning techniques. *Comput. Geotech.* **2024**, *169*, 106255. [CrossRef]
20. Nan, Z. Stability Analysis of Loess Slopes in Jingshan and Study of Management Options. Master's Thesis, Lanzhou University, Lanzhou, China, 2023.
21. Peng, J.; Wang, S.; Wang, Q.; Zhuang, J.; Huang, W.; Zhu, X.; Leng, Y.; Ma, P. Distribution and genetic types of loess landslides in China. *J. Asian Earth Sci.* **2019**, *170*, 329–350. [CrossRef]
22. Wei, L.; Zeng, Z.; Yan, J. Factors Affecting the Stability of Loess Landslides: A Review. *Appl. Sci.* **2024**, *14*, 2735. [CrossRef]
23. Zhang, Z.; Qi, W.; Zhao, Z.; Zheng, T. Influence of loess–mudstone strata structure on slope seismic stability of loess plateau in China. *Appl. Sci.* **2023**, *13*, 12854. [CrossRef]
24. Zeng, F.; Nait Amar, M.; Mohammed, A.S.; Motahari, M.R.; Hasanipanah, M. Improving the performance of LSSVM model in predicting the safety factor for circular failure slope through optimization algorithms. *Eng. Comput.* **2021**, *38*, 1755–1766. [CrossRef]
25. Liu, X.; Shao, S.; Shao, S. Landslide susceptibility zonation using the analytical hierarchy process (AHP) in the Great Xi'an Region, China. *Sci. Rep.* **2024**, *14*, 2941. [CrossRef]
26. Feng, W.; Liu, Y.; Chen, Z.; Li, Y.; Huang, Y. Theoretical and practical research into excavation slope protection for agricultural geographical engineering in the Loess Plateau: A case study of China's Yangjuangou catchment. *J. Rural Stud.* **2022**, *93*, 309–317. [CrossRef]

27. *GB 50025-2018*; Standard for Building Construction in Collapsible Loess Regions. Ministry of Housing and Urban-Rural Development of the People's Republic of China: Beijing, China, 2018.
28. *JTG D30-2015*; Specifications for Design of Highway Subgrades. Ministry of Transport of the People's Republic of China: Beijing, China, 2015.
29. Deris, A.M.; Solemon, B.; Taha, H.; Omar, R.C. Determination of influencing factors for slope stability using grey relational analysis (GRA) technique. *Syst. Rev. Pharm* **2020**, *11*, 202–206.
30. Zhao, X.; Yan, L.; Yang, L.; Chi, F.; Ning, Y. Deformation characteristics and influential factors of a toppling rock slope based on the grey relational analysis. *Eur. J. Environ. Civ. Eng.* **2023**, *27*, 2507–2518. [CrossRef]
31. Dou, F.; Xing, H.; Li, X.; Yuan, F.; Lu, Z.; Li, X.; Ge, W. 3D Geological suitability evaluation for urban underground space development based on combined weighting and improved TOPSIS. *Nat. Resour. Res.* **2022**, *31*, 693–711. [CrossRef]
32. Ye, S.; Huang, A.; Zhu, Y. Slope Stability Prediction Using Grey Entropy Correlation Analysis. *Soil Mech. Found. Eng.* **2024**, *61*, 419–425. [CrossRef]
33. Ye, S.H.; Huang, A.P. Sensitivity analysis of factors affecting stability of cut and fill multistage slope based on improved grey incidence model. *Soil Mech. Found. Eng.* **2020**, *57*, 8–17. [CrossRef]
34. Zhang, P.; Wang, Y.; Jiang, N.; Ge, L.; Ding, Y.; Tang, P. Screening method for evaluation index of distribution network regulation level based on improved grey correlation. *J. Power Syst. Autom.* **2018**, *30*, 64–69.
35. Hu, X.D.; Gao, J.Z.; Zhou, M.T.; Shen, J.; Zhao, T.H. Evaluating the success of engineering disturbed slope eco-restoration in the alpine region, southeast Qinghai-Tibet Plateau, China. *J. Mt. Sci.* **2021**, *18*, 2820–2832. [CrossRef]
36. Kim, J.; Kim, C.; Kim, G.; Kim, I.; Abbas, Q.; Lee, J. Probabilistic tunnel collapse risk evaluation model using analytical hierarchy process (AHP) and Delphi survey technique. *Tunn. Undergr. Space Technol.* **2022**, *120*, 104262. [CrossRef]
37. Yuan, X.; Hou, R.; Liu, Y.; Wei, X.; Wang, X.; Ying, Y.; Zhao, Y.; Yao, Y.; Chang, J. The evaluation of soil stability in loess hilly and gully region of Northern Shaanxi based on GIS. *Geol. J.* **2018**, *53*, 379–386. [CrossRef]
38. Xiao, H. Dynamic Evaluation Method and Application of Slope Stability in Small and Medium-Sized Open Pit Mines. Ph.D. Thesis, China University of Mining and Technology, Xuzhou, China, 2019.

applied sciences

MDPI

Article

Shear Properties and Failure Mechanism of Matched Discontinuities Between Two Different Rock Types Under Direct Shear

Xiaobo Zhang [1,2], Shu Ouyang [2], Chenglong Fan [2], Le Yi [2] and Da Liu [1,*]

1 Jiangxi Hydraulic Safety Engineering Technology Research Center, Jiangxi Academy of Water Science and Engineering, Nanchang 330029, China; zhangxb@ncu.edu.cn
2 School of Infrastructure Engineering, Nanchang University, Nanchang 330031, China; oys2468@163.com (S.O.); 412500220055@email.ncu.edu.cn (L.Y.)
* Correspondence: liud_sky@foxmail.com

Abstract: The shear mechanical properties of rock discontinuities with different joint wall compressive strengths are a practical basis for the stability analysis of layered rock mass. Shear tests on discontinuities possessing different joint wall strengths were carried out. The shear strength and failure characteristics were analyzed, and the influences of discontinuity morphology on its shear properties were investigated. Meanwhile, numerical tests were performed to study the shear mechanical behavior and dilation evolution of discontinuities possessing different joint wall compressive strengths. Results show that the shear process of discontinuities possessing different joint wall strengths can be divided into four stages: meshing and compacting, climbing wear of soft rock and crack formation of hard rock, shear of part of soft rock and crack expansion of hard rock, complete shearing of the rock discontinuity. Shear failure of discontinuities was mainly concentrated on the morphological structure facing the shear direction. The dilatancy evolution process of discontinuities was mainly affected by the roughness and normal stress. The magnitude of dilation, peak shear strength and residual shear strength of discontinuities possessing different joint wall strengths were between the discontinuities possessing identical joint wall strengths composed of soft and hard rock, under the same loading condition.

Keywords: rock discontinuity; different joint wall strength; shear property; failure mechanism; shear dilation

1. Introduction

Rock joint is a kind of geological discontinuity that widely exists in rock mass. The mechanical properties of rock joints are complex and are affected by many factors such as lithology, surface morphology, compressive strength of joint wall, filling materials and load actions [1–5]. Because the failure of rock masses is commonly manifested as shear slip along rock joints, the overall stability of the rock mass is largely controlled by the mechanical properties, especially the shear resistance behavior of rock joints. Therefore, the shear mechanical characteristics and the failure mechanism of rock joints have always been a research hotspot in the field of rock mechanics [6,7].

Over the years, a lot of researchers have carried out extensive research on the shear mechanism and shear strength of rock joints [2,8–10]. Barton and Choubey [11] systematically conducted shear tests on natural rock joints. A joint roughness coefficient (JRC) for estimating the roughness of rock joints was proposed and a JRC-JCS shear strength

https://doi.org/10.3390/app15105289

model of rock joints was developed, which have been of widespread interest for both researchers in laboratories and engineers in the field. Due to the limitation that JRC can only characterize joint roughness in two dimensions, Grasselli and Egger [12] introduced three-dimensional roughness-characterization parameters considering the contact area of rock joints, and developed a shear strength criterion based on three-dimensional roughness. Moradian et al. [13] carried out shear tests on joint specimens of three different materials and monitored the shear process using acoustic emission. The results showed that the peak value of acoustic-emission events appeared after the peak shear stress, which revealed the progressive shear failure of joints. Asadi et al. [14] conducted shear tests of sawtooth-shaped and wave-shaped mortar joints under different normal stresses, and adopted PFC2D to numerically study the mesoscopic shear mechanism of joints. The results showed that with the increase in joint roughness, the failure mode of the asperities changed from sliding friction failure to asperity degradation. Some researchers have conducted laboratory direct shear tests on the factors affecting the shear strength of rock joints. Taking into account the effects of the filling on the shear strength of joints, Oliveira and Indraratna [15] adopted the UDEC code to numerically study the shear behavior of soil infilled rock joints under constant normal stress, and a new shear strength model of soil infilled joints was proposed. In order to investigate the influence of joint materials on the shear strength of joints, Hossaini et al. [16] used plaster and concrete with different Young's modulus to make artificial joints, and shear test results showed that the deformability of asperity significantly affected the shear mechanical behavior of joints. Atapour and Moosavi [17] prepared flat and rough artificial joints made by gypsum and concrete and conducted direct shear tests at different shear rates. The results showed that the shear strength of flat and rough joints decreased as the shear rate increased. For the purpose of studying the effect of roughness on the apparent cohesion of joints, Rulliere et al. [18] made four types of artificial mortar joints with different roughness and carried out direct shear tests under low normal stress. The results showed that the apparent cohesion was positively correlated with the roughness, and the apparent cohesion was significantly reduced when the specimen was damaged or poorly interlocked. In order to assess the effects of material strength of joints on apparent cohesion, Rulliere et al. [19] carried out shear tests on three types of joints composed of different materials under low normal stress. The results showed that due to the conditions of normal stress, the impacts of material strength and types of joints on shear behavior, shear strength, apparent cohesion and friction angle was limited. The above studies, however, neglected the influence of the effective contact area between the lower and upper joint surfaces during the shearing process on the shear properties of rock joints. Therefore, Tang and Wong [20] considered different contact states of rock joints using different dislocations in the shear direction and proposed an empirical formula for the shear strength of rock joints under different contact states. Based on shear test results, Zhang et al. [21] introduced a two-dimensional roughness parameter considering the shear contact area, and developed an empirical formula for predicting the peak shear strength of rock joints. Ban et al. [22] carried out shear tests of rock joints with different contact area ratios, and deduced an anisotropic parameter AAHD for proposing a new peak shear strength criterion.

The current studies on shear properties of rock discontinuities are mostly aimed at the discontinuities with identical joint wall strength; that is to say, the lithology on both sides of rock discontinuities is the same. However, there are few studies on the shear properties of discontinuities possessing different joint wall strengths. Soft–hard interbedded rock masses composed of different lithologic combinations are pretty common in practical engineering [23]. The discontinuity possessing different joint wall strengths is prone to shear slip failure under load action, which is a critical weak part affecting the strength and overall stability of soft–hard interbedded rock masses. Therefore, it is essential to study the shear mechanical behavior of discontinuities possessing different joint wall strengths. For this reason, Lin et al. [24] produced joints with different strengths by pouring mortar with different sand–cement ratios at the top of rock-like mortar materials to form binary medium structural planes. Direct shear tests revealed that the peak and residual shear strength of the binary structural plane were much lower than those of the unitary structural plane. The difference of cohesion between the unitary and binary structural planes decreased with the increase in the sand–cement ratio, while the friction angle φ increased first and then decreased with the increase in the sand–cement ratio. Jiang et al. [25] carried out shear tests of discontinuities possessing different joint wall strengths with different JRC, and stated that the peak shear strength of discontinuities possessing different joint wall compressive strengths increased exponentially with the increase in JRC. Lin et al. [26] employed thePFC2D code to establish a binary medium shear model including flat joints. Numerical shear tests under different constant normal loads were carried out on models with different parallel bond modulus ratios. Based on the simulation results, an empirical formula of shear strength considering modulus difference of discontinuities possessing different joint wall strengths was proposed. Wu et al. [27] conducted direct shear tests and numerical simulations on 14 pairs of natural discontinuities possessing different joint wall strengths to evaluate surface damage through damage zone distribution and variations in JRC. The results showed that when the JRC of the upper and lower sides of the discontinuity were similar, the shear damage difference between the upper and lower blocks of the specimen was closely related to the strength difference.

The aforementioned studies mainly focused on laboratory tests and numerical simulations, but neglected theoretical studies for shear strength of discontinuities possessing different joint wall strengths. To address this issue, Ghazvinian et al. [28] used three types of plaster mortars to make saw-toothed discontinuities possessing different joint wall strengths. Based on shear test results on these joints, a peak shear strength criterion of discontinuities possessing different joint wall strengths was derived. Wu et al. [29] conducted shear tests of serrated discontinuities possessing different joint wall strengths and developed a prediction model of shear strength of discontinuities possessing different joint wall strengths through a neural network approach. Tang et al. [30] fabricated three types of rock-like discontinuities with different joint wall strengths and different roughnesses. On the basis of shear test results, a shear strength model considering three-dimensional morphology of rock joints was proposed.

Generally, shear tests of rock joints need a certain amount of specimens with different surface morphologies. The traditional method of preparing rock joint specimens is often time-consuming and laborious when accurately fabricating rough rock joints in batches. Because of the fabrication errors, there might be a certain deviation between the test results and the shear properties of natural rock joints. Fortunately, the three-dimensional engraving technology developed in recent years can effectively solve this problem [31]. Using this technique, rock joints with complex surface morphology can be directly carved on natural rock with high precision; thereby, a mass of duplicable tests can be carried out to explore the influence of various factors on the shear properties of rock joints [32,33]. This study adopted the Barton standard rock joint profiles as prototypes; discontinuities possessing different joint wall strengths and with different surface morphologies were fabricated with three-dimensional (3D) engraving technology. The shear mechanical properties and shear failure characteristics were experimentally investigated. Simultaneously, numerical simulations were performed to reveal the influence mechanism of the joint wall strength difference and joint morphology on the shear behavior.

2. Specimen Preparation

2.1. Preparation of Rock and Rock-like Materials

Sandstone and granite blocks were collected from Jinjiang of Fujian Province, southeast China. These rock blocks were taken to the laboratory and processed into test specimens with a size of 100 mm × 100 mm × 50 mm (length × width × height). A total of 20 complete sandstone specimens and 20 complete granite specimens for subsequent 3D carving of rough joints were prepared. In order to obtain two kinds of different discontinuities possessing different joint wall strengths with the above sandstone and granite specimens, 40 cement mortar specimens of the same size were fabricated. The mortar specimens were made of quartz sand, 32.5# Portland cement and water with a weight ratio of 4:2:1.2. The specific manufacturing process is introduced as follows.

(1) Quartz sand and cement were mixed in an electric mixer according to the designed weight ratio (as shown in Figure 1a). To ensure the initial strength and fluidity of the model material, early water reducing agent with 0.5% cement mass was added, and the mixture was fully stirred for 2 min.
(2) Water was poured into the mixer and fully stirred with the dry material for 2 min to obtain the mortar material.
(3) Mortar specimens were prepared using a triple cast iron mold with dimensions of 300 mm × 100 mm × 100 mm (length × width × height) (see Figure 1b). In order to facilitate the demolding of the sample after curing, a layer of demolding agent was sprayed on the inner surface of the mold. The mold filled with mortar was placed on the low-frequency electric vibration table (see Figure 1c).
(4) In order to obtain the basic physical and mechanical properties of the mortar material, cylindrical specimens with size of ϕ 50 mm × 100 mm (diameter × height) and size of ϕ 50 mm × 30 mm (diameter × height) were prepared. The mortar was slowly poured into the mold, and then the mold was vibrated on the vibration table for 3 min.
(5) After all mortar specimens were cured for 24 h, the specimens were extracted from the mold and placed at indoor temperature of about 25 °C for 4 weeks.

(a) (b) (c)

Figure 1. Apparatus for mortar specimen fabrication. (**a**) Concrete mixer; (**b**) triple cast iron mold; and (**c**) vibration table.

2.2. Three-Dimensional Carving of Rough Joints

A total of 12 sandstone specimens, 12 granite specimens and 24 mortar specimens were carved to form 12 pairs of sandstone–mortar rough discontinuities and 12 pairs of granite–mortar rough discontinuities. An HY-6060 stone mold engraving machine made by Jinan Heyi Machinery Equipment Company, Jinan, China, as shown in Figure 2, was adopted to carve the surface morphology of joint specimens. The joint morphology carving process is presented as follows.

Figure 2. HY-6060 stone mold engraving machine.

(1) Generation of JRC profiles. Three curves with JRC of 2.8, 10.8 and 18.7 from the Barton joint profiles were taken as the prototype. The coordinate points of the selected joint profiles given by Li and Zhang [34] were employed and were imported into AutoCAD 2016 software such that three JRC curves with length of 100 mm were plotted with the spline curve command and saved in R50 format.
(2) Three-dimensional stretching of Barton joint profiles. By employing JD Paint software, the three two-dimensional (2D) JRC curves obtained by step (1) were stretched 100 mm along the direction perpendicular to the extension direction. By doing that, three 3D surfaces with dimensions 100 mm × 100 mm were obtained.
(3) Curving parameter setting of the engraving machine. The vertex of the lower left corner of each 3D stretching surface generated by step (2) was selected as the coordinate origin. The path spacing of the carving tool was set to 0.06 mm and a taper ball-end cutter with a diameter of 0.6 mm was installed as the carving tool. The tool path obtained by JD Paint software was transformed into a format that can be recognized by the engraving machine.
(4) Surface morphology carving of joints. The prepared rock and mortar specimens with size 100 mm × 100 mm × 50 mm (length × width × height) were firmly fixed on the bench clamp of the engraving machine. The origin of the carving path and the carving origin were calibrated. Subsequently, the engraving machine was operated to automatically engrave the surface morphology of joints according to the carving path assigned by step (3). Once the carving for one specimen was completed, the cutting tool path was rotated 180° and the matching specimen was carved by repeating the above process to gain a pair of joint blocks with identical morphology. Finally, a pair of fully mated rough joint specimens was obtained.
(5) Evaluation of the joint-engraving precision. After completing the engraving work of all rough joint specimens, the morphological information of the rough joint was extracted to estimate the carving precision. The image of joints with different joint wall materials was binarized by the image-editing software ImageJ v1.8.0 to obtain the gray image of the carving acquisition line of joint. The joint image and the carving morphology of the joint were compared with the Barton joint profile, as shown in Figure 3. It can be realized from Figure 3 that the carving morphologies of the joint are in good agreement with the Barton joint profiles, indicating that the 3D carving method is applicable to carve the rock and rock-like materials to obtain rock joint specimens with specific morphology in batches.

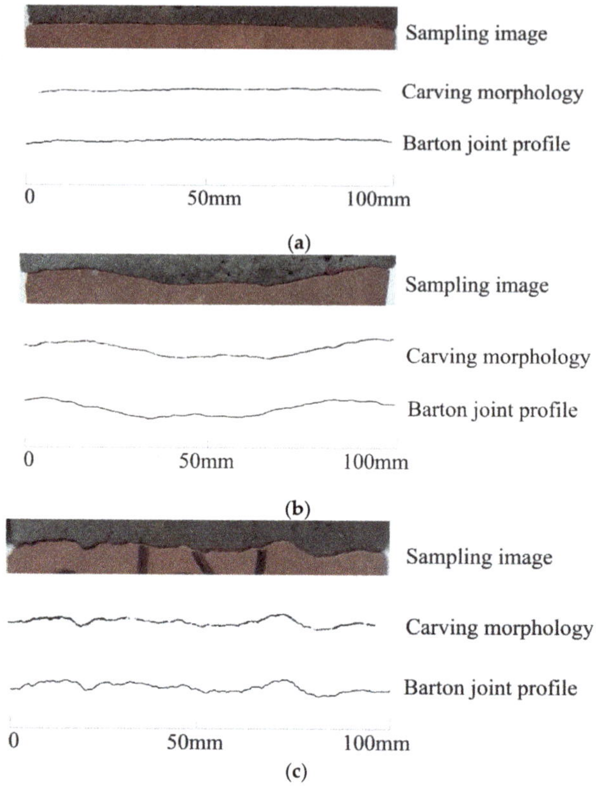

Figure 3. Comparison of sampling image, carving morphology and Barton joint profiles with JRC of (**a**) 2.8, (**b**) 10.8 and (**c**) 18.7.

3. Experiment Scheme

3.1. Physical and Mechanical Properties of Materials

In order to obtain the basic physical and mechanical properties of rock and cement mortar specimens, uniaxial compression tests, triaxial compression tests and Brazilian splitting tests were carried out using a temperature–stress–seepage coupling test system of rocks. Meanwhile, tilt tests were conducted on flat joint specimens by the three materials. The basic physical and mechanical properties obtained from tests are listed in Table 1. It can be seen from Table 1 that the three types of material used for matching joint specimens have different strength. Therefore, either the sandstone–mortar joint or the granite–mortar joint studied herein had different joint wall strengths.

Table 1. Physical and mechanical properties of materials used for making joint specimens.

Property	Cement Mortar	Sandstone	Granite
Density ρ (g/cm^3)	2.10	2.43	2.80
Uniaxial compression strength *UCS* (MPa)	32.60	95.56	261.55
Tensile strength σ_t (MPa)	1.80	2.13	6.48
Cohesion c (MPa)	7.00	12.00	23.00
Basic friction angle Φ_b (°)	31.50	29.50	35.00
Young's module E (GPa)	22.32	18.84	57.91
Poisson's ratio ν	0.30	0.28	0.22

3.2. Shear Test Method

After the joint specimen was installed in the shear box, the normal load was applied at a rate of 0.5 kN/s until it reached the specific normal stress level, which then remained constant. Subsequently, the shear load was applied at a rate of 0.5 mm/min. Each test was terminated when the shear displacement reached 10 mm. Figure 4 shows the joint specimens used in the test.

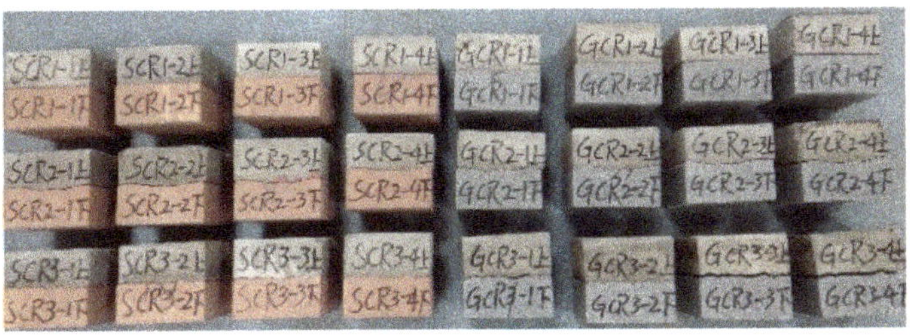

Figure 4. Joint specimens before shear tests.

All shear tests were performed on a YZW50 electro-hydraulic servo shear test system, as shown in Figure 5a. The shear apparatus has a servo control function, which can apply a normal load and a tangential load simultaneously. The maximum normal load and the maximum tangential load are 500 kN and 300 kN, respectively. The sensors equipped on the hydraulic cylinder can monitor the normal and tangential loads as well as displacements in real time. The shear box used in the test has semi-open structure, which is composed of four refined steel plates and four sets of high-strength bolts, as presented in Figure 5b.

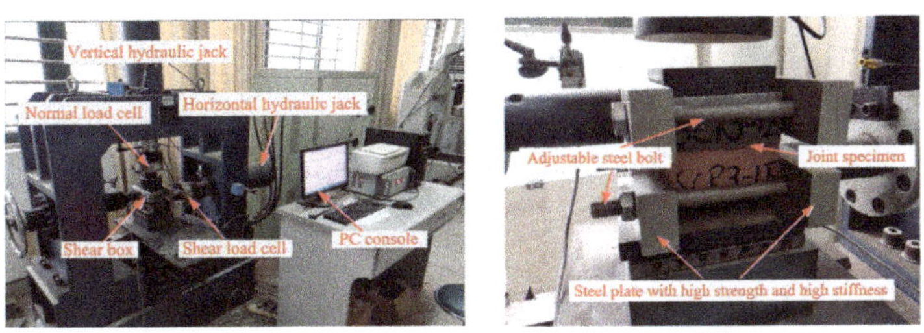

(a) (b)

Figure 5. YZW50 electro-hydraulic servo shear test system. (**a**) Overall view of the apparatus; (**b**) the local view of shear box.

The shear tests were conducted under normal stresses of 0.5, 1.0, 2.0 and 3.0 MPa. According to the type of joint materials, the sandstone–mortar joint and granite–mortar joint were named SC and GC, respectively. The specimens with JRC of 2.8, 10.8 and 18.7 were numbered R1, R2 and R3, respectively. For the sake of observing the failure characteristics of joint surfaces after test, the joint surface of all mortar blocks was coated with a layer of red ink and dried naturally for three days before shear test.

4. Experiment Results and Analyses

4.1. Shear Properties of Rough Discontinuities Possessing Different Joint Wall Strengths

Figure 6 presents the curves of shear stress vs. shear displacement of sandstone–mortar joints and granite–mortar joints with JRC of 2.8, 10.8 and 18.7. It can be found that with the increase in shear displacement, the shear stress increased approximately linearly until it reached the peak value. After the peak shear stress, with the continuous increase in shear displacement, the shear stress of joints with JRC of 2.8 remained almost stable, while the shear stress of joints with JRC of 10.8 and 18.7 appeared to show stress drop but eventually approached a stable value, namely the residual shear strength. In general, the shear stiffness and peak shear strength of the two types of rough discontinuities possessing different joint wall strengths increased significantly with the increase in normal stress.

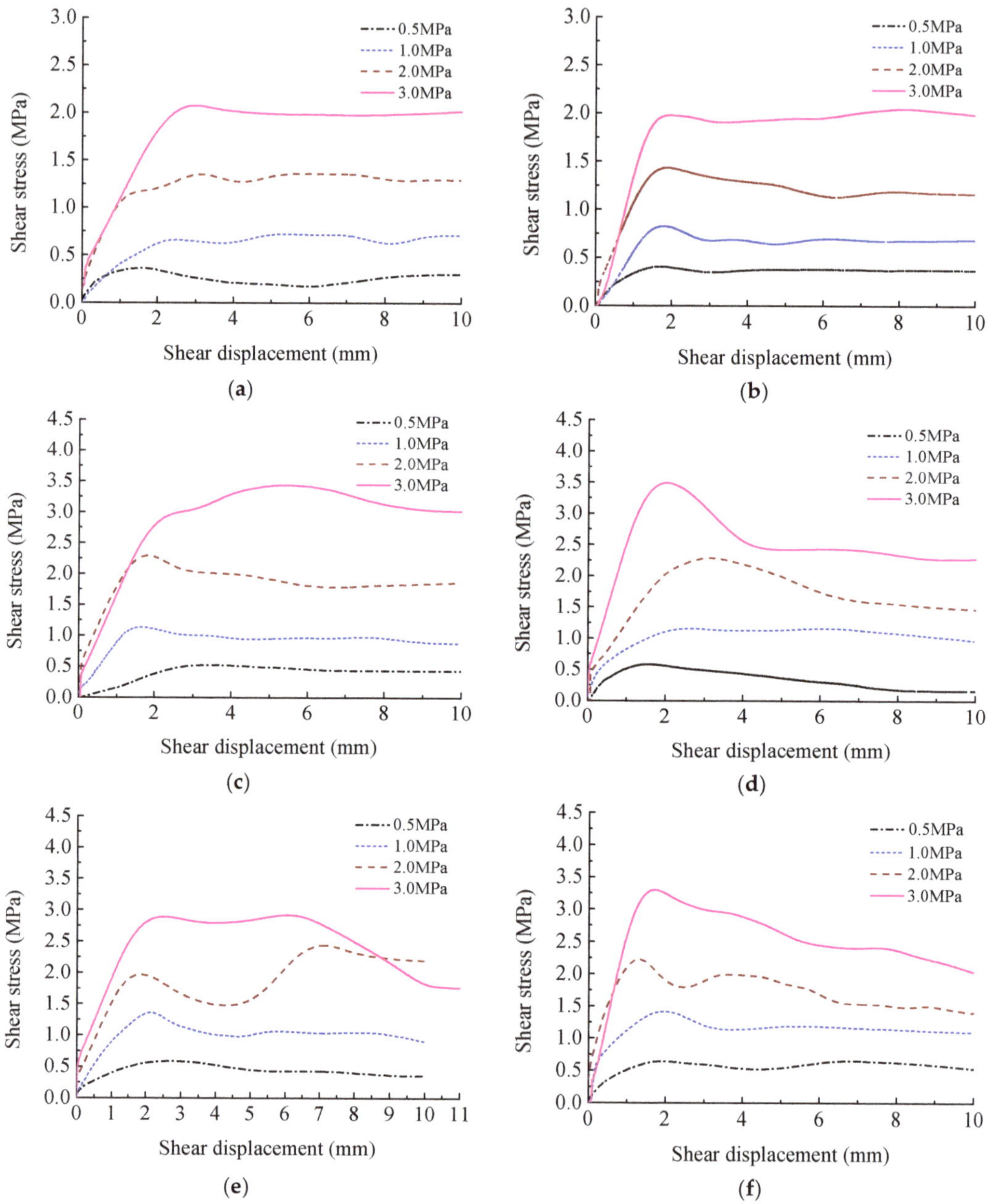

Figure 6. Shear curves of rough discontinuities. (**a**) GCR1; (**b**) SCR1; (**c**) GCR2; (**d**) SCR2; (**e**) GCR3; and (**f**) SCR3.

4.2. Failure Mechanism of Rough Discontinuities Possessing Different Joint Wall Strengths

Figure 7 presents the failure characteristics of discontinuities possessing different joint wall strengths with JRC of 2.8. It can be seen from Figure 7 that the failure of joint specimens with JRC of 2.8 mainly showed up as friction failure. For sandstone–mortar discontinuities, there were different degrees of wear on the surface of both sandstone and cement mortar, and the damage of cement mortar was more obvious. The damage area and damage depth of the joint surface increased with the increase in normal stress. For granite–mortar discontinuities, the damage was mainly concentrated on the side of the cement mortar. The asperity slightly climbed during the shear process. Due to the slip friction between the upper and lower joint blocks, the asperity on the surface of the mortar was ground and microcracks appeared on the asperity of granite. With the migration of ground mortar particles, the mortar particles were crushed and bonded on the surface of the granite. With the increase in normal stress, the degree of friction damage on the surface of mortar was strengthened and the number of ground mortar particles increased. Figure 7e presents the microscopic failure mechanism of the joint specimen with JRC of 2.8. Based on the strength of rock and mortar, granite and sandstone are named as hard rock, while mortar is named as soft rock. Slight dilation accompanied with the sliding friction between asperities appeared in the joint during the shearing process. Due to the strength difference of hard rock and soft rock, the asperity of soft rock was ground and the asperity of hard rock produced microcracks.

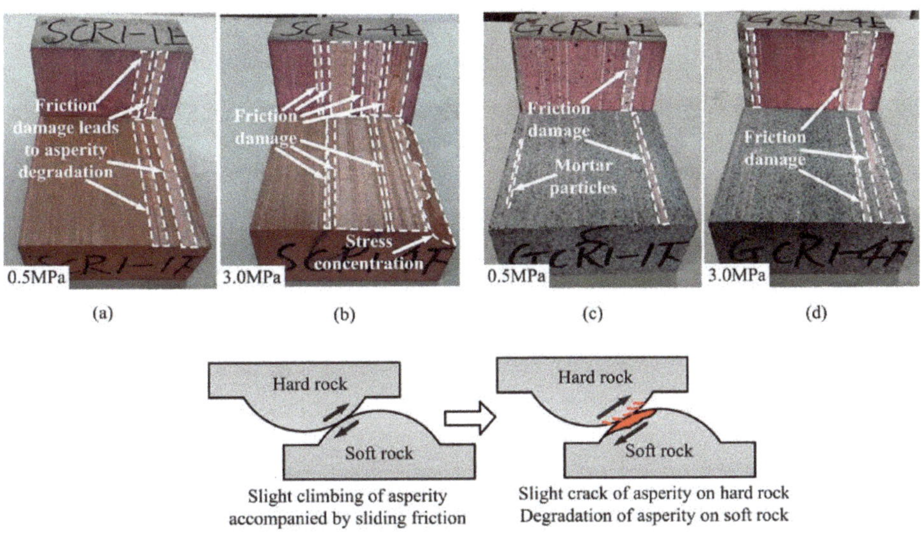

Figure 7. Failure characteristics of specimen SCR1 under normal stress of (**a**) 0.5MPa, (**b**) 3MPa, and specimen GCR1 under normal stress of (**c**) 0.5MPa, (**d**) 3MPa, and (**e**) sketch map for illustrating the failure mechanism of discontinuities possessing different joint wall strengths with JRC of 2.8.

Figure 8 shows the failure characteristics of joints with JRC of 10.8. Both types of discontinuities possessing different joint wall strengths exhibited shear failure after dilation. For sandstone–mortar discontinuities, sandstone and mortar only exhibited relative friction and wear between asperities after dilation under low normal stress, and slight scratches appeared on the surface of sandstone and mortar. Meanwhile, the edge of the sandstone surface was fractured due to stress concentration. With the increase in normal stress, the size of the damaged area increased gradually. Under high normal stress, both the asperities of sandstone and mortar exhibited climbing shear failure, and stress concentration occurred at the edge of mortar. For granite–mortar discontinuities, some asperities on the mortar surface were crushed and cut off under low normal stress, and only a small amount of friction damage appeared on the granite surface. In contrast, under high normal stress, the damage area of granite asperities increased significantly and some asperities were cut off. Meanwhile, the edge of the granite broke due to stress concentration. A great deal of asperities on the surface of cement mortar were cut off. Under the interaction of normal stress and shear stress, the generated moment led to tensile cracks in cement mortar. Figure 8e presents the microscopic failure mechanism of the joint specimen with JRC of 10.8. Discontinuities exhibited obvious dilation during the shearing process and the torque generated during the climbing process led to tensile cracks in both hard rock and soft rock. The shear damage of asperities was deeper due to the lower strength of soft rock.

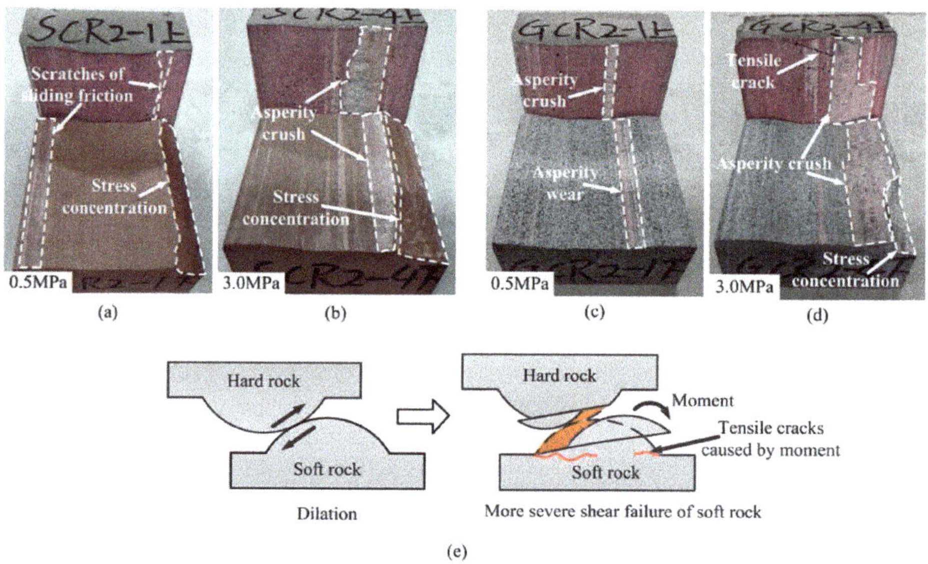

Figure 8. Failure characteristics of specimen SCR2 under normal stress of (**a**) 0.5 MPa, (**b**) 3.0 MPa, and specimen GCR2 under normal stress of (**c**) 0.5 MPa, (**d**) 3.0 MPa, and (**e**) sketch map for illustrating the failure mechanism of discontinuities possessing different joint wall compressive strengths with JRC of 10.8.

Figure 9 presents the failure characteristics of discontinuities possessing different joint strengths with JRC of 18.7. The damage depth and damage area of discontinuities with JRC of 18.7 were greater. Plenty of asperities were cut off on the surface of hard rock and soft rock, and massive fragmented rock particles were produced by extrusion and crushing of asperities on the discontinuity. Under high normal stress, multiple tensile cracks appeared in the mortar, and asperities of mortar were completely chewed off due to the interpenetration of tensile cracks. Figure 9e presents the microscopic failure mechanism of the joint specimen with JRC of 18.7. Under the effect of the torque generated during the shear process, asperities of hard rock were partially sheared due to the higher strength of the hard rock. Meanwhile, tensile cracks inside the soft rock were interpenetrated and asperities of soft rock were completely chewed off.

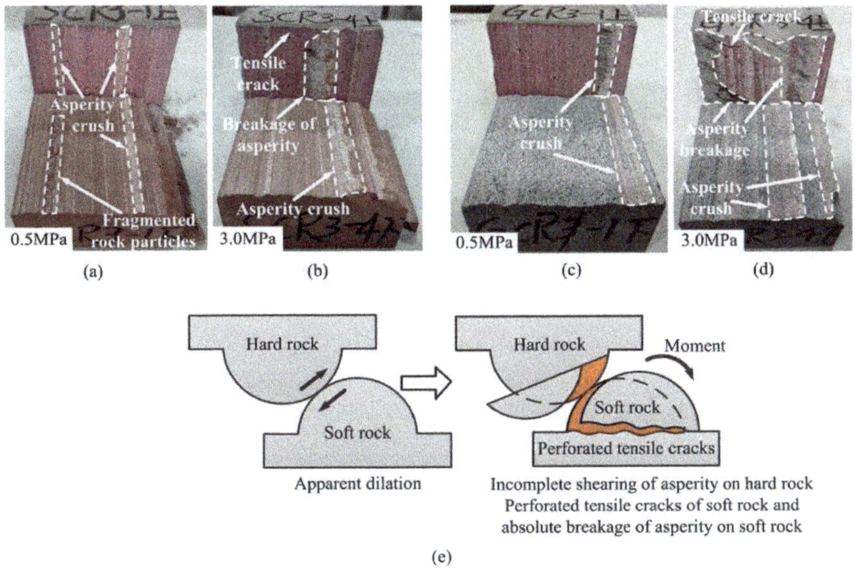

Figure 9. Failure characteristics of specimen SCR3 under normal stress of (**a**) 0.5 MPa, (**b**) 3.0 MPa, and specimen GCR3 under normal stress of (**c**) 0.5 MPa, (**d**) 3.0 MPa, and (**e**) sketch map for illustrating the failure mechanism of discontinuities possessing different joint wall compressive strengths with JRC of 18.7.

It can be realized from the above analysis that discontinuities possessing different joint wall strengths needed to overcome the dilation and friction resistance of the surface asperity under low normal stress, and the damage was mainly manifested as the wear and shear of asperities on the soft rock. The shear strength of discontinuities was mainly influenced by the strength of the soft rock. The upper and lower blocks of the discontinuities came into close contact with the increase in normal stress. Plenty of asperities on the soft rock were sheared and crushed, while asperities of hard rock were partially embedded in the soft rock and tightly engaged with the soft rock subjected to higher normal stress. Moreover, asperities of the hard rock also successively experienced dilation and shear failure under sustained shear effect, which greatly improved the peak shear strength and shear stiffness of the discontinuities. The shear strength of discontinuities possessing different joint wall strengths was mainly governed by the shear strength of the hard rock.

On the basis of the above analysis, the shear process of discontinuities possessing different joint wall strengths can be divided into four stages combined with the shear process of the GCR3 specimen, as illustrated in Figure 10. (a) Stage of tight matching: The upper and lower blocks of the discontinuities possessing different joint wall strengths came into light contact and coupled compaction under the effect of normal stress. (b) Stage of climbing wear of soft rock and crack formation of hard rock: As the shear proceeds, the joint walls of discontinuities dislocate relatively accompanied by dilation. Meanwhile, part of the asperities of the soft rock is worn down and cracks appears in the asperity of the hard rock. (c) Stage of partial shear of soft rock and crack propagation of hard rock: Asperities of the soft rock are partially cut off when the shear stress endured by the soft rock reaches its shear strength. Asperities of the soft rock are completely cut off and cracks in asperities of the hard rock further expands under high normal stress. (d) Stage of entire cutting of asperities: As the shear stress of discontinuities increases continually, asperities of the hard rock are partially cut off. Meanwhile, obvious tensile cracks occurred due to the tensile stress in the soft rock exceeding the tensile strength. Interpenetration of tension cracks causes the soft rock to be completely chewed and crushed; hence, the discontinuity was completely broken. The summary of these four stages is shown in Figure 10e.

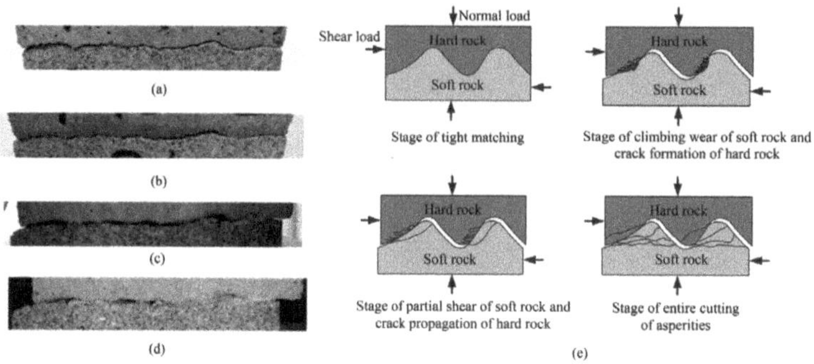

Figure 10. Shear failure process of rough discontinuities possessing different joint wall strengths. (**a**) Stage of tight matching; (**b**) stage of climbing wear of soft rock and crack formation of hard rock; (**c**) stage of partial shear of soft rock and crack propagation of hard rock; (**d**) stage of entire cutting of asperities; (**e**) sketch map of the shearing process.

5. Numerical Simulation of Shear Test of Discontinuities Possessing Different Joint Wall Strengths

5.1. Fundamental Introduction of UDEC

The Universal Distinct Element Code (UDEC) is a two-dimensional block discrete element software for solving discontinuous medium problems, which adopts continuum mechanics theory to analyze rock mechanics problems. Moreover, the UDEC is a calculation program guided by discrete element theory, which satisfies the basic requirements of dealing with engineering problems of a discontinuous medium. For discontinuities of structures (such as structural planes in rock mass, cracks and joints), the UDEC handles them as an internal boundary interface between divided blocks and assigns values to the mechanical parameters of rock joints. The new contact between the joint walls during shearing can also be automatically identified according to the calculation process. As for the divided discrete blocks, the UDEC can handle them as rigid blocks or deformed blocks. The deformed block is covered by the grid element, and the mechanical effect inside the rock block is simulated by the given constitutive criterion between the grids. The deformed block can also generate certain displacement, rotation and deformation along the rock joint. Compared with the traditional numerical software of geotechnical engineering which sets the grid unit as the rigid block, the UDEC is more appropriate for the simulation of mechanical properties of real rock materials.

5.2. Determination of Constitutive Model and Its Parameters

In this study, the Mohr–Coulomb model is adopted to simulate the material of joint blocks. The parameters used for numerical simulation include material density (ρ), bulk modulus (K), shear modulus (G), cohesion (c), internal friction angle (Ø) and tensile strength (T). The bulk modulus (K) and shear modulus (G) are calculated from the following equation,

$$K = \frac{E}{3(1-2V)},\ G = \frac{E}{2(1+V)} \quad (1)$$

where V is the Poisson's ratio, E is the Young's modulus.

The constitutive model of discontinuities possessing different joint wall strengths was a continuously yielding model (CY model). The required parameters for this model include shear stiffness (Ks), normal stiffness (Kn), roughness (R), initial peak friction angle (Øim) and basic friction angle (Øb). The shear stiffness (Ks) and normal stiffness (Kn) of the discontinuities were calculated according to the following equation.

$$K_n = a_n \sigma_n{}^{e_n},\ K_s = a_s \sigma_n{}^{e_s} \quad (2)$$

where a_n and e_n are the parametric factors of the normal stiffness, a_s and e_s are the parametric factors of the shear stiffness, σ_n is the normal stress. Based on the laboratory test results, the normal and shear stiffness parameter factors were calibrated by the trial and error method. The parameters used in numerical simulation are listed in Table 2.

Table 2. Mechanical properties of joints.

Numerical Parameters	a_n	e_n	a_s	e_s	R	$Ø^i{}_m$ (°)	Ø (°)
SCR1	0.815	0.738	0.574	0.832	0.0008	32.530	30.270
SCR2	0.932	0.796	0.865	0.894	0.0023	45.280	30.270
SCR3	1.348	0.951	1.097	0.875	0.0047	46.840	30.270
GCR1	0.946	0.803	0.681	1.073	0.0008	33.190	33.490
GCR2	1.245	0.992	1.157	0.926	0.0023	43.450	33.490
GCR3	1.578	1.179	1.437	0.841	0.0047	44.620	33.490

5.3. Establishment of Numerical Model of Shear Tests

A rectangular block with a size of 100 mm × 40 mm was established, and discontinuities possessing different joint wall strengths with three types of morphology were imported into the middle of the rectangle to construct numerical models corresponding to the joint specimens tested in the laboratory. The initial rectangular block was divided into the triangular mesh element. The smaller the size of the mesh element, the more realistic the shear mechanics regularity of discontinuities. However, the computational load of the computer will also increase, which reduces the computational efficiency of the numerical model. Therefore, in order to elaborately investigate the shear mechanical properties of discontinuities with inclusion of the computational efficiency, the block region within 6 mm from the discontinuity was finely divided into grids with size 1 mm. In contrast, the grid size of the remaining block region was 1.5 mm. With reference to the laboratory shear test, the upper block of the model was set as sandstone or granite, while the lower block of the model was set as mortar. The numerical model of discontinuities possessing different joint wall strengths is shown in Figure 11. The loading scheme in numerical simulation was consistent with laboratory tests. The numerical analysis program of discontinuities was compiled with Fish, which monitored the shear stress, shear displacement and normal displacement.

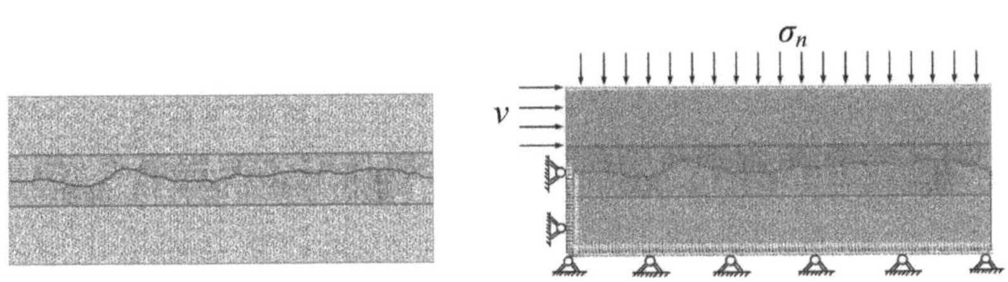

(a) (b)

Figure 11. Numerical model of discontinuities possessing different joint wall strengths. (**a**) Two-dimensional grid model, and (**b**) numerical model with boundary conditions.

6. Results and Analysis of Numerical Simulation

6.1. Numerical Shear Characteristic of Discontinuities Possessing Different Joint Wall Strengths

After each shear simulation was completed, the numerical test data were derived. Figure 12 presents the comparison between the experimental and numerical results of the shear curves of joint specimens.

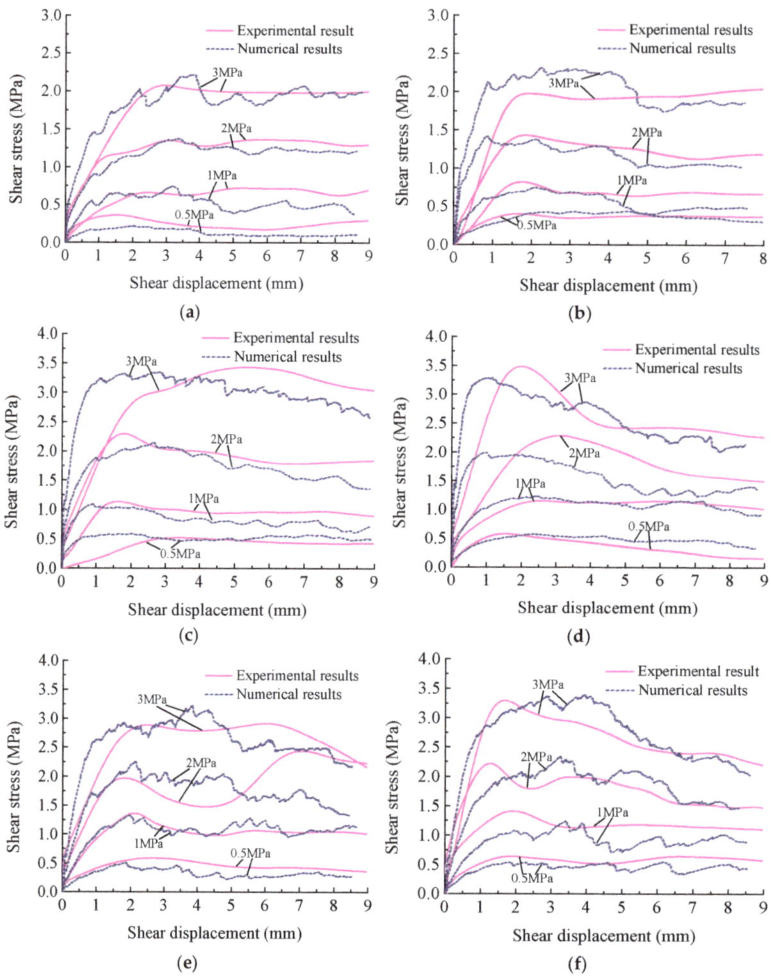

Figure 12. Comparison results of numerical curve and test curve of discontinuities possessing different joint wall strengths. (**a**) GCR1; (**b**) SCR1; (**c**) GCR2; (**d**) SCR2; (**e**) GCR3; and (**f**) SCR3.

It is visible from Figure 12 that the numerical results are basically consistent with the experimental results. Both the test curve and the numerical curve had obvious peak values and showed a progressive failure process. The peak shear strength, residual shear strength and shear stiffness in numerical results increased significantly with the gradual increase in normal stress. Under high normal stress, the shear stress of discontinuities with JRC of 18.7 decreased steeply after reaching the peak value, which means the brittle failure was caused by the great difference in strength between the hard rock and the soft rock. That is to say, for discontinuities possessing different joint wall strengths with JRC of 18.7, the convex structure of the discontinuity would suddenly break under the effect of high normal stress. Therefore, the overall failure mode of discontinuity was manifested as brittle shear failure, which was consistent with the test results. It can be seen that the numerical results fully verified the accuracy of the test results, and further that the CY model in UDEC is completely applicable to the shear numerical simulation of discontinuities possessing different joint wall strengths. Figure 13 presents the shear stress contours of the discontinuities possessing different joint wall strengths when the shear displacement reached 5 mm.

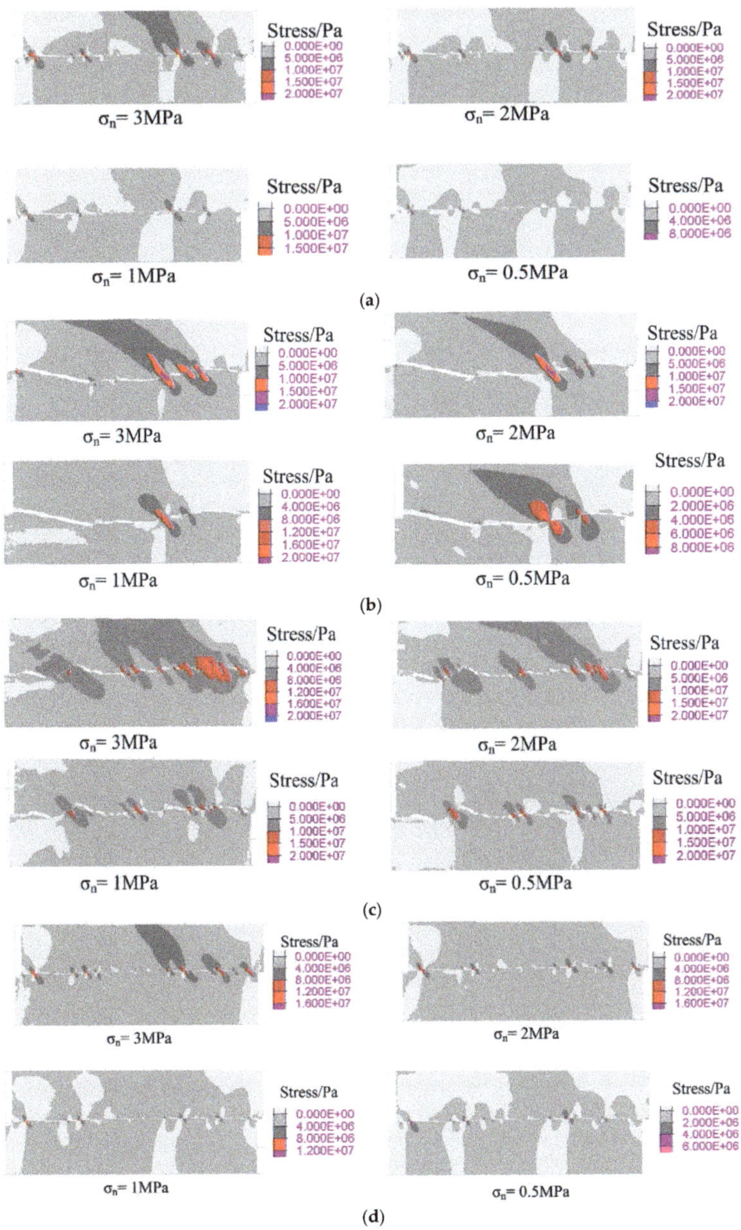

Figure 13. *Cont.*

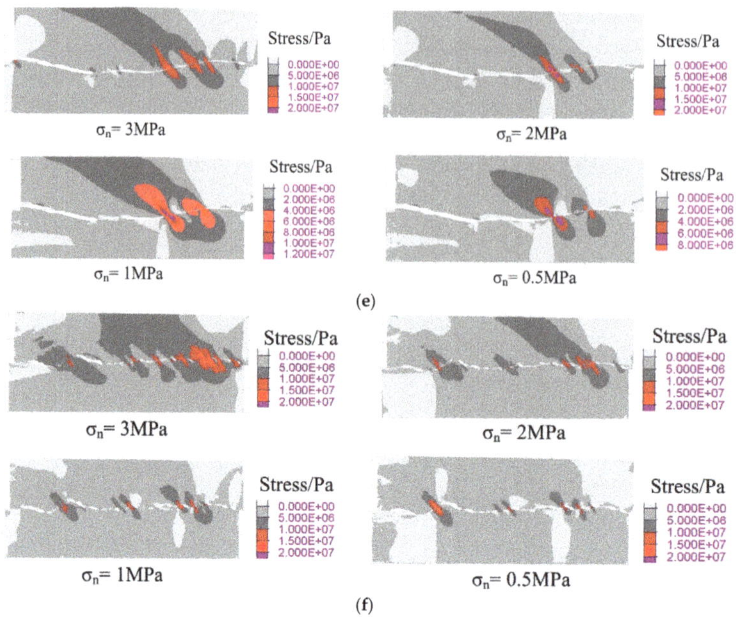

Figure 13. Shear stress contours of discontinuities possessing different joint wall strengths with shear displacement of 5 mm. (**a**) GCR1; (**b**) GCR2; (**c**) GCR3; (**d**) SCR1; (**e**) SCR2; and (**f**) SCR3.

It is visible from Figure 13 that in the case of equal roughness, the shear stress of discontinuities increased with the increase in normal stress. In the case of the same normal stress, the shear stress endured by discontinuities with large roughness was significantly greater than that of discontinuities with small roughness. Simultaneously, the shear stress endured by discontinuities was mostly concentrated on the asperity facing the shear direction. Comparing with the test results, the friction damage and cutting phenomena of asperities mostly occurred in the morphology facing the shear direction. Meanwhile, it can be realized from the shear stress contour that the shear stress endured by granite–mortar discontinuities was greater than shear stress endured by sandstone–mortar discontinuities under the same normal stress. This is owing to the strength difference between granite and mortar being greater, and the embedding effect and furrow effect of asperity being more significant in the shear process. Therefore, the shear stress endured by granite–mortar discontinuities increased obviously.

6.2. Dilation Law of Discontinuities Possessing Different Joint Wall Strengths

The numerical dilation data of discontinuities possessing different joint wall compressive strengths were derived, and curves of normal displacement vs. shear displacement were plotted, as shown in Figure 14.

The dilation displacement of discontinuities increased with the increase in shear displacement. In the initial stage of the numerical test, the dilation displacement increased rapidly. However, there was a turning point of increasing rate when the peak shear displacement was reached, the increasing rate of dilation displacement decreased and the dilation curve tended to be gentle. Dilation displacement eventually reached a stable value with the continuous increasing of shear displacement. This was because asperities of discontinuities were cut off in quantities after the shear stress reached the peak value. The dilation effect was significantly weakened until asperities entered the grinding and migration stage, and the dilation displacement then tended to be stable. In conclusion, the dilation displacement of discontinuities was concerned with normal stress. With the gradual increase in normal stress, the cutting effect between asperities of discontinuities was significantly enhanced, thus weakening the climbing dilation effect between asperities and decreasing the dilation displacement gradually. In addition, the increasing rate of dilation displacement also gradually decreased. Dilation values (mm) of discontinuities possessing different joint wall strengths with different morphologies are listed in Table 3.

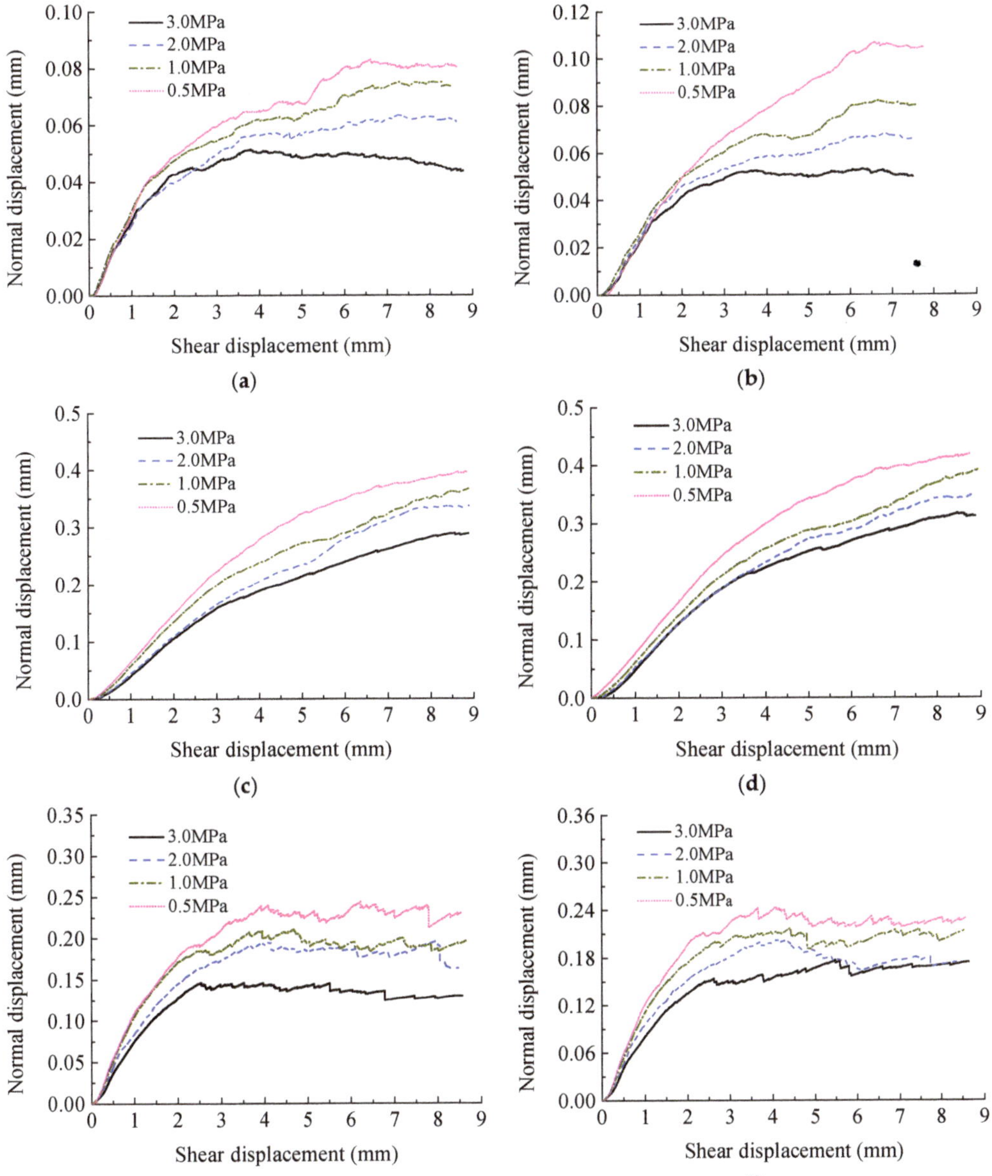

Figure 14. Shear dilation curves of discontinuities possessing different joint wall strengths. (**a**) GCR1; (**b**) SCR1; (**c**) GCR2; (**d**) SCR2; (**e**) GCR3; and (**f**) SCR3.

Table 3. Dilation displacement of discontinuities possessing different joint wall strengths (mm).

σ_n (MPa)	SCR1	SCR2	SCR3	GCR1	GCR2	GCR3
0.5	0.107	0.419	0.244	0.083	0.397	0.243
1	0.082	0.391	0.217	0.060	0.367	0.211
2	0.068	0.353	0.204	0.056	0.338	0.196
3	0.053	0.317	0.177	0.051	0.289	0.147

It is visible from Table 3 that under the effect of the same normal stress, the dilation displacement of discontinuities with JRC of 10.8 was the maximum, the dilation displacement of discontinuities with JRC of 18.7 was second, while the dilation displacement of discontinuities with JRC of 2.8 was the minimum. The reason is that the undulation of discontinuities with JRC of 10.8 was the largest; thus, the climbing effect was the most significant correspondingly. As for discontinuities with JRC of 2.8, the number of rough construction was the least, which is why the dilation effect was basically weakened and the dilation displacement was the minimum. Comparing two types of discontinuities possessing different joint wall compressive strengths with the same roughness, asperities of granite had more obvious embedding effects on mortar due to the compressive strength of granite being approximately three times that of sandstone. Therefore, the damage degree of granite–mortar discontinuities during the shearing process was more severe as a result of the meshing and shearing effects of asperities being greatly enhanced, which weakened the climbing dilation effect correspondingly. Eventually, the dilation value of granite–mortar discontinuities was smaller than that of sandstone–mortar discontinuities.

6.3. Shear and Dilation Properties of Rock–Mortar Discontinuities and Rock–Rock Discontinuities

In order to further investigate the difference of shear properties between discontinuities possessing identical joint wall strengths and discontinuities possessing different joint wall strengths, numerical shear tests of granite discontinuities, sandstone discontinuities and mortar discontinuities with JRC of 10.8 were performed. The identifier of granite discontinuities was set as GGR2, the identifier of sandstone discontinuities was set as SSR2 and the identifier of mortar discontinuities was set as CCR2. Shear curves of discontinuities possessing identical joint wall strengths and discontinuities possessing different joint wall strengths under different normal stresses are shown in Figures 15 and 16.

After comparing the numerical results of discontinuities possessing identical joint wall strengths and discontinuities possessing different joint wall strengths, it can be realized that the peak shear strength and residual shear strength of granite–mortar discontinuities were between those of granite discontinuities and mortar discontinuities. Meanwhile, the peak shear strength and residual shear strength of sandstone–mortar discontinuities were also between those of sandstone discontinuities and mortar discontinuities. Therefore, it can be concluded that the peak shear strength and residual shear strength of discontinuities possessing different joint wall strengths are between those of discontinuities possessing identical joint wall compressive strengths composed of hard rock and discontinuities possessing identical joint wall compressive strengths composed of soft rock. Comparing the numerical results of three types of discontinuities possessing identical joint wall compressive strengths, the results showed that the peak shear strength of granite discontinuities was greater than that of sandstone discontinuities, and the peak shear strength of mortar discontinuities was the smallest. Moreover, the residual shear strength of the three types of discontinuities possessing identical joint wall strengths had identical regularity. Figures 17 and 18 present the dilation curves of the discontinuities possessing identical joint wall strengths and discontinuities possessing different joint wall strengths.

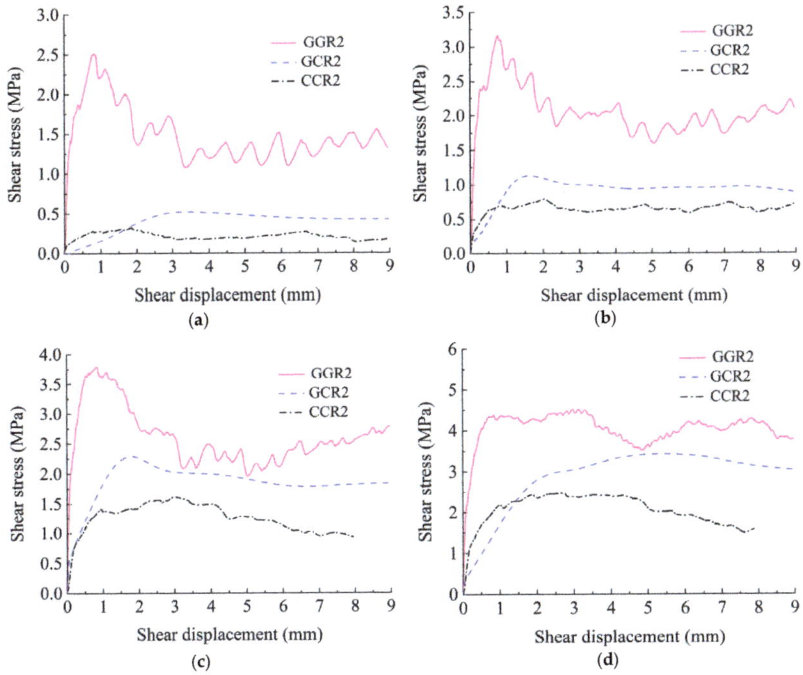

Figure 15. Shear curves of GGR2, GCR2 and CCR2 under normal stresses are (**a**) 0.5 MPa, (**b**) 1.0 MPa, (**c**) 2.0 MPa and (**d**) 3.0 MPa.

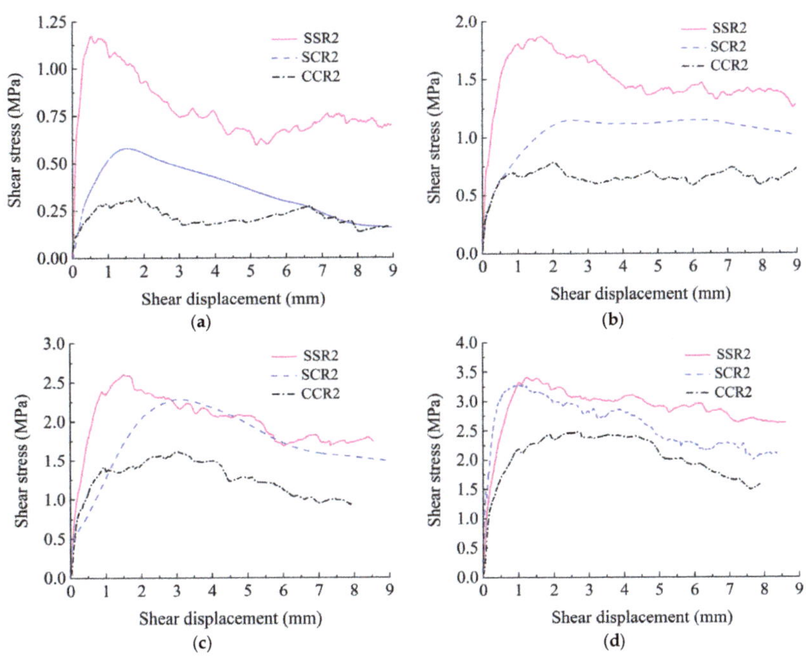

Figure 16. Shear curves of SSR2, SCR2 and CCR2 under normal stresses are (**a**) 0.5 MPa, (**b**) 1.0 MPa, (**c**) 2.0 MPa and (**d**) 3.0 MPa.

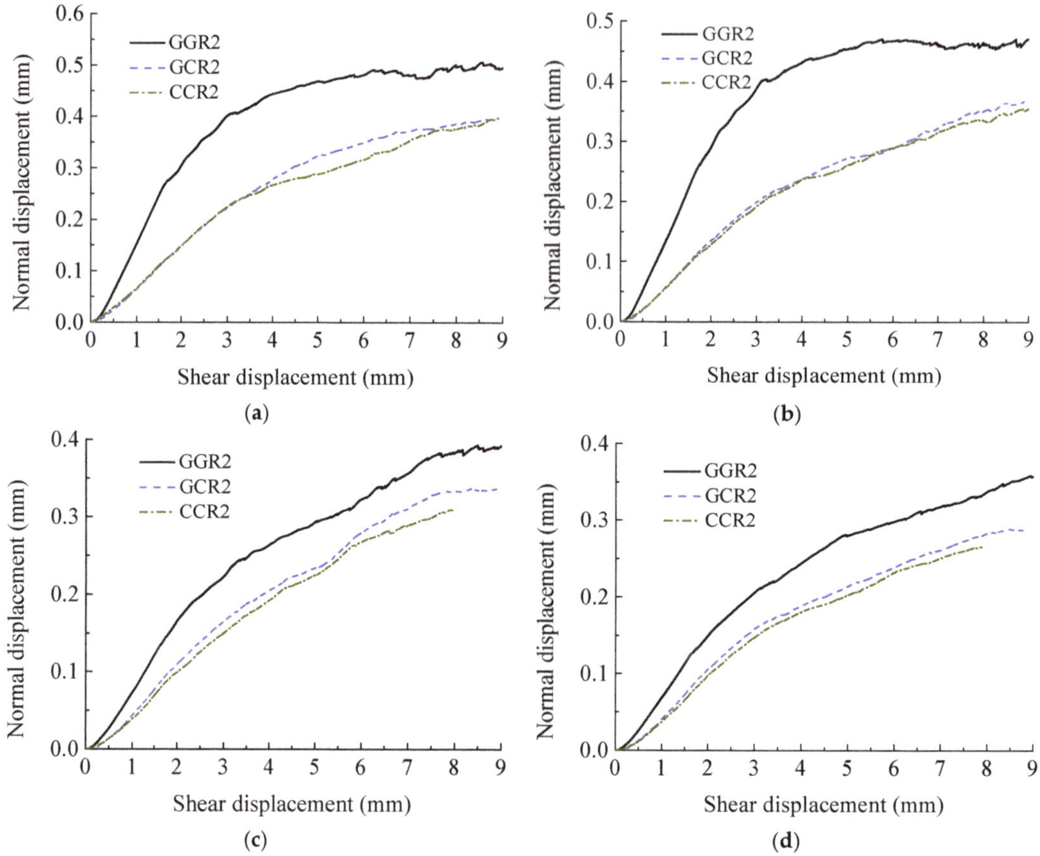

Figure 17. Dilation curves of GGR2, GCR2 and CCR2 under normal stresses are (**a**) 0.5 MPa, (**b**) 1.0 MPa (**c**) 2.0 MPa and (**d**) 3.0 MPa.

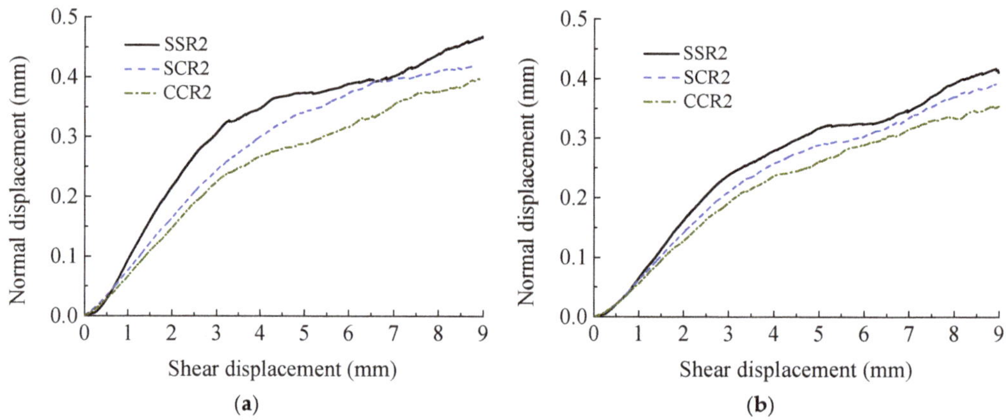

Figure 18. *Cont.*

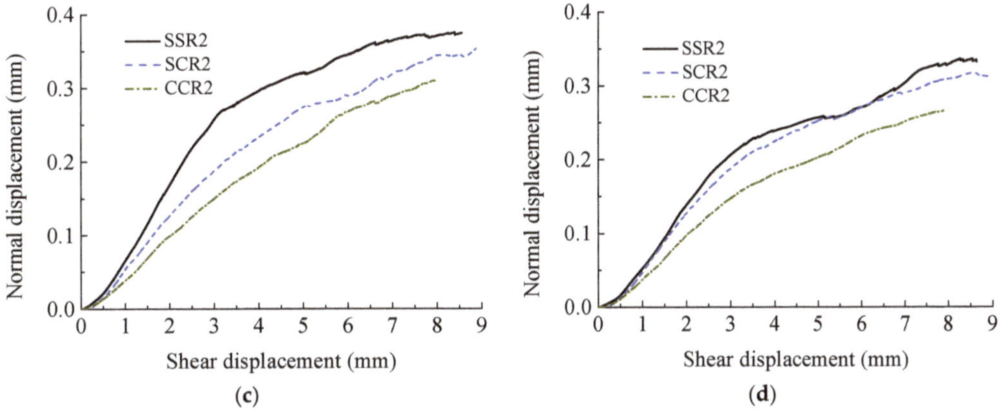

Figure 18. Dilation curves of SSR2, SCR2 and CCR2 under normal stresses are (**a**) 0.5 MPa, (**b**) 1.0 MPa, (**c**) 2.0 MPa and (**d**) 3.0 MPa.

Figure 17 shows that the dilation displacement of granite–mortar discontinuities was between that of granite discontinuities and mortar discontinuities. Figure 18 shows that the dilation displacement of sandstone–cement mortar discontinuities was between sandstone discontinuities and mortar discontinuities. Simultaneously, the climbing dilation effect was strengthened due to the increase in joint wall strength on both sides of discontinuities possessing identical joint wall compressive strengths. Therefore, the dilation displacement of granite discontinuities was the largest, the dilation displacement of sandstone discontinuities was second and the dilation displacement of mortar discontinuities was the smallest. The specific dilation values of rock–rock discontinuities and rock–mortar discontinuities are listed in Table 4.

Table 4. Dilation values of rock–rock discontinuities and rock–mortar discontinuities (mm).

Normal Stress (MPa)	The Identifier of Specimens				
	SSR2	SCR2	CCR2	GCR2	GGR2
0.5	0.468	0.419	0.398	0.397	0.506
1	0.418	0.391	0.355	0.367	0.471
2	0.376	0.353	0.311	0.338	0.398
3	0.337	0.317	0.265	0.289	0.362

7. Conclusions

(1) Three-dimensional engraving technology can effectively fabricate joint specimens that are applicable to laboratory tests. The joint specimen fabricated by this method has the morphological characteristics of being highly consistent with the prototype, which is available to analyze the shear properties and morphological characteristics of the rock joint with different joint wall strengths made by various rock and rock-like materials.

(2) The shear curve of the numerical simulation is basically consistent with the test curve, which verifies the accuracy of the CY constitutive model in simulating progressive failure of discontinuities possessing different joint wall strengths. Integrating the shear stress contour of the numerical simulation and the test results to analyze, the damage area in the shearing process of discontinuities possessing different joint wall strengths is mainly concentrated on the asperity facing the shear direction. The shear wear and cutting destruction of discontinuities possessing different joint wall strengths are mainly distributed in the soft rock.

(3) The dilation displacement of discontinuities possessing different joint wall compressive strengths decreases with the increase in normal stress. Due to the cooperative effect of roughness and undulation, the dilation displacement of discontinuities possessing different joint wall strengths with JRC of 10.8 is the largest, discontinuities possessing different joint wall strengths with JRC of 18.7 is second and discontinuities possessing different joint wall strengths with JRC of 2.8 is the smallest. The embedding effect of asperities is enhanced and the climbing dilation effect is weakened as a result of the great difference in joint wall strength on both sides of granite–mortar discontinuities. Therefore, the dilation displacement of granite–mortar discontinuities is smaller compared with sandstone–mortar discontinuities.

(4) On basis of the numerical results, the peak shear strength and residual shear strength of discontinuities possessing different joint wall strengths are between those of discontinuities possessing identical joint wall strengths composed of hard rock and discontinuities possessing identical joint wall strengths composed of soft rock. Meanwhile, the dilation displacement of discontinuities possessing different joint wall strengths is also between discontinuities possessing identical joint wall strengths composed of hard rock and discontinuities possessing identical joint wall strengths composed of soft rock.

Author Contributions: All authors made substantial contributions to the conception and design of the study. X.Z.: conceptualization, methodology, funding acquisition and writing—review & editing. S.O.: investigation, data curation and writing original draft. C.F.: methodology and writing—review & editing. L.Y.: data curation and writing—review & editing. D.L.: supervision, methodology and writing—review & editing. All authors have read and agreed to the published version of the manuscript.

Funding: Open Foundation of Jiangxi Hydraulic Safety Engineering Technology Research Center, Jiangxi Academy of Water Science and Engineering (Grant No. 2022SKSG06), Young Elite Scientists Sponsorship Program by JXAST (Grant No. 2023QT06), Jiangxi Provincial Natural Science Foundation for Distinguished Young Scholars (Grant No. 20242BAB23047), National Natural Science Foundation of China (Grant Nos. 52369019).

Institutional Review Board Statement: Not applicable.

Informed Consent Statement: Not applicable.

Data Availability Statement: The data presented in this study are available on request from the corresponding author. The data is not publicly available due to privacy.

Conflicts of Interest: The authors declare no conflicts of interest.

References

1. Liu, G.; Yang, Y. Deep Soft Rock Tunnel Perimeter Rock Control Technology and Research. *Appl. Sci.* **2025**, *15*, 278. [CrossRef]
2. Liu, D.; Yan, W.; Yan, S.; Kang, Q. Study on the effect of axial and hydraulic pressure coupling on the creep behaviors of sandstone under multi-loading. *Bull. Eng. Geol. Environ.* **2021**, *80*, 6107–6120. [CrossRef]
3. Dang, W.; Tao, K.; Fu, J.; Wu, B. Experimental investigation on unloading-induced sliding behavior of dry sands subjected to constant shear force. *Appl. Sci.* **2025**, *15*, 401. [CrossRef]
4. Xie, S.; Jiang, Z.; Lin, H.; Ma, T.; Peng, K.; Liu, H.; Liu, B. A new integrated intelligent computing paradigm for predicting joints shear strength. *Geosci. Front.* **2024**, *15*, 40111884. [CrossRef]
5. Zhang, X.; Cao, Z.; Yao, C.; Yang, J.; Ma, Y.; Ye, Z.; Zhou, C. Effect of nonstationarity on shear properties of rock joints and a new peak shear strength criterion. *Int. J. Geomech.* **2023**, *23*, 04023213. [CrossRef]
6. Tang, Z.C.; Zhang, Z.F.; Zuo, C.Q.; Jiao, Y.Y. Peak shear strength criterion for mismatched rock joints: Revisiting JRC-JMC criterion. *Int. J. Rock Mech. Min.* **2021**, *147*, 104894. [CrossRef]

7. Jiang, F.; Wang, G.; He, P.; Hou, B.; Zhang, S.B.; Sun, S.Q.; Zheng, C.C.; Wu, Y. Mechanical failure analysis during direct shear of double-joint rock mass. *Bull. Eng. Geol. Environ.* **2022**, *81*, 410. [CrossRef]
8. Pirzada, M.A.; Bahaaddini, M.; Moradian, O.; Roshan, H. Evolution of contact area and aperture during the shearing process of natural rock fractures. *Eng. Geol.* **2021**, *291*, 106236. [CrossRef]
9. Zhang, X.B.; Yi, L.; Yao, C.; Ma, Y.L.; Yang, J.H.; Jiang, Q.H.; Zhou, C.B. Comparative study on shear mechanical properties and failure mechanisms of rough joints under CNL and DNL boundary conditions. *Chin. J. Rock Mech. Eng.* **2025**, *44*, 865–880. (In Chinese)
10. Pratomo, F.Y.; Wei, X.; Zou, C.; Zhao, G.F. Investigation of the shear failure of rock joints using the four-dimensional lattice spring model. *Int. J. Rock Mech. Min.* **2022**, *152*, 105070. [CrossRef]
11. Barton, N.; Choubey, V. The shear strength of rock joints in theory and practice. *Rock Mech. Rock Eng.* **1977**, *10*, 1–54. [CrossRef]
12. Grasselli, G.; Egger, P. Constitutive law for the shear strength of rock joints based on three-dimensional surface parameters. *Int. J. Rock Mech. Min.* **2003**, *40*, 25–40. [CrossRef]
13. Moradian, Z.A.; Ballivy, G.; Rivard, P.; Gravel, C.; Rousseau, B. Evaluating damage during shear tests of rock joints using acoustic emissions. *Int. J. Rock Mech. Min.* **2010**, *47*, 590–598. [CrossRef]
14. Asadi, M.S.; Rasouli, V.; Barla, G. A laboratory shear cell used for simulation of shear strength and asperity degradation of rough rock fractures. *Rock Mech. Rock Eng.* **2013**, *46*, 683–699. [CrossRef]
15. Oliveira, D.; Indraratna, B. Comparison between models of rock discontinuity strength and deformation. *J. Geotech. Geoenviron.* **2010**, *136*, 864–874. [CrossRef]
16. Hossaini, K.A.; Babanouri, N.; Nasab, S.K. The influence of asperity deformability on the mechanical behavior of rock joints. *Int. J. Rock Mech. Min.* **2014**, *70*, 154–161. [CrossRef]
17. Atapour, H.; Moosavi, M. The influence of shearing velocity on shear behavior of artificial joints. *Rock Mech. Rock Eng.* **2014**, *47*, 1745–1761. [CrossRef]
18. Rullière, A.; Rivard, P.; Peyras, L.; Breul, P. Influence of roughness on the apparent cohesion of rock joints at low normal stresses. *J. Geotech. Geoenviron.* **2020**, *146*, 04020003.1–04020003.15. [CrossRef]
19. Rullière, A.; Rivard, P.; Peyras, L.; Breul, P. Influence of material strength on the apparent cohesion of unbounded gravity dam joints under low normal stress. *J. Geotech. Geoenviron.* **2021**, *147*, 06021012.1–06021012.8. [CrossRef]
20. Tang, Z.C.; Wong, L.N.Y. New criterion for evaluating the peak shear strength of rock joints under different contact states. *Rock Mech. Rock Eng.* **2016**, *49*, 1191–1199. [CrossRef]
21. Zhang, X.; Jiang, Q.; Chen, N.; Wei, W.; Feng, X. Laboratory investigation on shear behavior of rock joints and a new peak shear strength criterion. *Rock Mech. Rock Eng.* **2016**, *49*, 3495–3512. [CrossRef]
22. Ban, L.; Qi, C.; Chen, H.; Yan, F.; Ji, C. A new criterion for peak shear strength of rock joints with a 3D roughness parameter. *Rock Mech. Rock Eng.* **2023**, *53*, 1755–1775. [CrossRef]
23. Yang, J.; Xia, Y.; Chen, W.; Zhang, L.; Li, L. Shear behavior of silty clay–concrete interface based on large-scale direct shear test. *Int. J. Geomech.* **2023**, *23*, 04023084. [CrossRef]
24. Lin, H.; Wang, M.; Xu, W.Z. Shear behaviors of the binary structural plane. *Geotech. Geol. Eng.* **2018**, *36*, 939–948. [CrossRef]
25. Jiang, Y.; Wang, Y.; Yan, P.; Luan, H.; Chen, Y. Experimental investigation on the shear properties of heterogeneous discontinuities. *Geotech. Geol. Eng.* **2019**, *37*, 4959–4968. [CrossRef]
26. Lin, H.; Han, L.; Wang, Y.; Cao, R.; Zhao, Y.; Jiang, C. Influence of the micro-deformation characteristics of binary media on the shear behavior of structural plane. *Geotech. Geol. Eng.* **2020**, *39*, 347–358. [CrossRef]
27. Wu, Q.; Jiang, Y.; Tang, H.; Luo, H.; Wang, X.; Kang, J.; Zhang, S.; Yi, X.; Fan, L. Experimental and numerical studies on the evolution of shear behaviour and damage of natural discontinuities at the interface between different rock types. *Rock Mech. Rock Eng.* **2020**, *53*, 3721–3744. [CrossRef]
28. Ghazvinian, A.H.; Taghichian, A.; Hashemi, M.; Mar'Ashi, S.A. The shear behavior of bedding planes of weakness between two different rock types with high strength difference. *Rock Mech. Rock Eng.* **2010**, *43*, 69–87. [CrossRef]
29. Wu, Q.; Xu, Y.; Tang, H.; Fang, K.; Jiang, Y.; Liu, C.; Wang, X. Peak shear strength prediction for discontinuities between two different rock types using neural network approach. *Bull. Eng. Geol. Environ.* **2019**, *78*, 2315–2329. [CrossRef]
30. Tang, Z.C.; Zhang, Z.F.; Jiao, Y.Y. Three-dimensional criterion for predicting peak shear strength of matched discontinuities with different joint wall strengths. *Rock Mech. Rock Eng.* **2021**, *54*, 3291–3307. [CrossRef]
31. Zhang, X.; Zhu, X.; Yao, C.; Ouyang, S.; Yang, J.; Jiang, Q. Investigation on the shear behavior of rough rock joints prepared by three- dimensional engraving technique. *J. Test. Eval.* **2022**, *50*, 1425–1440. [CrossRef]
32. Jiang, Q.; Feng, X.; Gong, Y.; Song, L.; Ran, S.; Cui, J. Reverse modelling of natural rock joints using 3D scanning and 3D printing. *Comput. Geotech.* **2016**, *73*, 210–220. [CrossRef]

33. Jaber, J.; Conin, M.; Deck, O.; Moumni, M.; Godard, O.; Kenzari, S. Investigation of the mechanical behavior of 3D printed polyamide-12 joints for reduced scale models of rock mass. *Rock Mech. Rock Eng.* **2020**, *53*, 2687–2705. [CrossRef]
34. Li, Y.; Zhang, Y. Quantitative estimation of joint roughness coefficient using statistical parameters. *Int. J. Rock Mech. Min.* **2015**, *77*, 27–35. [CrossRef]

applied sciences

MDPI

Article

Application of a Dynamic Step Size Iterative Method for Parameter Inversion in the Unified Hardening Models

Binglong Zhu [1,2,3], Degou Cai [3,4,*], Hongye Yan [2,3], Zongqi Bi [2,3], Mingzhe Ouyang [2,3] and Junkai Yao [2,3]

1 China Academy of Railway Sciences Corporation Limited, Beijing 100081, China; blzhucars@163.com
2 Railway Engineering Research Institute, China Academy of Railway Sciences Corporation Limited, Beijing 100081, China; yanhongye_2005@163.com (H.Y.); bizongqitky@163.com (Z.B.); ouyangmingzhe_hit@163.com (M.O.); yao_junkai@126.com (J.Y.)
3 State Key Laboratory of Highspeed Railway Track System, China Academy of Railway Sciences Corporation Limited, Beijing 100081, China
4 Infrastructure Inspection Research Institute, China Academy of Railway Sciences Corporation Limited, Beijing 100081, China
* Correspondence: caidegou@126.com

Abstract: The unified hardening model for clays and sands (CSUH) can adequately represent the stress–strain characteristics of various soil types. However, being an incremental elastoplastic constitutive model, the CSUH model requires extensive iterative computations during parameter identification, resulting in significant computational time. To improve computational efficiency, this study derives the elastoplastic compliance matrix and stress–strain incremental relationships under different stress paths, eliminating the repeated solving of equations typically required during iterative processes. Furthermore, a dynamic step size iterative method is proposed based on the changing slope characteristics of the stress–strain curves. This method divides the total axial strain into two segments: in the initial segment (approximately the first 30% of total strain), where the curve slope is steep, smaller step sizes with arithmetic progression distribution are employed, while in the latter segment (approximately the remaining 70%), characterized by a gentle curve slope, larger and uniformly distributed step sizes are adopted. Comparative analyses between the proposed dynamic step size method and the traditional constant-step iterative method demonstrate that, under the premise of ensuring calculation accuracy, the dynamic step size method significantly reduces the iteration steps from 3000 to 50, thus decreasing the computational time by approximately 47 times. Finally, the proposed method is applied to parameter identification of Fujinomori clay, calcareous sand, and Changhe dam rockfill materials using the CSUH model. The predictions closely match experimental results, confirming the CSUH model's capability in accurately describing the mechanical behaviors of different soils under various stress paths. The dynamic step size iterative approach developed in this study also provides valuable insights for enhancing computational efficiency and parameter identification of other elastoplastic constitutive models.

Keywords: CSUH model; dynamic step size; flexibility matrix; incremental relationship; soil mechanical properties

1. Introduction

The constitutive relationship of soil, also known as the constitutive model, focuses on the stress–strain behavior of soil, providing a mathematical expression that reflects the complex stress–strain characteristics of soil. It serves as a fundamental theoretical basis in geotechnical engineering and has significant applications in dam construction, railways, and other engineering projects [1–4]. The Modified Cam-Clay (MCC) model [5,6], a classical elastoplastic constitutive framework, has been widely successful in capturing the mechanical behavior of normally consolidated clay. However, the MCC model shows limitations in capturing the behavior of overconsolidated clays, such as abrupt transitions in stress–strain curves between elastic and plastic responses and an overestimation of peak strength [7]. To better characterize the strain softening and dilatancy behavior of overconsolidated saturated clay, Yao et al. [8–10] introduced the Unified Hardening (UH) model based on the MCC model. By modifying only the hardening parameter of the MCC model, the UH model can uniformly describe strain hardening and softening, as well as contraction and dilatancy, in clays with different densities (both normally and overconsolidated clays), demonstrating broad applicability.

In engineering practice, not only clays but also granular soils such as sands, rockfill materials, and railway ballast are encountered. However, the UH model, which is designed for clays, cannot adequately describe the stress–strain behavior of granular soils. To address this, Yao et al. [11] further developed the Unified Hardening model for Clays and Sands (CSUH) based on the UH model. The CSUH model is capable of describing both clays and granular soils and can degenerate into the UH and MCC models, representing a theoretical enhancement and innovative development of the MCC model [12].

Before applying the CSUH model to laboratory experiments and engineering simulations, it is essential to accurately calibrate its parameters. When using inverse methods for parameter calibration, it is necessary to compute the error between the theoretical predictions of the constitutive model and the experimental observations, which requires numerical integration to obtain the theoretical response curves. POPP et al. [13] imported experimental data into Isight and used its optimization algorithms to automatically adjust the model parameters in the finite element software ABAQUS 2025, thereby minimizing the error between the simulated and experimental stress–strain curves for parameter calibration. In recent years, adaptive step-size control in the field of finite element computational methods has received considerable attention in the field of computational mechanics. For instance, Sloan et al. [14,15] proposed an adaptive step-size scheme based on local truncation error estimation, where each substep is computed using both a formula of order p and another of order $p + 1$. The difference between the results is used to estimate the local error, which then guides the automatic adjustment of the step size. Such methods are known for their robustness and general applicability, especially in finite element integration where the shape of the stress–strain curve is not known a priori. However, these approaches typically require multiple computations and error analyses at each step, resulting in increased computational effort. Other adaptive schemes, such as embedded Runge–Kutta or arc-length methods, are also widely used in related fields.

An alternative approach for computing theoretical constitutive model curves during parameter inversion is explicit iteration [12,16]. Common iterative methods include the forward Euler method and the fourth-order Runge–Kutta method. The fourth-order Runge–Kutta method provides high computational accuracy and stability, but it is computationally expensive. In contrast, the forward Euler method is simple to implement, yet only first-order accurate; when the curves exhibit steep slopes or strong nonlinearities, significant integration errors may occur unless a very small step size is used. Consequently, improving accuracy with the forward Euler method typically requires increasing the number of iterations, which reduces computational efficiency in practical applications.

Among these approaches, the uniform-step forward Euler method is the most commonly used. However, its limited accuracy often necessitates finer discretization, further increasing computational costs. For instance, Yin et al. [17] developed the ErosOpt software for parameter calibration, which can conveniently call the user material subroutine (UMAT) in ABAQUS to compute theoretical response curves for various constitutive models. Nevertheless, because ErosOpt employs the uniform-step forward Euler method and requires assembling the element matrix and solving the system of equations based on boundary conditions at each incremental step, the process consumes significant computational resources, resulting in relatively slow parameter inversion.

Soil stress–strain curves typically exhibit nonlinear characteristics, with significant variations in slope at different stages. Existing methods for solving theoretical constitutive model curves primarily rely on FEM or explicit iteration methods with uniform step sizes, which do not fully consider the influence of stress–strain curve slope variations on the required iteration step size. To improve the computational efficiency of the CSUH model, this study proposes a dynamic iteration method with adaptive step size adjustment. This method adjusts iteration step sizes according to variations in the stress–strain curve slope, significantly reducing the number of iterations while maintaining computational accuracy.

This paper first introduces the flexibility matrix of general elastoplastic constitutive models and presents the stress–strain incremental iteration equations under different stress paths. Subsequently, the fundamental theoretical framework of the CSUH model is outlined, followed by a detailed explanation of the proposed dynamic iteration method. The effectiveness of the dynamic step size iteration method is verified through comparisons with traditional uniform step size iteration methods. Finally, this method is employed for parameter inversion of the CSUH model using Fujinomori clay, calcareous sand, and Changhe dam rockfill material. The prediction results demonstrate that the CSUH model accurately describes the mechanical behavior of different soil types under various stress paths. The findings of this study provide valuable insights for improving the computational efficiency of the CSUH model and other elastoplastic constitutive models.

2. Elastoplastic Constitutive Relationship Theory

2.1. General Elastoplastic Constitutive Relationship

In the elastoplastic constitutive relationship of soil, the elastic part is generally described by Hooke's Law, while the plastic part consists of the yield condition, flow rule, and hardening law.

The yield condition can be quantitatively described using the yield surface equation, f, which determines the stress condition at which plastic deformation begins. The yield surface equation, f, is a function of stress, σ, and the hardening parameter, H:

$$f(\sigma, H) = 0 \tag{1}$$

The flow rule can be quantitatively described using the plastic potential surface equation g, which defines the direction of plastic strain increments. The plastic potential surface equation, g, is generally expressed as follows:

$$g(\sigma, C) = 0 \tag{2}$$

where C is a constant.

The hardening law essentially determines the hardening parameter, H, which governs the magnitude of plastic strain increments. The hardening parameter, H, is generally a function of plastic strain:

$$H(\varepsilon^{\mathrm{p}}) \tag{3}$$

where ε^{p} represents the plastic strain. Different constitutive models adopt different definitions of plastic strain. For example, the MCC model uses the plastic volumetric strain, $\varepsilon_{\mathrm{v}}^{\mathrm{p}}$, as the hardening parameter, i.e., $H(\varepsilon^{\mathrm{p}}) = \varepsilon_{\mathrm{v}}^{\mathrm{p}}$. The CSUH model, on the other hand, adopts a plastic volumetric strain incorporating a stress ratio coefficient as the hardening parameter, expressed as $H\left(\varepsilon_V^p\right) = \int \frac{\left(M_Y^4 - \eta^4\right)}{\left(M_c^4 - \eta^4\right)} d\varepsilon_V^p$, where η represents the stress ratio, while M_Y^4 and M_c^4 denote the potential failure stress ratio and characteristic state stress ratio, respectively. Further details will be introduced later.

2.2. Elastoplastic Stress–Strain Relationship

According to the classical small-strain elastoplastic theory, the total strain, ε, is decomposed into an elastic strain part, ε^e, and a plastic strain part, ε^p. The elastic strain is calculated using Hooke's Law, while the plastic strain is determined using plasticity theory. When adopting an incremental formulation, we have the following:

$$\{d\varepsilon\} = \{d\varepsilon^e\} + \{d\varepsilon^p\} \tag{4}$$

Assuming that the constitutive relationship of the material in the elastic region is given by the following:

$$\{d\varepsilon^e\} = [C^e]\{d\sigma\} \tag{5}$$

where $[C^e]$ is the elastic compliance matrix. If the elastic strain increment vector, $\{d\varepsilon^e\}$, and the stress increment vector, $\{d\sigma\}$, are principal strain increments and principal stress increments, respectively, then $[C^e]$ is a third-order compliance matrix. When both vectors have dimensions of 6×1, $[C^e]$ is a sixth-order compliance matrix.

Since plastic strain increments do not contribute to stress increments, from Equation (5), the relationship between the total stress increment and the total strain increment is given by the following:

$$\{d\varepsilon - d\varepsilon^p\} = [C^e]\{d\sigma\} \tag{6}$$

Alternatively, this can be rewritten as follows:

$$\{d\varepsilon\} = ([C^e] + [C^e])\{d\sigma\} = [C^e]\{d\sigma\} \tag{7}$$

where $[C^e]$ is the plastic compliance matrix. Using incremental plasticity theory, for non-associated flow rules, we can express the formulation as follows:

$$\{d\varepsilon^p\} = \Lambda\left\{\frac{\partial g}{\partial \sigma}\right\} \tag{8}$$

where Λ is a scalar factor representing the magnitude of the plastic strain increment, referred to as the plastic multiplier.

During **plastic loading**, the **consistency condition** $df = 0$ must be maintained, leading to the following:

$$df = \frac{\partial f}{\partial \sigma} : \{d\sigma\} + \frac{\partial f}{\partial H} dH = 0 \tag{9}$$

Since the hardening parameter, H, is often considered a function of plastic strain or accumulated plastic strain, we have the following:

$$dH = \frac{\partial H}{\partial \varepsilon^p} : \{d\varepsilon^p\} \tag{10}$$

Substituting Equation (8) into Equation (10):

$$dH = \Lambda\left[\frac{\partial H}{\partial \varepsilon^p} : \left\{\frac{\partial g}{\partial \sigma}\right\}\right] \tag{11}$$

Substituting Equation (11) into Equation (9) to obtain Λ:

$$\Lambda = -\frac{\frac{\partial f}{\partial \sigma} : \{d\sigma\}}{\frac{\partial f}{\partial H}\left[\frac{\partial H}{\partial \varepsilon^p} : \left\{\frac{\partial g}{\partial \sigma}\right\}\right]} \tag{12}$$

Substitute Equations (8) and (12) into Equation (6):

$$\{d\varepsilon\} = [C^e]\{d\sigma\} + \left(-\frac{\frac{\partial f}{\partial \sigma} : \{d\sigma\}}{\frac{\partial f}{\partial H}\left[\frac{\partial H}{\partial \varepsilon^p} : \left\{\frac{\partial g}{\partial \sigma}\right\}\right]}\right)\left\{\frac{\partial g}{\partial \sigma}\right\} \tag{13}$$

Equation (13) can also be rewritten in terms of the elastoplastic compliance matrix, forming the stress–strain increment equation:

$$\{d\varepsilon\} = \left\{[C^e] + \left(-\frac{\left\{\frac{\partial g}{\partial \sigma}\right\}\left\{\frac{\partial f}{\partial \sigma}\right\}^T}{\frac{\partial f}{\partial H}\left[\frac{\partial H}{\partial \varepsilon^p} : \frac{\partial g}{\partial \sigma}\right]}\right)\right\}\{d\sigma\} \tag{14}$$

2.3. *Iterative Formula of Stress and Strain Increment Under Different Boundary Conditions*

The principal stress and principal strain can be used to simulate common variables in laboratory true triaxial and conventional triaxial tests. The approach to solving the constitutive model's stress–strain increment iteration equations under different boundary conditions, also known as different stress paths, involves solving a system of equations consisting of the third-order compliance matrix form of the stress–strain increment Equation (16), the intermediate principal stress coefficient Equation (18), and the stress path equation. The maximum principal strain increment, $d\varepsilon_1$, is used to express the intermediate principal strain increment, $d\varepsilon_2$, the minimum principal strain increment, $d\varepsilon_3$, the maximum principal stress increment, $d\sigma_1$, the intermediate principal stress increment, $d\sigma_2$, and the minimum principal stress increment, $d\sigma_3$. From Equation (14), the stress–strain equations expressed in the third-order compliance matrix form can be obtained as follows:

$$\{d\varepsilon_i\} = [C_{ij}]\{d\sigma_j\}, i = 1,2,3, \quad j = 1,2,3 \tag{15}$$

Of course, Equation (15) can also be expressed in the following form:

$$\begin{Bmatrix} d\varepsilon_1 \\ d\varepsilon_2 \\ d\varepsilon_3 \end{Bmatrix} = \begin{bmatrix} C_{11} & C_{12} & C_{13} \\ C_{21} & C_{22} & C_{23} \\ C_{31} & C_{32} & C_{33} \end{bmatrix} \begin{Bmatrix} d\sigma_1 \\ d\sigma_2 \\ d\sigma_3 \end{Bmatrix} \tag{16}$$

In experiments, the maximum principal strain increment, $d\varepsilon_1$, is generally a known quantity, representing the axial strain. To express the remaining five variables in terms of $d\varepsilon_1$, two additional equations are needed. One equation can be derived from the stress path, while the other can be obtained using the intermediate principal stress coefficient, b, or the stress Lode angle, θ. The definition of the intermediate principal stress coefficient, b, is given as follows:

$$b = \frac{\sigma_2 - \sigma_3}{\sigma_1 - \sigma_3} \tag{17}$$

where b is a constant, provided as an initial condition. This relationship remains valid throughout the increment process:

$$b = \frac{d\sigma_2 - d\sigma_3}{d\sigma_1 - d\sigma_3} \tag{18}$$

The increment formulation is independent of the constitutive model, and different models use different compliance matrices. During loading and unloading, the compliance matrix, $[C_{ij}]$, is taken as the elastoplastic compliance matrix and the elastic compliance matrix, respectively. The following sections present the incremental formulations for some commonly used stress paths.

2.3.1. Isotropic Compression (ISO) Stress Path

In the isotropic compression stress path, hydrostatic pressure is applied to compress the specimen uniformly in all directions. The stress increments in the three principal stress directions are equal, $d\sigma_1 = d\sigma_2 = d\sigma_3$, and the strain increments are also equal, $d\varepsilon_1 = d\varepsilon_2 = d\varepsilon_3$. Using the condition that the three stress increments are identical, it follows that the intermediate principal stress coefficient $b = 0$, $d\sigma_1 = d\sigma_3$. By combining this condition with Equation (16), a system of equations consisting of five equations and six variables is obtained:

$$\begin{cases} d\sigma_1 = d\sigma_3 \\ b = \frac{d\sigma_2 - d\sigma_3}{d\sigma_1 - d\sigma_3} = 0 \\ d\varepsilon_1 = C_{11}d\sigma_1 + C_{12}d\sigma_2 + C_{13}d\sigma_3 \\ d\varepsilon_2 = C_{21}d\sigma_1 + C_{22}d\sigma_2 + C_{23}d\sigma_3 \\ d\varepsilon_3 = C_{31}d\sigma_1 + C_{32}d\sigma_2 + C_{33}d\sigma_3 \end{cases} \quad (19)$$

where b is a known quantity determined from the experiment, and the compliance matrix elements are provided by the constitutive model. Consequently, $d\varepsilon_2$, $d\varepsilon_3$, $d\sigma_1$, $d\sigma_2$, and $d\sigma_3$ can all be expressed in terms of $d\varepsilon_1$:

$$\begin{cases} d\sigma_1 = \frac{d\varepsilon_1}{C_{11} + C_{12} + C_{13}} \\ d\sigma_2 = d\sigma_1 \\ d\sigma_3 = d\sigma_1 \\ d\varepsilon_2 = (C_{21} + C_{22} + C_{23})d\sigma_1 \\ d\varepsilon_3 = d\varepsilon_2 \end{cases} \quad (20)$$

If the initial shear state (initial stress p_0 and initial void ratio e_0) is known, the MCC and CSUH models can perform iterative calculations based on Equation (20).

2.3.2. Oedometer Compression (K_0) Stress Path

The oedometer compression test involves placing a soil sample in a rigid ring that restricts lateral deformation while allowing vertical drainage and compression. The boundary condition for this test is $d\varepsilon_2 = d\varepsilon_3 = 0$, which implies that the lateral stresses, $d\sigma_2$ and $d\sigma_3$, are not equal, but their relationship can be determined. This leads to $b = 0$. By formulating and solving the system of equations in a manner similar to the isotropic compression stress path, we obtain the following:

$$\begin{cases} d\sigma_1 = \frac{(C_{22} + C_{23})d\varepsilon_1}{C_{11}(C_{22} + C_{23}) - C_{21}(C_{12} + C_{13})} \\ d\sigma_2 = \frac{C_{21}d\varepsilon_1}{C_{21}(C_{12} + C_{13}) - C_{11}(C_{22} + C_{23})} \\ d\sigma_3 = d\sigma_2 \\ d\varepsilon_2 = 0 \\ d\varepsilon_3 = 0 \end{cases} \quad (21)$$

2.3.3. Constant σ_3 Drained Shear Stress Path

A constant σ_3 shear stress path maintains σ_3 unchanged throughout the shear process. To obtain different values of σ_3, a consolidation (drained compression) test is typically required first. Common consolidation tests include isotropic compression path consolidation tests and K_0 consolidation tests. Different consolidation paths lead to different initial stresses and initial void ratios during the shear stage. After consolidation, the pressure conditions and void ratio can be determined, and the shear test can then proceed. In the equation of the constant σ_3 shear stress path, $d\sigma_3 = 0$. By formulating and solving the system of equations in a manner similar to the isotropic compression stress path, we obtain the following:

$$\begin{cases} d\sigma_1 = \frac{d\varepsilon_1}{C_{11}+bC_{12}} \\ d\sigma_2 = \frac{bd\varepsilon_1}{C_{11}+bC_{12}} \\ d\sigma_3 = 0 \\ d\varepsilon_2 = (C_{21} + bC_{22})d\sigma_1 \\ d\varepsilon_3 = (C_{31} + bC_{32})d\sigma_1 \end{cases} \tag{22}$$

Conventional triaxial consolidated drained shear (CD) stress path testing is the most common constant σ_3 shear test. To solve for the stress–strain relationship in a CD test, one only needs to set $b = 0$ in Equation (22) and iterate. When $b = 1$, Equation (22) can be used to calculate a triaxial extension test. It is important to distinguish between triaxial extension tests and triaxial extrusion tests. In a triaxial extension test, the confining pressure σ_3 remains unchanged, whereas in a triaxial extrusion test, the axial stress remains unchanged while the confining pressure increases.

2.3.4. Undrained Shear Stress Path with Constant Confining Pressure

The undrained shear stress path is unique as it is a volumetric strain-controlled path. The boundary condition is that the volumetric strain increment satisfies $d\varepsilon_v = d\varepsilon_3 + d\varepsilon_3 + d\varepsilon_3 = 0$, By formulating and solving the system of equations similarly to the isotropic compression stress path, we obtain the following:

$$\begin{cases} d\sigma_1 = \frac{d\varepsilon_1}{(C_{11}+bC_{12}) - \frac{A\cdot D}{B}} \\ d\sigma_3 = -\frac{A\cdot d\sigma_1}{B} \\ d\sigma_2 = bd\sigma_1 + (1-b)d\sigma_3 \\ d\varepsilon_2 = (C_{21} + bC_{22})d\sigma_1 + ((1-b)C_{22} + C_{23})d\sigma_3 \\ d\varepsilon_3 = (C_{31} + bC_{32})d\sigma_1 + ((1-b)C_{32} + C_{33})d\sigma_3 \end{cases} \tag{23}$$

In Equation (23), A, B, and D are defined as follows:

$$\begin{cases} A = C_{11} + C_{21} + C_{31} + b(C_{12} + C_{22} + C_{32}) \\ B = C_{13} + C_{23} + C_{33} + (1-b)(C_{12} + C_{22} + C_{32}) \\ D = (1-b)C_{12} + C_{13} \end{cases} \tag{24}$$

The terms A, B, and D have no physical significance and are only used to simplify the length of Equation (23). When $b = 0$, Equation (23) reduces to the conventional triaxial consolidated undrained (CU) stress path.

2.3.5. Constant Mean Effective Stress (p) Drained Shear (p0) Stress Path

The constant p drained shear stress path maintains the effective mean principal stress, p, constant during shearing. The condition for this stress path is $dp = \frac{d\sigma_1+d\sigma_2+d\sigma_3}{3} = 0$. By formulating and solving the system of equations, similar to the isotropic compression stress path, we obtain the following:

$$\begin{cases} d\sigma_1 = \frac{(b-2)d\varepsilon_1}{(b-2)C_{11}+(1-2b)C_{12}+(1+b)C_{13}} \\ d\sigma_2 = \frac{(1-2b)d\varepsilon_1}{(b-2)C_{11}+(1-2b)C_{12}+(1+b)C_{13}} \\ d\sigma_3 = \frac{(1+b)d\varepsilon_1}{(b-2)C_{11}+(1-2b)C_{12}+(1+b)C_{13}} \\ d\varepsilon_2 = \frac{((b-2)C_{21}+(1-2b)C_{22}+(1+b)C_{23})d\varepsilon_1}{(b-2)C_{11}+(1-2b)C_{12}+(1+b)C_{13}} \\ d\varepsilon_3 = \frac{((b-2)C_{31}+(1-2b)C_{32}+(1+b)C_{33})d\varepsilon_1}{(b-2)C_{11}+(1-2b)C_{12}+(1+b)C_{13}} \end{cases} \tag{25}$$

At this point, the iterative stress–strain increment equations for the five commonly used stress paths have been fully derived.

3. CSUH Model: Framework and Theory

3.1. Three-Dimensional Constitutive Model Framework

Yao Yangping et al. [11] proposed a unified hardening (CSUH) model based on the Modified Cam-Clay (MCC) model, which can describe both overconsolidated clay and sand. The CSUH model effectively captures the stress–strain behavior of various soil types, including clay, sand, and rockfill, under different initial conditions and stress paths. It represents a theoretical advancement and innovative development of the MCC model [12].

To apply the constitutive model to true triaxial laboratory test predictions or engineering practice, it is necessary to extend the model to three dimensions. The Transformed Stress (TS) method, proposed by Yao et al. [18], is a three-dimensional extension approach for constitutive models. Its principal stress formulation is expressed as follows:

$$\tilde{\sigma}_i = p + \frac{q_c}{q}(\sigma_i - p), i = 1,2,3 \tag{26}$$

where σ_i, p, and q represent the principal stresses, mean principal stress, and generalized shear stress in the real stress space, respectively; $\tilde{\sigma}_i$ and q_c denote the principal stresses and generalized shear stress in the transformed stress space, respectively. The q_c varies based on different strength criteria. The mean principal stress ppp in the real stress space is equal to that in the transformed stress space.

Since this method can be formulated based on different strength criteria (such as the SMP criterion, Mohr–Coulomb criterion, Lade criterion, and generalized nonlinear strength criteria), it is applicable to materials with varying strength characteristics. Figure 1 illustrates the transformed stress method based on the SMP strength criterion. Figure 1a shows the SMP strength criterion and the generalized Mises strength criterion in the real stress space. The SMP criterion in the real stress space is transformed into the generalized Mises strength criterion in the transformed stress space, assuming that the constitutive model is established in the transformed stress space. The real stress–strain relationship is obtained by differentiating the yield surface in the transformed stress space with respect to the real stress space. Figure 1b presents the transformation illustration in the π-plane. Figure 1c depicts the yield surface of the CSUH model in both the real principal stress space and the transformed principal stress space. Figure 1d shows the p-q cross-section of Figure 1c, where M_e represents the critical state stress ratio under triaxial extension.

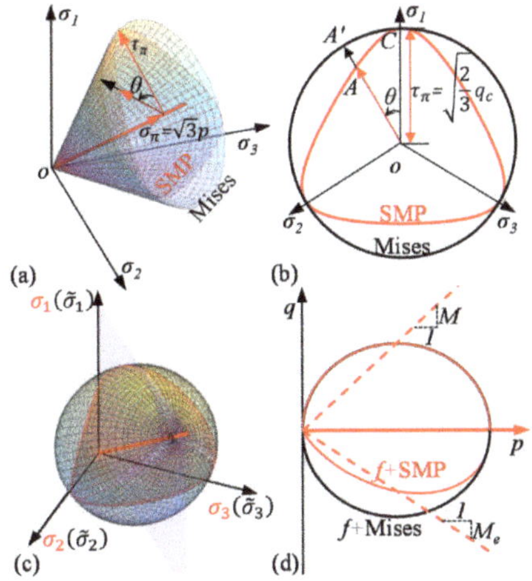

Figure 1. Transformed stress method: (**a**) SMP and von Mises Yield Criteria in True Stress Space; (**b**) Transformation Illustration on the π-Plane; (**c**) CSUH Model Yield Surface(χ = 0) in Principal Stress Space; (**d**) Yield Surface in the p–q Plane.

When applying the TS method to extend the CSUH model into three dimensions, it is assumed that the CSUH model is formulated in the transformed stress space. The plastic potential surface equation, g, of the CSUH model in the transformed stress space is expressed as follows:

$$\tilde{g} = \ln \tilde{p}^2 + \ln\left(1 + \frac{\tilde{q}^2}{M_c^2 \tilde{p}^2}\right) = 0 \tag{27}$$

where $\tilde{p}$ and $\tilde{q}$ denote the effective mean principal stress and shear stress in the transformed stress space, respectively. The tilde symbol (~) indicates variables in the transformed stress space. The parameter p_y represents the horizontal coordinate of the plastic potential surface's intersection with the p-axis. The characteristic state stress ratio, $\tilde{M}_c$, has a geometric interpretation as the stress ratio at the upper quadrant point of the ellipse.

The yield surface equation, f, of the CSUH model in the transformed stress space is given by the following:

$$\tilde{f} = \ln\left[\left(1 + \frac{(1+\chi)\tilde{q}^2}{M^2\tilde{p}^2 - \chi\tilde{q}^2}\right)\tilde{p} + \tilde{p}_s\right] - \ln(\tilde{p}_{x0} + \tilde{p}_s) - \frac{\tilde{H}}{c_p} = 0 \tag{28}$$

where M is the critical state stress ratio, representing the slope of the conventional triaxial compression critical state line (CSL) in the p-q space. The critical state parameter, χ, adjusts the shape of the yield surface. When χ = 0, the function, f, degenerates into an elliptical yield surface, as shown in Figure 1d. The parameter, p_s, represents the hardening parameter. The plastic compression coefficient is defined as $c_p = (\lambda - \kappa)/(1 + e_0)$, where λ and κ are the compression and rebound indices from isotropic compression tests, respectively. The variable p_{x0} denotes the horizontal coordinate of the initial yield surface intersection with the p-axis under the initial conditions. The term H represents the hardening parameter:

$$\tilde{H} = \int \frac{\tilde{M}_Y^4 - \tilde{\eta}^4}{\tilde{M}_c^4 - \tilde{\eta}^4} d\varepsilon_v^p \tag{29}$$

In this equation, $\tilde{\eta} = \frac{\tilde{q}}{\tilde{p}}$ represents the current stress ratio in the transformed stress space, $\tilde{M}_Y$ is the potential failure stress ratio, and $d\varepsilon_v^p$ is the increment of plastic volumetric strain. During the shearing process, the characteristic state stress ratio, $\tilde{M}_c$, and the potential failure stress ratio, $\tilde{M}_Y$, change with the state parameter, $\tilde{\xi}$, which indicates the current degree of compaction, as shown in Equations (30) and (31). From these equations, it is evident that $\tilde{H}$ is also state-dependent.

$$\tilde{M}_c = M \cdot \exp(-m \cdot \tilde{\xi}) \tag{30}$$

$$\tilde{M}_Y = 6\left[\sqrt{\frac{12(3-M)}{M^2}\exp\left(-\frac{\tilde{\xi}}{\lambda-\kappa}\right)+1}+1\right]^{-1} \tag{31}$$

In these equations, the dilatancy parameter, m, scales the influence of $\tilde{\xi}$ and thereby affects $\tilde{M}_c$. The state parameter, $\tilde{\xi}$, is defined as follows:

$$\tilde{\xi} = Z - \lambda \ln\left(\frac{\tilde{p}+p_s}{1+p_s}\right) - (\lambda-\kappa)\ln\left(\frac{\left(\frac{M^2+\tilde{\eta}^2}{M^2-\chi\tilde{\eta}^2}\right)\tilde{p}+p_s}{\tilde{p}+p_s}\right) - e \tag{32}$$

where p_s is the hardening parameter:

$$p_s = \exp\left(\frac{N-Z}{\lambda}\right) - 1 \tag{33}$$

The Unified Hardening Model for Clays and Sands (CSUH) has eight fundamental parameters: M, ν, κ, λ, N, Z, χ and m. For the same type of soil, regardless of different stress paths and initial states (initial confining pressure $p0$ and initial void ratio $e0$), the same set of parameters is used.

3.2. Third-Order Flexibility Matrix of the Model

Based on Equation (14), the incremental relationship is expressed in the form of a third-order flexibility matrix:

$$\{d\varepsilon\}_{3\times1} = \left\{\left[C^e\right]_{3\times3} + \left[C^p\right]_{3\times3}\right\}\{d\sigma\}_{3\times1} \tag{34}$$

For isotropic materials:

$$[C^e]_{3\times3} = \frac{1}{E}\begin{bmatrix} 1 & -\nu & -\nu \\ -\nu & 1 & -\nu \\ -\nu & -\nu & 1 \end{bmatrix} \tag{35}$$

where E is the elastic modulus and ν is Poisson's ratio.

In Equation (34), the third-order plastic flexibility matrix $[C^e]_{3\times3}$ is given as follows:

$$\left[C^p\right]_{3\times3} = -\frac{\left\{\frac{\partial\tilde{g}}{\partial\tilde{\sigma}}\right\}_{3\times1}\left\{\frac{\partial\tilde{f}}{\partial\sigma}\right\}^T_{1\times3}}{\frac{\partial\tilde{f}}{\partial\tilde{H}}\left[\left\{\frac{\partial\tilde{H}}{\partial\varepsilon^p}\right\}^T_{1\times3}\left\{\frac{\partial\tilde{g}}{\partial\tilde{\sigma}}\right\}_{3\times1}\right]} \tag{36}$$

where the yield surface equation, $\tilde{f}$, is differentiated with respect to the true stress, σ, and the hardening parameter, $\tilde{H}$, the plastic potential surface equation, $\tilde{g}$, is differentiated with respect to the transformed stress, $\tilde{\sigma}$, and the hardening parameter, $\tilde{H}$, is differentiated with respect to the true strain, ε^p. These differentiation processes require the application of the chain rule. The yield surface equation, $\tilde{f}$, and plastic potential surface equation, $\tilde{g}$, are generally expressed as functions of $\tilde{p}$ and $\tilde{q}$, while $\tilde{p}$ and $\tilde{q}$ are functions of stress, σ_i.

Based on the SMP strength criterion and the transformed stress (TS) method, we obtain the following:

$$\begin{cases} \tilde{p} = p = (\sigma_1 + \sigma_2 + \sigma_3)/3 \\ \tilde{q} = q_c = \\ \frac{I_1 I_2 - 9I_3 + 3\sqrt{(I_1 I_2 - I_3)(I_1 I_2 - 9I_3)}}{4I_2} \end{cases} \quad (37)$$

where I_1, I_2, and I_3 are the first, second, and third invariants of the principal stress tensor, respectively, and q_c is the transformed generalized shear stress based on the SMP strength criterion. The derivation results for Equation (36) of the CSUH model are provided in [11] and will not be reiterated here. Equation (8) can also be expressed as follows:

$$\left\{ d\varepsilon_v^p \right\} = \tilde{\Lambda}\left\{ \frac{\partial \tilde{g}}{\partial \tilde{p}} \right\} \quad (38)$$

The derivative of the hardening parameter with respect to plastic strain can also be expressed as the derivative with respect to plastic volumetric strain. Therefore, for the CSUH model, Equation (36) can be rewritten as follows:

$$\left[C_{ij}^p\right] = c_p \frac{\tilde{M}_c^4 - \tilde{\eta}^4}{\tilde{M}_Y^4 - \tilde{\eta}^4} \frac{\frac{\partial \tilde{f}}{\partial \sigma_j}}{\frac{\partial \tilde{g}}{\partial \tilde{p}}} \frac{\partial \tilde{g}}{\partial \tilde{\sigma}_i}, i = 1,2,3,\ j = 1,2,3 \quad (39)$$

The loading and unloading criteria are defined as follows:

$$df_{trial} = \frac{\partial \tilde{f}}{\partial \sigma_1} d\sigma_1 + \frac{\partial \tilde{f}}{\partial \sigma_2} d\sigma_2 + \frac{\partial \tilde{f}}{\partial \sigma_3} d\sigma_3 \quad (40)$$

If the strain increment dεd\varepsilondε calculated using the elastic flexibility matrix in Equation (34) indicates loading, the total flexibility matrix, which includes both the elastic and plastic flexibility matrices, should be used to recalculate the stress–strain increment in Equation (34). If the strain increment calculated using only the elastic flexibility matrix indicates unloading, the calculation is considered correct, and the next step of elastic prediction can proceed.

If the trial value, f_{trial}, calculated using the elastic compliance matrix, $[C^e]$, in Equation (34) is greater than zero, it indicates loading. In this case, the elastoplastic compliance matrix, which is the sum of the elastic compliance matrix and the plastic compliance matrix, should be used to recalculate the stress–strain increments in Equation (34). Conversely, if the trial value, df_{trial}, calculated using the elastic compliance matrix is less than zero, it indicates unloading, meaning the trial calculation is correct. In this scenario, the stress–strain increments in Equation (34) can be accurately computed using only the elastic compliance matrix, and the next step of the elastic trial calculation can proceed.

4. Dynamic Iteration Method

4.1. Common Iteration Methods

The inversion of constitutive model parameters using optimization algorithms requires extensive computations, with the most time-consuming part being the iterative stress–strain increment calculation, similar to Equation (20). The process of determining constitutive model parameters through optimization involves providing different parameter sets to the constitutive model. The model then computes the theoretical stress–strain curve corresponding to the given parameters and initial conditions from the experiment. The computed theoretical stress–strain curve is compared with the experimental stress–strain curve, and the error is fed back to the optimization algorithm. Based on this error feedback, the optimization algorithm adjusts and attempts to provide improved model parameters. Consequently, the inversion process requires frequent evaluation of stress–strain increments to generate the stress–strain curve.

As previously discussed, different constitutive models can adopt a unified iterative stress–strain increment formulation, similar to Equation (20), for stress–strain curve calculations. Several numerical methods can be applied for iteration, including the forward Euler method, the fourth-order Runge–Kutta (RK4) method, and implicit iteration methods. The RK4 method offers higher computational accuracy and better stability but faces challenges in practical applications due to its complex implementation and high computational cost. Implicit iteration methods, such as the backward Euler method and the Newton-Raphson method, allow for larger incremental steps but require solving nonlinear equations iteratively, which may lead to convergence issues. The semi-implicit return mapping method performs well in handling elastoplastic problems, but its implementation is complex, and its computational efficiency is relatively low.

Due to its simplicity, intuitiveness, and ease of implementation, the forward Euler method is widely used. This method is based on a first-order Taylor expansion, where the stress state of the next step is directly computed using the tangent modulus of the current step. The core concept is to discretize the continuous stress–strain relationship into an incremental form. According to the forward Euler method, the stress–strain increment in Equation (20) and subsequent stress paths can be simplified as $C_n d\varepsilon_1$:

$$[\sigma, \varepsilon]_{n+1} = [\sigma, \varepsilon]_n + C_n \mathrm{d}\varepsilon_1 \quad (41)$$

where $[\sigma, \varepsilon]_{n+1}$ represents the stress and strain at step n + 1, $[\sigma, \varepsilon]_n$ represents the stress and strain at step n, C_n is the compliance matrix computed based on the stress state at step n, and $d\varepsilon_1$ is the incremental maximum principal strain.

Figure 2 illustrates the accuracy comparison between the RK4 method and the forward Euler method. While RK4 provides higher accuracy, its computational cost is significantly higher than the forward Euler method. However, it was observed that when the total axial strain is uniformly divided into 3000 increments, the integration curves obtained from both methods with 3000 uniform steps are nearly identical, as shown in Figure 2. This indicates that using 3000 uniform steps can yield an accurate stress–strain curve. Since RK4 requires solving the compliance matrix at four different substep increments, whereas the forward Euler method only requires one calculation, the computational cost of the forward Euler method is significantly lower. Therefore, when the iteration step count reaches 3000, the forward Euler method not only reduces computational effort but also provides satisfactory results, making it the preferred choice for computing the constitutive model stress–strain curve.

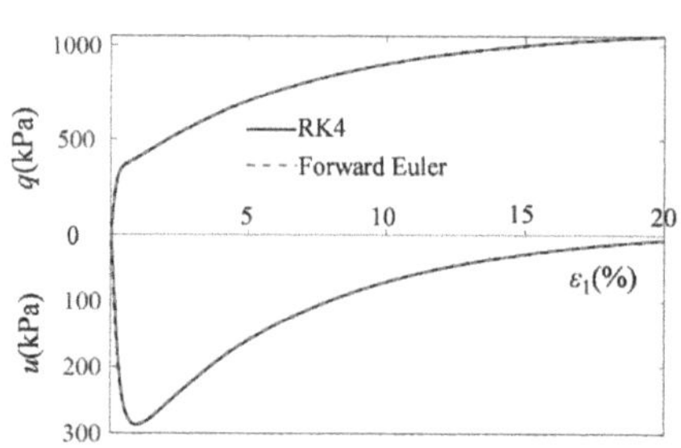

Figure 2. Stress–strain curves for the CU stress path obtained using different integration methods with 3000 uniform step sizes.

Compared with the computational cost of the optimization algorithm itself, the iterative evaluation of the constitutive model's stress–strain increment equation is the most time-consuming part of the entire inversion process. Even with 3000 iteration steps using the forward Euler method, the computation time remains excessively long.

4.2. Proposed Dynamic Iterative Method

The slope of the stress–strain curve of soil is not uniform. As shown in Figure 3, the slope of the curve is steep in the front part and gentle in the latter part. By segmentally adjusting the model's iterative increment step size, the total number of iterations can be reduced, thereby accelerating the speed of parameter inversion.

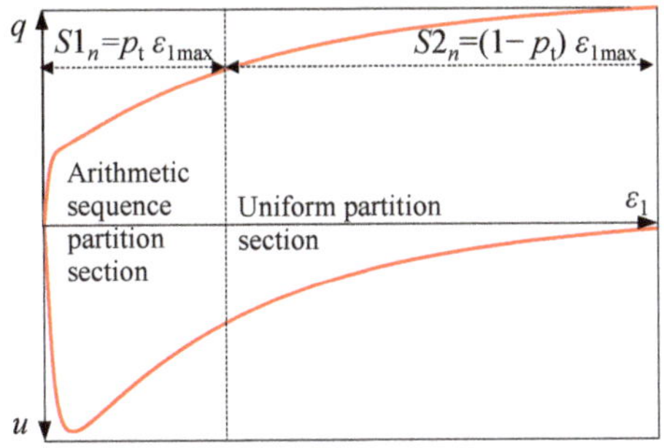

Figure 3. Segmented adjustment of the iterative increment step size $\Delta\varepsilon_1$ in the constitutive model.

When the curve slope is large, a small axial strain increment, $\Delta\varepsilon_1$, can cause a significant stress increment, whereas when the curve slope is small, a larger axial strain increment, $\Delta\varepsilon_1$, induces only a minor stress increment. If the same iterative increment step size, $\Delta\varepsilon_1$, is used regardless of the curve slope, $\Delta\varepsilon_1$ must be set sufficiently small to ensure accuracy when the slope is large, which results in a high number of iterations and slow computation.

To enhance the numerical computation efficiency of the constitutive model, a dynamic iterative method that adjusts the iterative increment step size in segments is proposed to reduce the total number of iterations. A smaller iterative step size is adopted when the stress–strain curve slope is large, while a larger step size is used when the slope is small. This approach ensures accuracy while reducing the number of iterations, thereby accelerating the computation speed. Generally, the axial strain distributed in an arithmetic sequence accounts for about 30% of the total axial strain, while the uniformly distributed axial strain accounts for about 70%. The segmentation position dynamically changes with the total axial strain.

As shown in Figure 3, the total axial strain, ε_{1max}, is divided into two parts. The first part adopts an arithmetic sequence distribution, with its axial strain, S_{1n}, accounting for a proportion, p_t, of ε_{1max}, where the strain increment, $\Delta\varepsilon_1$, follows an arithmetic sequence distribution. The second part adopts a uniform distribution, with axial strain S_{2n} accounting for a proportion $(1-p_t)$ of ε_{1max}, where $\Delta\varepsilon_1$ is uniformly distributed.

Given a total number of iterations, N_s, the number of iterations in the uniform distribution part is $N_{s2}=p_tN_s$, and the corresponding axial strain increment, $\Delta\varepsilon_1$, is as follows:

$$\Delta\varepsilon_1=\frac{S2_n}{N_{s2}}=\frac{(1-p_t)\varepsilon_{1max}}{p_tN_s} \quad (42)$$

Assuming that the initial iterative step size, a_1, in the arithmetic sequence part is 1/10 of the uniform distribution step size $\Delta\varepsilon_1$, we have:

$$a_1 = \frac{\Delta\varepsilon_1}{10} \tag{43}$$

Given the total number of iterations N_s, the number of iterations in the arithmetic sequence part is $N_{s1} = (1 - p_t)N_s$. The common difference d of the arithmetic sequence and the iterative increment step size, a_n, can be expressed as follows:

$$d = \frac{2(S1_n - a_1 N_{s1})}{N_{s1}(N_{s1} - 1)} \tag{44}$$

$$a_n = a_1 + (n-1)d \tag{45}$$

By selecting an appropriate proportion, p_t, and a reasonable total iteration count, N_s, a highly accurate stress–strain curve can be obtained with a reduced number of iterations. When ε_{1max} =20%, p_t = 0.3, and N_s = 50, the step sizes of different partitions obtained using Equations (42) to (45) are shown in Figure 4.

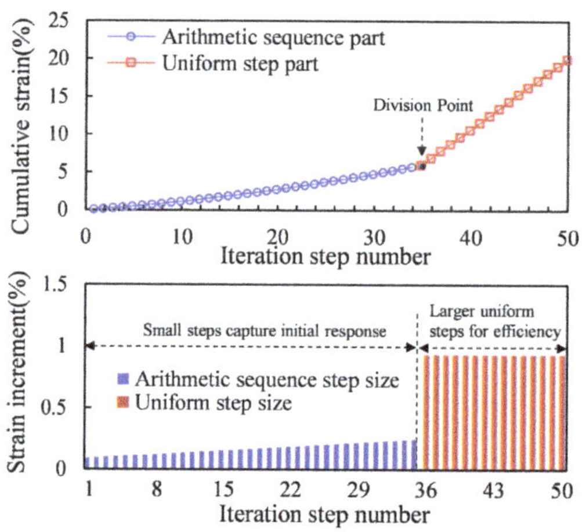

Figure 4. Step Size for Different Step Partitions.

As shown in Figure 5, the results of 50 steps arithmetic sequence step size (dynamic iteration method parameters p_t = 0.3, N_s = 50) are similar to those of 3000 steps uniform step size, indicating that the dynamic iteration method effectively reduces the number of iterations to 1/60 of the original while maintaining accuracy.

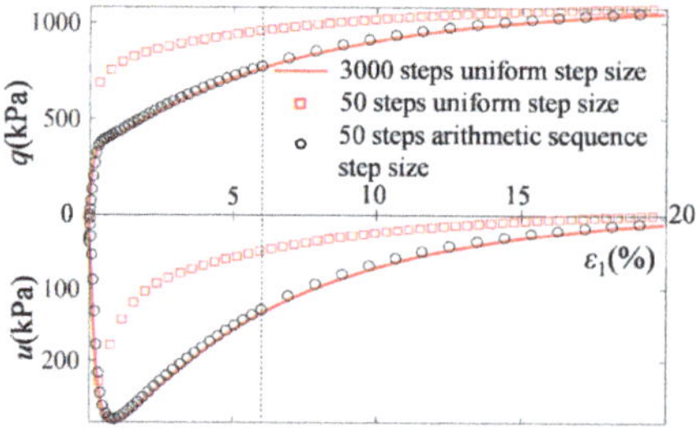

Figure 5. Comparison of Uniform and Arithmetic Sequence Distributed Iterative Increment Step Sizes.

Figures 2, 5 and 6 present the stress–strain curve of the CSUH model using Toyoura Sand parameters from Table 1 under a confining pressure of 500 kPa and a void ratio of 0.84 in CU tests. According to Equation (16), theoretically, the numerical computation iterative increment step size, $d\varepsilon_1$, should be infinitely small. However, practical calculations reveal that 3000 steps are generally sufficient for high accuracy (as shown in Figure 2). Since the step size is not infinitely small in actual computation, $d\varepsilon_1$ is denoted as $\Delta\varepsilon_1$, and the same applies to other principal strain and stress increments.

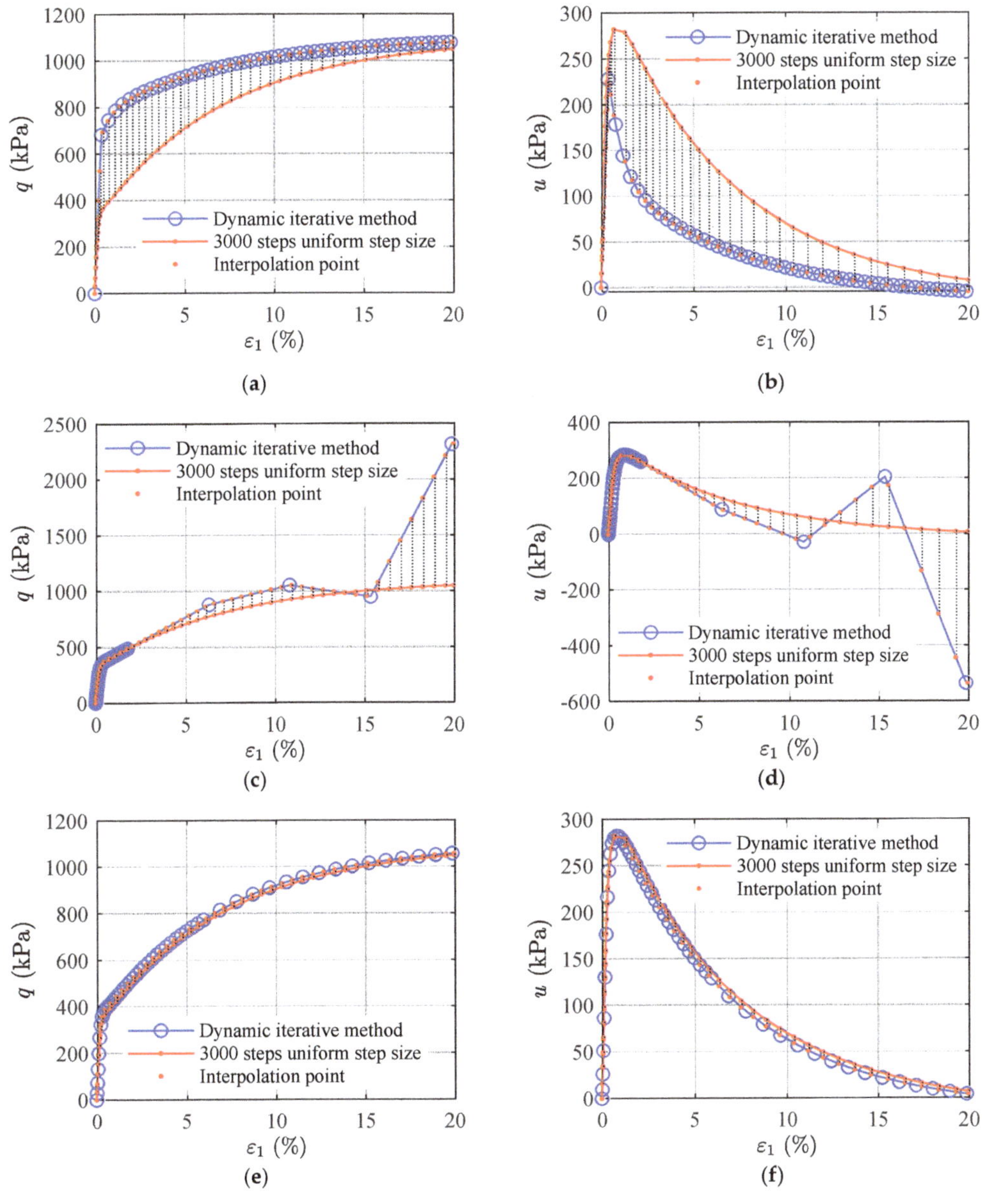

Figure 6. *Cont.*

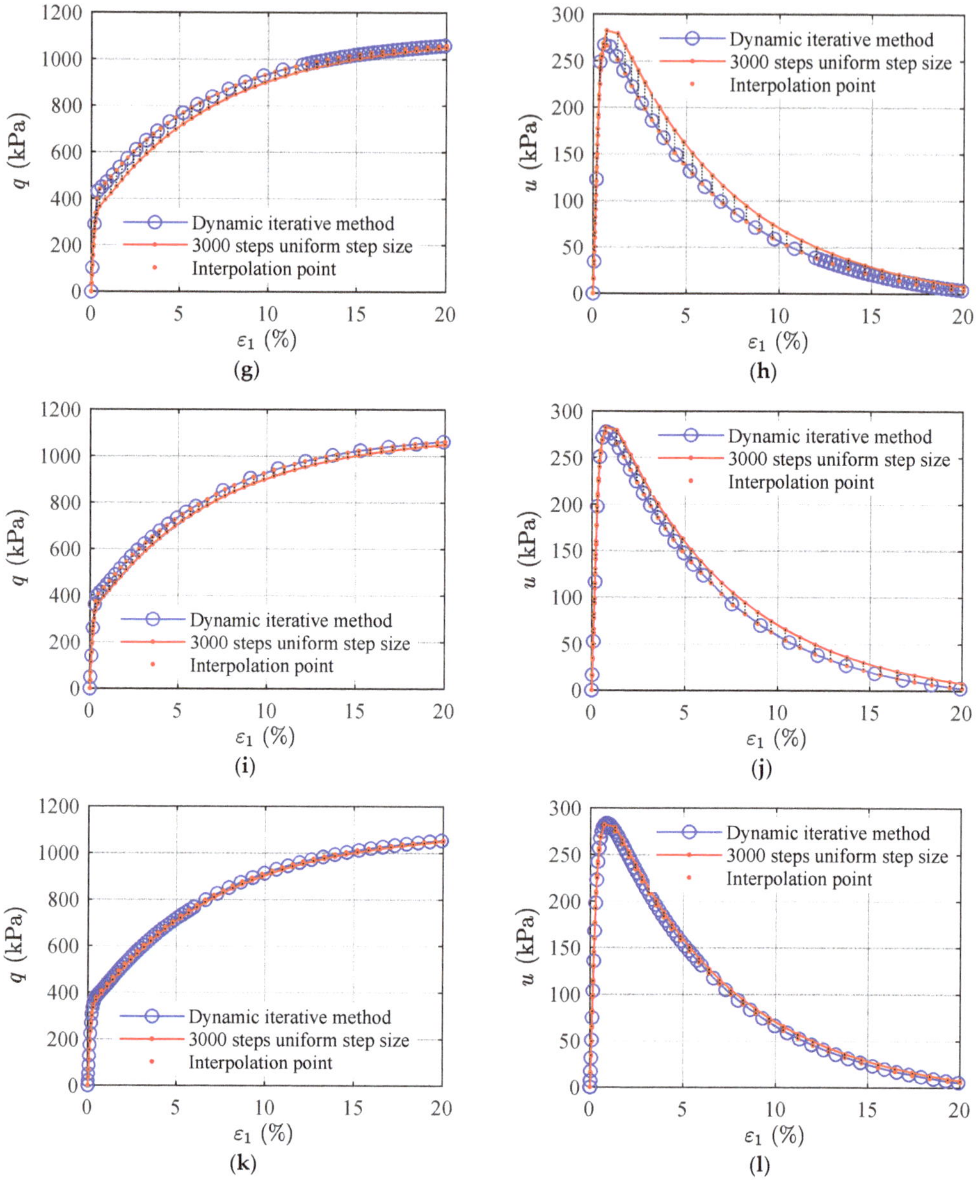

Figure 6. Differences between dynamic iteration method curves and 3000 uniform step size curves for different p_t and N_s values: (**a**) ε_1-q curve at $p_t = 0$, $N_s = 50$, with $MRE = 0.2495$; (**b**) ε_1-u curve at $p_t = 0$, $N_s = 50$, with $MRE = 0.5378$; (**c**) ε_1-q curve at $p_t = 0.09$, $N_s = 50$, with $MRE = 0.1512$; (**d**) ε_1-u curve at $p_t = 0.09$, $N_s = 50$, with $MRE = 3.4802$; (**e**) ε_1-q curve at $p_t = 0.3$, $N_s = 50$, with $MRE = 0.0229$; (**f**) ε_1-u curve at $p_t = 0.3$, $N_s = 50$, with $MRE = 0.07622$; (**g**) ε_1-q curve at $p_t = 0.6$, $N_s = 50$, with $MRE = 0.0619$; (**h**) ε_1-u curve at $p_t = 0.6$, $N_s = 50$, with $MRE = 0.1534$; (**i**) ε_1-q curve at $p_t = 0.3$, $N_s = 30$, with $MRE = 0.0401$; (**j**) ε_1-u curve at $p_t = 0.3$, $N_s = 30$, with $MRE = 0.1331$; (**k**) ε_1-q curve at $p_t = 0.3$, $N_s = 70$, with $MRE = 0.0153$; (**l**) ε_1-u curve at $p_t = 0.3$, $N_s = 70$, with $MRE = 0.0528$.

Table 1. Parameters used for calculating theoretical curve values of the CSUH model [11].

Parameters	M	ν	κ	λ	N	Z	χ	m
Values	1.25	0.3	0.04	0.135	1.973	0.934	0.4	1.8

Figure 6 shows the computational results of the dynamic iteration method with different values of p_t and N_s. The red line represents the calculation curve with 3000 uniform steps. Fifty interpolation points are taken at equal arc lengths on the curve to represent the experimental values. Corresponding points to the interpolation points are found on the dynamic iteration method calculation curve. The error between the dynamic iteration method calculation curve and the curve with 3000 uniform steps is measured using mean relative error (*MRE*) from reference [12]. The dynamic iteration method with p_t = 0 and N_s = 50 is equivalent to using 50 uniform step iterations. As shown in Figure 6a,b, the dynamic iteration method (blue dots) differs significantly from the 3000 uniform steps calculation curve (red), with an *MRE* of 0.2495 for q and 0.5378 for u. The dynamic iteration method with p_t = 0.09 and N_s = 50 yields the worst computational results, as shown in Figures 6c,d and 7a.

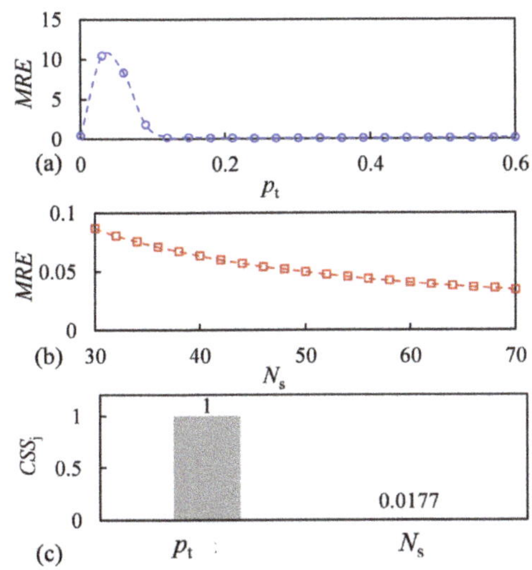

Figure 7. Sensitivity analysis of p_t and N_s: (**a**) Average *MRE* of q and u corresponding to different p_t when N_s = 50; (**b**) average *MRE* of q and u corresponding to different N_s when p_t = 0.3; (**c**) composite scaled sensitivity (CSS) of p_t and N_s.

Figure 7 shows the variation in MRE and parameter sensitivity analysis for different p_t and N_s values. The sensitivity analysis method adopts a local sensitivity analysis method, namely Composite Scale Sensitivity (CSS) [19]. Figure 7c shows that pt has a larger sensitivity while N_s has a smaller sensitivity, indicating that the choice of pt is the most important in the dynamic iteration method. From Figure 7a, it can be seen that when p_t is greater than 0.12, the MRE does not change significantly. Therefore, for this typical stress–strain curve with a slope that is initially large and then small, p_t = 0.3 is appropriate. As shown in Figure 7, under the extreme condition where the total number of iteration steps N_s is reduced to 30, the proposed dynamic iteration algorithm still demonstrates excellent numerical stability, with the generated stress–strain curves maintaining a high degree of agreement with the full 3000-step iteration results (*MRE*s of q and u are 0.041 and 0.1331, respectively). In contrast, the traditional numerical scheme with equal step-size iteration encounters complex numbers due to the large initial step size, and both the fourth-order Runge–Kutta method and Euler method exhibit severe numerical divergence. The dynamic iteration method, with a smaller initial step size, improves iteration efficiency by two orders of magnitude while ensuring computational accuracy.

Taking the inversion of the CSUH model parameters for the Changhe dam rockfill in Section 5.3 as an example, the inversion process includes one isotropic compression test and four CD tests, totaling 414 data points. Based on the dynamic iterative method with segmented adjustment of the model iteration increment step size, this paper developed parameter inversion software using VBA language on the Excel platform (website: https://github.com/longbiscuit/ModelParaOpt2.0_Excel_STADE, accessed on 29 April 2025.). The software runs on a Dell Inspiron 13 7000 series laptop (Dell Inc., Austin, TX, USA). Experimental results show that when the number of iterations is 50, the computation time is about 7 min; when the number of iterations increases to 3000, the required time is 332 min. Compared with the traditional method, the computation time is reduced to 1/47 of the original, as shown in Figure 8.

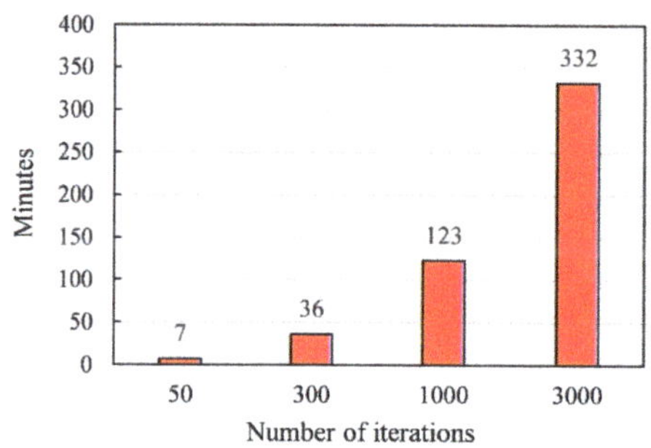

Figure 8. Computational Time for Inversion with Different Iteration Counts.

In conclusion, the proposed dynamic iterative method with segmented iterative increment step sizes significantly reduces the number of iterations while ensuring prediction accuracy, greatly enhancing the efficiency of parameter inversion.

5. CSUH Model Iterative Inversion Parameters

5.1. Fujinomori Clay

In all parameter inversion calculations presented in Section 5, the dynamic iterative method is consistently used, with parameter $p_t = 0.3$ and $N_s = 50$. The parameter inversion process refers to references [12,16].

To verify the capability of the CSUH model in describing the stress–strain characteristics of different types of soil, a dynamic iterative method was used to predict the CSUH model parameters for Fujinomori clay. The initial conditions are shown in Table 2.

Table 2. Initial Shear State of Fujinomori Clay.

OCR	1	2	4	8
p_0	196	196	196	98
e_0	0.769	0.719	0.668	0.684

For the sample with OCR = 8.0, where p_0 = 98 kPa, the process involved first consolidating to the preconsolidation pressure σ_c = 784 kPa, then unloading to p_0 = 98 kPa, and finally performing a constant p shear test at p_0 = 98 kPa. The loading and unloading processes for other overconsolidation ratios were similar. The CSUH model parameters are shown in Table 3.

Table 3. CSUH Model Parameters for Fujinomori Clay.

Parameters	M	ν	κ	λ	N	Z	χ	m
Values	1.36	0.0	0.02	0.093	1.26	1.26	0.05	5

5.1.1. Constant p Compression Test

As shown in Figure 9, the points represent the experimental results, while the lines represent the CSUH model predictions. The minimum overconsolidation ratio (OCR) is 1.0, corresponding to normally consolidated soil with the lowest density. The maximum OCR is 8, corresponding to heavily overconsolidated soil with the highest density.

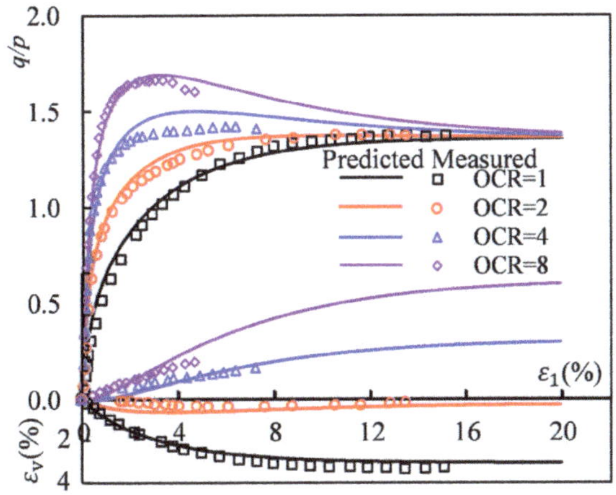

Figure 9. Comparisons between the constant p compression test results of the Fujinormori clay and the CSUH predictions [20].

Figure 9 shows that normally consolidated clay exhibits strain hardening and shear contraction, while heavily overconsolidated clays display strain softening and volumetric dilation. As the OCR increases, the peak stress ratio gradually increases, and the volumetric deformation transitions from shear contraction to initial shear contraction followed by shear dilation. When OCR = 8, the peak stress ratio is the highest, and the amount of volumetric dilation is the largest. Moreover, the CSUH model predictions align well with the experimental results for clays with different overconsolidation ratios, indicating that the CSUH model can reasonably quantify the differences in dilatancy behavior among soils with varying densities.

5.1.2. True Triaxial Test

The true triaxial test for Fujinomori clay follows the basic procedure of first undergoing isotropic consolidation to p = 196 kPa, followed by a constant p (or constant σ_m) shear test at an axial strain rate of 1% per day.

As shown in Figure 10, the CSUH model predictions agree well with the experimental results. This further demonstrates that, after incorporating the transformed stress method, the CSUH model can quantitatively describe the differences in the frictional behavior of soils under different stress Lode angles (θ).

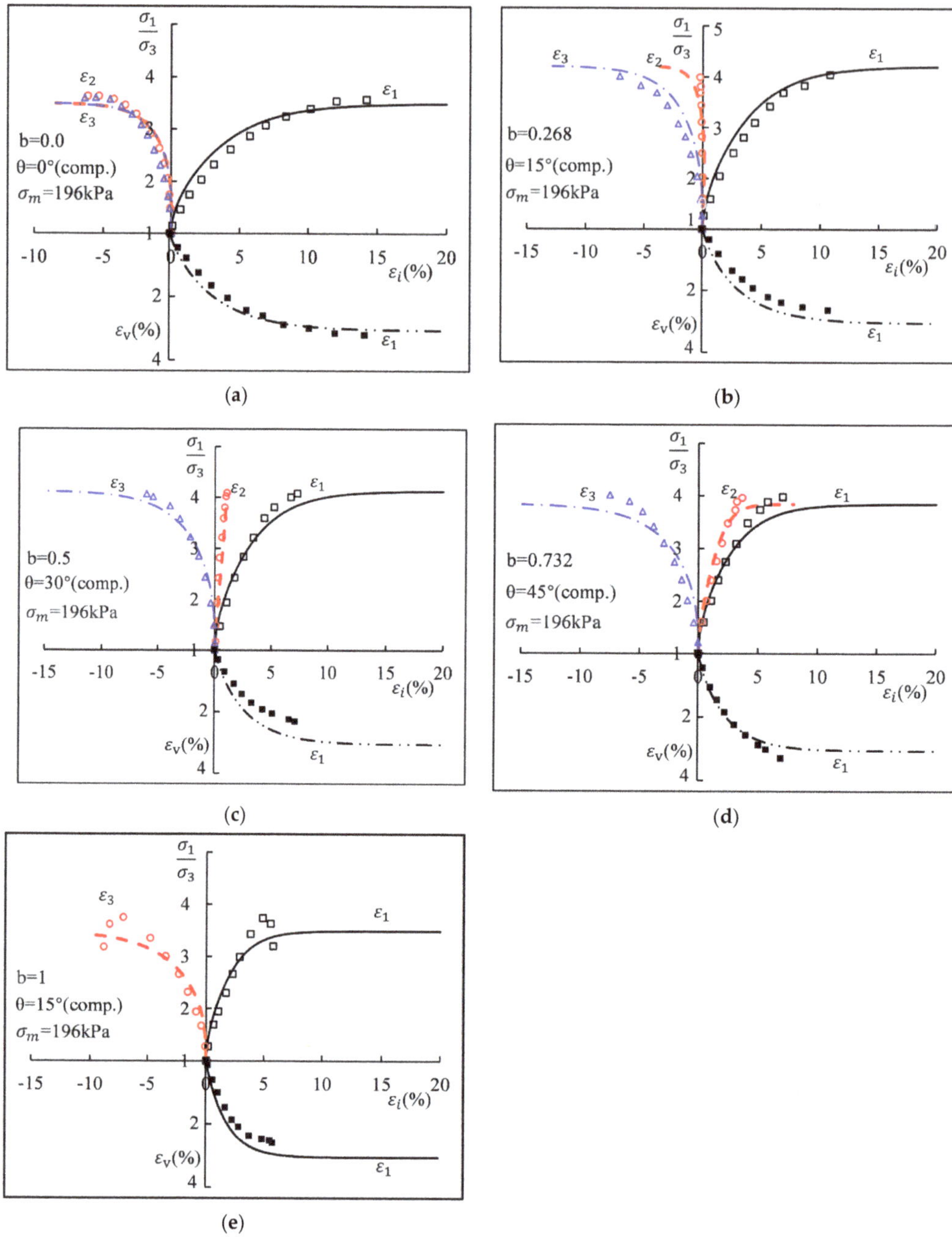

Figure 10. Comparisons between the true triaxial test measured results under constant p stress path with different medium principal stress coefficient b of the Fujinormori clay and the CSUH predictions [21] (dots: measurements; lines: predictions): (**a**) Medium principal stress ratio $b = 0$, equivalent to stress Lode angle $\theta = 0°$; (**b**) Medium principal stress ratio $b = 0.268$; (**c**) Medium principal stress ratio $b = 0.5$; (**d**) Medium principal stress ratio $b = 0.732$; (**e**) Medium principal stress ratio $b = 1$.

As shown in Figure 10, the horizontal axis represents ε_i, where i may denote 1, 2, or 3. The upper part of the vertical axis corresponds to σ_1 and the ratio σ_1/σ_3, while the lower part represents the volumetric stress σ_v. In Figure 10, lines indicate theoretical calculation results, whereas symbols denote experimental data. Specifically, the black solid square points represent ε_1-ε_v, the black open square points denote ε_1-σ_1/σ_3, the red open circles indicate ε_2-σ_1/σ_3, and the blue open triangles correspond to ε_3-σ_1/σ_3. The relevant explanations have been added to the figure caption.

In summary, the predictions for conventional triaxial and true triaxial tests of saturated Fujinomori clay indicate that the CSUH model can not only quantitatively distinguish the differences in dilatancy behavior among clays with different densities but also accurately capture the variations in frictional characteristics under different stress Lode angles.

5.2. Calcareous Sand

Weng Yiling [22] conducted CD and CU tests on dredged calcareous sand (also known as coral reef sand) from the Nansha Islands. The calcareous sand foundation mainly consists of gravelly sand or gravel soil, which is an uncemented loose material. In this study, experimental data for calcareous sand particles smaller than 0.075 mm were used to validate the model. The CSUH model parameters are listed in Table 4, and the comparison results are shown in Figures 11 and 12.

Table 4. CSUH model parameters for calcareous silt.

Parameters	M	v	κ	λ	N	Z	χ	m
Values	1.473	0.337	0.023	0.057	1.332	1.21	0.99	0.145

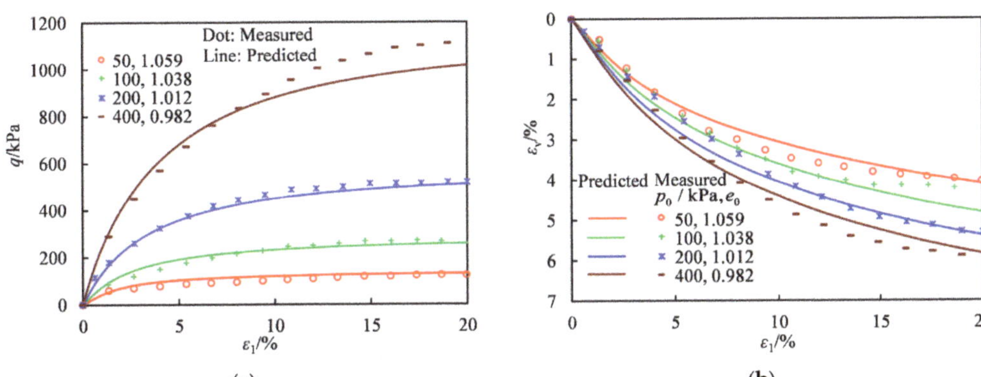

Figure 11. Comparisons between the CD test measured results of the calcareous silt and the CSUH predictions [22]: (**a**) Axial Strain vs. Deviatoric Stress; (**b**) Axial Strain vs. Volumetric Strain.

From Figure 11, it can be observed that under different confining pressures, the calcareous sand exhibits strain hardening and shear contraction in the CD tests.

As shown in Figure 12, the calcareous sand reaches peak strength at a relatively small strain, followed by strain softening until the strength nearly approaches zero. Meanwhile, pore pressure rises to a level almost equal to the axial stress, indicating that the soil sample is approaching a liquefied state. Therefore, when analyzing the effective stress ratio, it can be observed that the measured effective stress ratio exhibits anomalies, which were also noted in the literature [22].

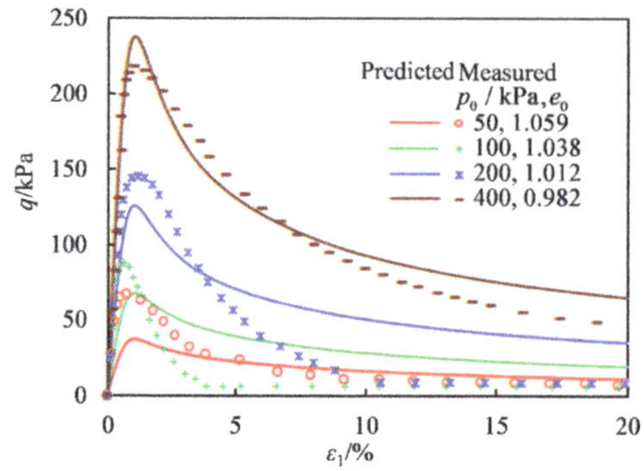

(**a**)

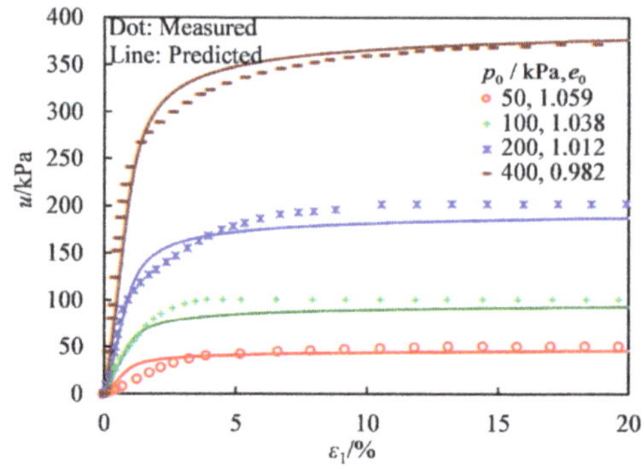

(**b**)

Figure 12. Comparisons between the CU test measured results of the calcareous silt and the CSUH predictions [22]. (**a**) Axial Strain vs. Deviatoric Stress; (**b**) Axial Strain vs. Excess Pore Pressure.

5.3. Rockfill Material

Liu et al. [23] conducted isotropic compression, CD, and CU tests on the main rockfill material (composed of hard diorite) from the Changhe dam. The CSUH model parameters for the rockfill material are listed in Table 5.

Table 5. CSUH model parameters for the rockfill material from the Changhe dam.

Parameters	M	ν	κ	λ	N	Z	χ	m
Values	1.678	0.272	0.021	0.087	1.125	0.742	0.385	1.716

5.3.1. Isotropic Compression Test

Figure 13 presents the results of isotropic compression tests on the rockfill material from the Changhe dam, along with the corresponding CSUH model predictions. The comparison between the predicted and experimental results indicates that the CSUH model can effectively describe the compression hardening behavior of large-grained granular soils such as rockfill. Compression hardening refers to the characteristic where the modulus of the soil increases with increasing density during compression, often described as "the more it compresses, the harder it gets" [24].

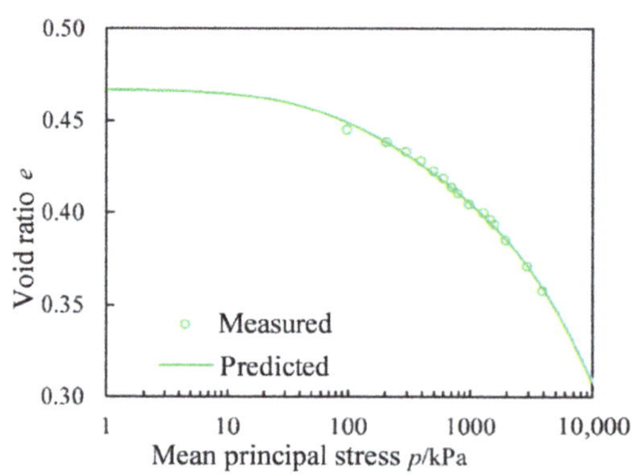

Figure 13. Comparisons between the isotropic compression test measured results of the rockfill from the Changhe dam and the CSUH predictions [23].

5.3.2. CD Triaxial Compression Test

Figure 14 compares the CD test results of the rockfill material under different confining pressures with the CSUH model predictions. The experimental results indicate that at lower confining pressures, the material exhibits slight strain softening and volume dilation, whereas at higher confining pressures, it exhibits strain hardening with volume contraction throughout. This behavior is primarily attributed to particle breakage in the rockfill material.

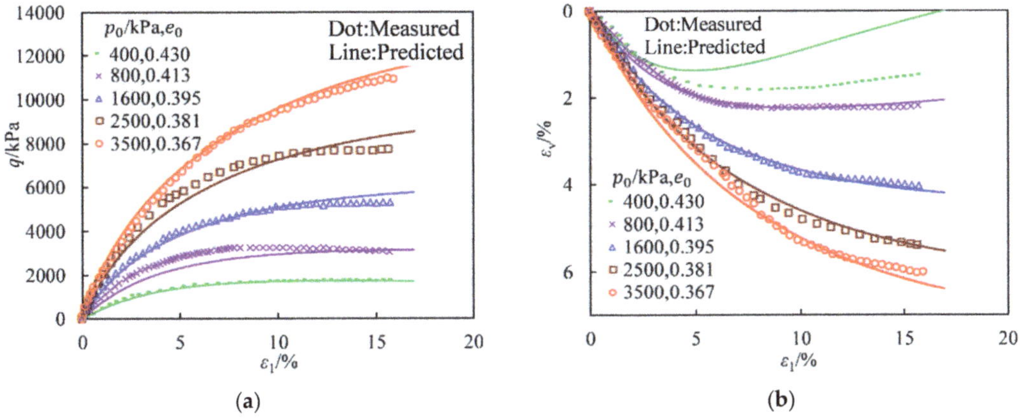

Figure 14. Comparisons between the CD test measured results of the rockfill from the Changhe dam and the CSUH predictions [23]: (**a**) Axial Strain vs. Deviatoric Stress; (**b**) Axial Strain vs. Volumetric Strain.

Since both rockfill and sand exhibit breakage characteristics, their qualitative mechanical behaviors are similar. However, rockfill is more prone to breakage than sand, leading to quantitative differences in test results. For example, in Figure 11, the calcareous sand with an initial state of (400 kPa, 0.982) and in Figure 14, the rockfill with an initial state of (3500 kPa, 0.367) both exhibit a final volumetric strain of approximately 6% after shearing. This suggests that if the initial conditions were the same, the final volumetric strains of the calcareous sand and rockfill would likely be different.

5.3.3. CU Triaxial Compression Test

Figure 15 compares the CU test results of the rockfill material under different confining pressures with the CSUH model predictions. From Figure 15a, it can be observed that under different confining pressures, the rockfill material exhibits strain hardening. Figure 15b shows that as axial strain increases, pore pressure initially rises rapidly and then gradually decreases, indicating a tendency for the rockfill material to undergo initial shear contraction followed by shear dilation in undrained conditions. Additionally, higher confining pressures result in greater pore pressure buildup, which offsets part of the confining pressure. As a result, for the same initial confining pressure, p_0, the undrained shear strength, q_f, is lower than the drained shear strength.

A comparison between Figures 14b and 15b reveals that under an initial confining pressure of 3500 kPa, the rockfill material undergoes continuous shear contraction in the drained test, whereas in the undrained test, it also exhibits dilation. The good agreement between the experimental results and the CSUH model predictions demonstrates that the CSUH model can quantitatively describe the differences in dilation behavior of the same material (rockfill) under different drainage conditions.

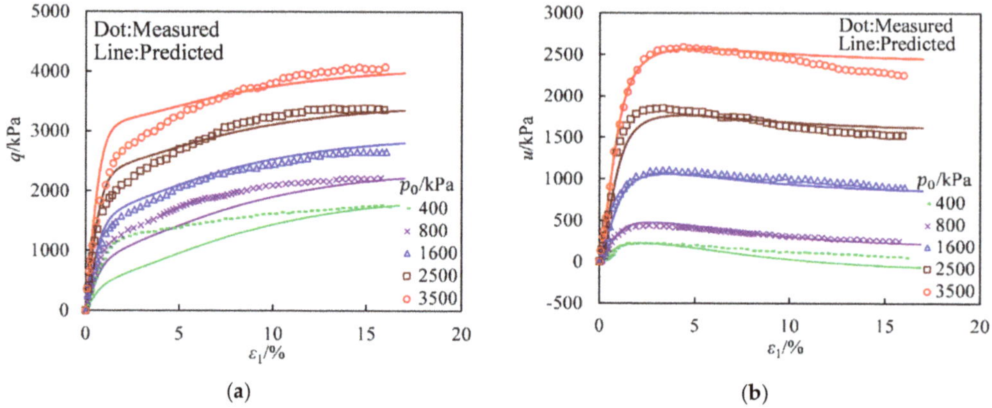

Figure 15. Comparisons between the CU test measured results of the rockfill from the Changhe dam and the CSUH predictions [23]: (**a**) Axial Strain vs. Deviatoric Stress; (**b**) Axial Strain vs. Excess Pore Pressure.

6. Conclusions

This study proposes a dynamic iterative method with segmented adjustment of incremental step sizes to improve the efficiency of parameter inversion for the CSUH model. Through theoretical analysis and experimental validation, the following key conclusions are obtained:

(1) The elastoplastic flexibility matrix of the elastoplastic constitutive model and the incremental stress–strain relationships under different boundary conditions are derived. This eliminates the need to solve equation systems when calculating theoretical model curves, thereby accelerating computation speed.

(2) Based on the variation characteristics of the stress–strain curve slope, the iterative step size is distributed using an arithmetic sequence over 30% of the total axial strain range, while a uniform distribution is applied over the remaining 70%. This effectively balances computational accuracy and efficiency. The study shows that this method reduces the number of iterations from the traditional 3000 steps to just 50 steps, decreasing computation time by approximately 47 times while maintaining accuracy comparable to the fourth-order Runge–Kutta method.

(3) The CSUH model accurately describes the stress–strain characteristics of different soil types (such as clay, sand, and rockfill) under various stress paths, including isotropic compression, constant p drained conditions with different intermediate principal stress ratios, CD, and CU tests.

The dynamic iterative method proposed in this study significantly enhances the efficiency of parameter inversion for elasto-plastic models, creating favorable conditions for the practical engineering application of the CSUH model. However, the CSUH model exhibits limitations in characterizing stress–strain behaviors of structured soils, anisotropic formations, and environments influenced by variable water content, prolonged creep processes, or thermal variations. Specialized constitutive formulations are required for these complex scenarios, necessitating tailored parameter recalibration and iterative scheme redevelopment.

In addition, it should be noted that the current validation is primarily performed on datasets used for model calibration and has not yet been extended to independent datasets or new, unseen stress paths. Although the dynamic iterative method demonstrates clear advantages in computational efficiency and accuracy in the present cases, further comprehensive cross-validation and testing on independent datasets and more complex loading conditions are needed in future research to fully establish the generalizability and robustness of the proposed approach.

Author Contributions: Conceptualization, B.Z. and H.Y.; methodology, B.Z.; software, B.Z. and J.Y.; validation, D.C., H.Y. and B.Z.; formal analysis, B.Z.; investigation, Z.B.; resources, D.C.; data curation, M.O.; writing—original draft preparation, B.Z.; writing—review and editing, B.Z. and Z.B.; visualization, B.Z.; supervision, D.C.; project administration, H.Y.; funding acquisition, D.C. All authors have read and agreed to the published version of the manuscript.

Funding: This research was funded by the National Natural Science Foundation of China (Grant No. U2268216); the Science and Technology Research and Development Program of China State Railway Group Co., Ltd. (Grant No. L2023G009); research projects of China Academy of Railway Sciences Corporation Limited (Grant No. 2023YJ384); and the 2022 Tencent XPLORER PRIZE (Research on Prefabricated Railway Subgrade Structure and Fundamental Theory).

Institutional Review Board Statement: Not applicable.

Informed Consent Statement: Not applicable.

Data Availability Statement: All data that support the findings of this study are included within the article.

Acknowledgments: We give thanks to Yangping Yao for his guidance on the unified hardening constitutive model.

Conflicts of Interest: All authors were employed by the company China Academy of Railway Sciences Corporation Limited. The authors declare that the research was conducted in the absence of any commercial or financial relationships that could be construed as a potential conflict of interest.

References

1. Yao, Y.; Luo, T.; Hou, W. *Soil Constitutive Models*; People's Transportation Publishing House Co., Ltd.: Beijing, China, 2018.
2. Zhang, Y.; Chen, Y. A constitutive relationship for gravelly soil considering fine particle suffusion. *Materials* **2017**, *10*, 1217. [CrossRef] [PubMed]
3. Cong, S.; Ling, X.; Li, X.; Geng, L.; Xing, W.; Li, G. Elastoplastic model framework for saturated soils subjected to a freeze–thaw cycle based on generalized plasticity theory. *Materials* **2021**, *14*, 6485. [CrossRef] [PubMed]
4. Dong, L.; Tian, S.; Yao, C.; Han, X.; Wang, K. A nonlinear constitutive model for remoulded fine-grained materials used under the Qinghai–Tibet railway line. *Materials* **2022**, *15*, 5119. [CrossRef] [PubMed]
5. Roscoe, K.H.; Schofield, A.N.; Thurairajah, A. Yielding of clays in states wetter than critical. *Géotechnique* **1963**, *13*, 211–240. [CrossRef]
6. Roscoe, K.H.; Burland, J.B. On the generalized stress–strain behaviour of 'wet clay'. In *Engineering Plasticity*; Cambridge University Press: Cambridge, UK, 1968.
7. Hou, W.; Yao, Y. Analysis in behaviors of over-consolidated clays described by cam-clay model and unified hardening model. *Ind. Constr.* **2011**, *41*, 18–23+140. [CrossRef]
8. Yao, Y.P.; Gao, Z.W.; Zhao, J.D.; Wan, Z. Modified UH model: Constitutive modeling of overconsolidated clays based on a parabolic Hvorslev envelope. *J. Geotech. Geoenviron. Eng.* **2012**, *138*, 860–868. [CrossRef]
9. Yao, Y.P.; Hou, W.; Zhou, A. Constitutive model for overconsolidated clays. *Sci. China Ser. E-Technol. Sci.* **2008**, *51*, 179–191. [CrossRef]
10. Yao, Y.P.; Hou, W.; Zhou, A. UH model: Three-dimensional unified hardening model for overconsolidated clays. *Géotechnique* **2009**, *59*, 451–469. [CrossRef]
11. Yao, Y.P.; Liu, L.; Luo, T.; Tian, Y.; Zhang, J.M. Unified hardening (UH) model for clays and sands. *Comput. Geotech.* **2019**, *110*, 326–343. [CrossRef]
12. Zhu, B.; Chen, Z. Calibrating and validating a soil constitutive model through conventional triaxial tests: An in-depth study on CSUH model. *Acta Geotech.* **2022**, *17*, 3407–3420. [CrossRef]
13. Ilie-Octavian, P.; Mihai-Octavian, P.; Mihaela, O.; Gabriela-Petruța, R.; Lucia, C.A. Calibrating DC01 material properties for finite element analysis with Abaqus and Isight. *Methodology* **2023**, *7*, 8. [CrossRef]
14. Sloan, S.W.; Abbo, A.J.; Sheng, D. Refined explicit integration of elastoplastic models with automatic error control. *Eng. Comput.* **2001**, *18*, 121–194. [CrossRef]

15. Sloan, S.W. Substepping schemes for the numerical integration of elastoplastic stress–strain relations. *Int. J. Numer. Methods Eng.* **1987**, *24*, 893–911. [CrossRef]
16. Zhu, B.; Su, X.; Cao, Y.; Yan, F. Determining parameters of the CSUH constitutive model by genetic algorithm. *Jpn. Geotech. Soc. Spec. Publ.* **2020**, *8*, 188–193. [CrossRef]
17. Yin, Z.; Jin, Y. Development of Geotechnical Optimization Platform EROSOPT. In *Practice of Optimisation Theory in Geotechnical Engineering*; Springer: Singapore, 2019; pp. 243–292. [CrossRef]
18. Yao, Y.P.; Wang, N. Transformed stress method for generalizing soil constitutive models. *J. Eng. Mech.* **2014**, *140*, 614–629. [CrossRef]
19. Kadlíček, T.; Janda, T.; Šejnoha, M.; Najser, J.; Beneš, Š. Automated calibration of advanced soil constitutive models. Part I: Hypoplastic sand. *Acta Geotech.* **2022**, *17*, 3421–3438. [CrossRef]
20. Nakai, T.; Hinokio, M. A simple elastoplastic model for normally and over consolidated soils with unified material parameters. *Soils Found.* **2004**, *44*, 53–70. [CrossRef] [PubMed]
21. Nakai, T.; Matsuoka, H.; Okuno, N.; Tsuzuki, K. True triaxial tests on normally consolidated clay and analysis of the observed shear behavior using elastoplastic constitutive models. *Soils Found.* **1986**, *26*, 67–78. [CrossRef] [PubMed]
22. Weng, Y. Study on the Shear Strength of Calcareous Soil and Its Influencing Mechanism. Ph.D. Thesis, Guangxi University, Nanning, China, 2017.
23. Liu, E.; Chen, S.; Li, G.; Zhong, Q.M. Critical state of rockfill materials and a constitutive model considering grain crushing. *Rock Soil Mech.* **2011**, *32*, 148–154. [CrossRef]
24. Yao, Y.; Zhang, B.; Zhu, J. Behaviors, constitutive models and numerical simulation of soils. *China Civ. Eng. J.* **2012**, *45*, 127–150. [CrossRef]

applied sciences

Article

Numerical Modeling and Theoretical Analysis of Deformation Characteristics of Non-Equal-Width Retained Walls

Kai Cui [1], Zheng Yang [2,*] and Jing Li [3,*]

[1] College of Construction Engineering, Sanmenxia Polytechnic, Sanmenxia 472000, China
[2] School of Resources and Safety Engineering, Central South University, Changsha 410083, China
[3] School of Resources and Environmental Engineering, Wuhan University of Science and Technology, Wuhan 430081, China
* Correspondence: zhengyang.yz@outlook.com (Z.Y.); lijing1994@wust.edu.cn (J.L.)

Abstract: The width of the pillar is an important factor in the stability of the underground space and the efficiency of resource recovery. This study aims to model the performance of retained walls in panel barrier pillar stopes. By simplifying the three-dimensional problem based on the mining operation, a two-dimensional mechanical model of non-equal-width retained walls was established, and the stress and deflection were solved analytically. The calculated deformation characteristics of equal-width and non-equal-width retained walls were analyzed and compared with numerical simulations. The results indicated that the deformation of retained walls is mainly influenced by the roof loads, the uniaxial compressive strength, and the internal friction angle of backfill materials. For equal-width retained wall design, corresponding to the areas of pillar stopes where the uniaxial compressive strength and internal friction angle of backfill materials are low, great lateral pressure will be created on the retained walls. This results in significant flexural wall deformations in this area, increasing the risk of wall collapses. In comparison, for non-equal-width retained walls, the width is defined based on the surrounding backfill materials, which could greatly reduce the risk of potential damage. For the mining operation at the actual mine, the non-equal-width design with 2.5 m and 4.0 m intervals was adopted for the panel barrier pillar stopes, and the final displacement of the roof of the stope after the completion of the mining is 34 mm, and the two sides of the mine wall remain in good integrity with no significant peeling or cracking identified. This design improves the recovery rate of mineral resources and the stability of mining.

Keywords: optimization of retained walls; pillar stability; room and pillar mining; mine backfill effects

1. Introduction

With the depletion of shallow mineral resources, mining inevitably goes deeper, where the operation faces significantly more additional challenges [1–5]. At shallower depths, the ground stresses around the ore body are low, so only a minimal number of pillars need to be retained as necessary ground supports to ensure the safety of the underground operation [6,7]. However, for deep mining, more in-depth geological investigations are required to help select the best mining method and design the most suitable pillar layout, to make the mining process safe and efficient [8,9].

In deep mining, in situ stresses around the ore body are much higher and it is difficult to effectively control the pressure of surrounding rocks using only pillars. Therefore, in mining practice, a large number of mines try to fill the mined areas with tailings and other mining wastes to bear the stresses of surrounding rocks together with pillars [10]. Back-filling using tailings also brings additional environmental benefits by addressing the issues of accumulated tailings on the ground surface, hence helping promote the restoration of ecological environments after mining [11,12]. At the early stage of the backfilling process, mine pillars are still the main stress-bearing structures and the backfill mainly plays the role of limiting the lateral displacement of mine pillars and improving their bearing capacity [13–16]. With the development of backfill technology, the strength of backfilled material has been enhanced and the contact area between the backfill and the roof has been gradually increased. Therefore, the overall bearing capacity of the backfill has been increased significantly to improve its ability to limit the movement of the roof strata [17–19]. This potentially allows the mining of previously retained pillars to improve the recovery of mineral resources [20]. The original pillars are left temporarily to divide the mining area and to ensure that multiple stopes are mined simultaneously without interfering with each other. Subsequently, these temporary pillars are mined, with only a few thin walls (thin pillars) left permanently underground. These thin walls are mainly used to prevent the mixing of backfill materials and ores at each stope. At the same time, these permanent thin walls can also limit the lateral displacement of backfill materials and build a load-bearing system together with the backfill to maintain the stability of the goaf [21]. There are a significant number of published studies on the load-bearing mechanism of the backfill and pillars in underground space [22–24]. Wang et al. [25] investigated the joint bearing mechanism between the backfill and the surrounding rocks using indoor uniaxial compression tests, and the results indicated that pillars could enhance the uniaxial compressive strength and the elastic modulus of the backfill. Zhu et al. [26] used particle flow code numerical simulation software to investigate the load-bearing effects of support pillars consisting of backfill and coal walls under different conditions, and the results indicated that the width of the residual coal pillars had the greatest influence on the stability of the support pillars relative to factors such as mining depth, mining thickness, filling material, and filling ratio. However, most published research focuses on the interaction behavior of a single backfill material with the pillar, and there are few studies on the case of two backfill materials. In practice, more than one backfill material exists because of the economic benefits.

In this study, based on elastic mechanics, the stress and deformation characteristics of permanently retained thin walls in panel barrier pillar stopes are solved when different backfill materials are used in adjacent stopes. The theoretical findings are verified by numerical simulations and collected monitoring data. The results obtained from this study will help design the optimal width of the retained walls to maximize the recovery of mineral resources.

2. Project Overview and Mining Methods

The study area is a mine in China, where direct mining is risky due to the thickness of the ore body and the depth of its burial. The mining method is sublevel open stoping with backfilling. In this operation, the ore body is firstly divided into different panels along the trend; then, a fixed-width barrier pillar is designed between the panels, and the room stopes and pillar stopes are alternately distributed along the panels perpendicular to the ore body trend, as shown in Figure 1. The mining of the ore body is divided into three steps: The first step is to mine the room stopes under the primary support provided by ore pillars on both sides and then backfill the mined stopes using the cemented tailings backfill material. The second step is to mine the pillar stopes under the support of cemented backfills on both sides and then backfill the mine stopes using the full-tailings backfill material. After completing the previous two steps, most of the mineral resources have been recovered. To further improve the recovery of mineral resources, the third step is to mine the panel barrier pillar stopes, and finally backfill the mined area also using the full-tailings backfill material.

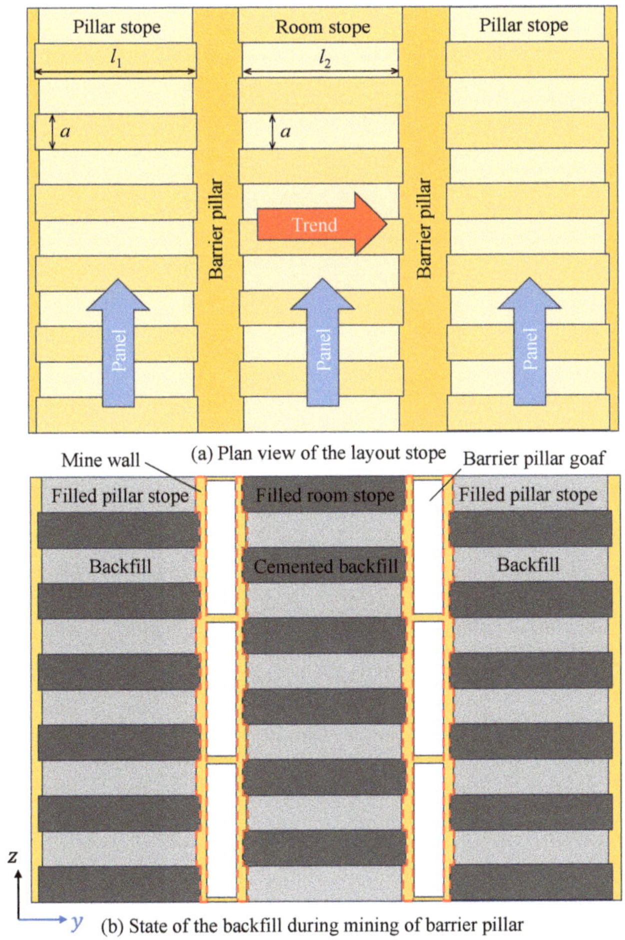

Figure 1. Schematic diagram of mining and filling sequence.

Although this step-by-step mining process is more complex, it can effectively control the regional stresses and maintain the stability of the stopes throughout the gradual mining and backfilling, and ultimately achieve the goal of less ore loss.

In Figure 1, l_1 is the length of the room stope; l_2 is the length of the pillar stope; and a is the width of the room stope and pillar stope.

3. Mechanical Modeling of the Permanent Thin Wall

At the third mining step, the thin walls on both sides of the barrier pillar stope are permanently retained in order to prevent mixing of the backfill material with the ore inside the stope. The thin wall has a small direct contact area with the roof compared with the backfill materials, which are the main bearers of the roof loads at this stage. On the horizontal plane, the wall has a large contact area with the backfill materials, so the forces acting on the wall can be simplified to only the lateral loads from the backfill materials. The basic assumptions for the thin walls are as follows: the wall is a continuous, homogeneous, isotropic, and perfectly elastic material, and the magnitude of the wall displacement is much smaller than its own size.

3.1. Mechanical Model of the Thin Wall When Using One Backfill Material

For the case of the same backfill material being used in both the room stopes and pillar stopes, if $l_1 = l_2$ (see Figure 1), the width of the thin wall does not change in the z-axis direction, as shown in Figure 2. In this case, it can be assumed that the lateral load q applied by the backfill materials to the thin wall does not change perpendicular to the wall. Therefore, the three-dimensional spatial problem can be simplified to a two-dimensional plane problem.

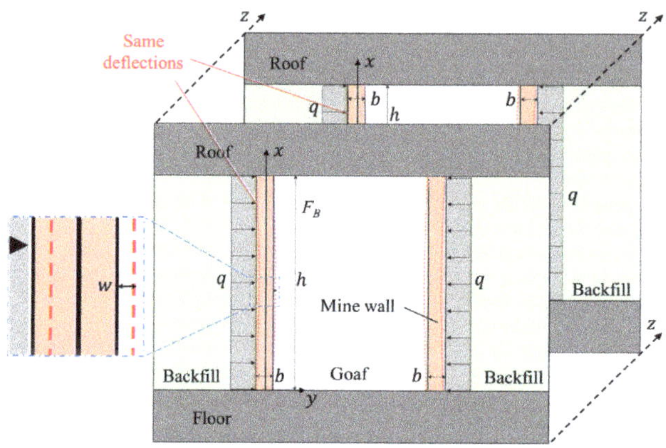

Figure 2. Mechanical model of the thin wall when using one backfill material.

As can be seen from Figure 2, the thin wall is only bearing lateral loads from backfill materials, and it can be regarded as a beam structure undergoing pure bending deformation; therefore, its deflection and stresses satisfy Equations (1) and (2), respectively.

$$E_0 I_z w'' = -M(x) \tag{1}$$

$$\sigma = -\frac{M(x)}{I_z} y \tag{2}$$

$$I_z = \frac{ab^3}{12} \tag{3}$$

where E_0 is the elastic modulus of the thin wall; I_z is the inertia moment of the wall cross-section about the neutral axis; b is the width of the wall; w is the deflection of the wall; and $M(x)$ is the bending moment of the wall under the lateral load q.

The two ends of the wall are parts of the surrounding rocks, so fixed constraints are applied. From structural mechanics, under uniform load, the bending moment and shear force at the two end points of the beam are as follows:

$$\begin{cases} F_{x=0} = F_{x=h} = qh/2 \\ M_{x=0} = M_{x=h} = -qh^2/12 \end{cases} \tag{4}$$

From Equation (4), the bending moment along the x direction can be derived as follows:

$$M(x) = -\frac{qh^2}{12} + \frac{qhx}{2} - \frac{qx^2}{2} \tag{5}$$

By combining Equations (1) and (5), the deflection equation of the wall can be solved by integration:

$$E_o I_z w = -\frac{q}{24}x^4 + \frac{qh}{12}x^3 - \frac{qh^2}{24}x^2 + C_1 x + C_2 \tag{6}$$

where C_1 and C_2 are undetermined coefficients.

Using Equation (6) and the boundary conditions of

$$\begin{cases} w|_{x=0} = 0 \\ w|_{x=h} = 0 \end{cases} \tag{7}$$

the deflection of the wall can be derived as follows:

$$w = -\frac{q}{2E_o ab^3}\left(x^4 - 2hx^3 + h^2x^2\right) \tag{8}$$

As expected, the deflection of the wall is directly proportional to the roof load and inversely proportional to the length of the mine wall, the elastic modulus, and the cube of the width.

3.2. Mechanical Model of the Thin Wall When Using Two Backfill Materials

In practice, pillar stopes are mined under the condition that both sides are backfilled at the first mining step, which uses the cemented tailings backfill material, which has sufficient strength to ensure that the backfill will not fail during the mining process of pillar stopes. After the recovery of pillar stopes, to reduce costs, full-tailings backfill instead of cemented tailings materials are used for the backfill. This results in different lateral loads being applied to the thin walls at different locations. Assuming that the lateral loads applied to the wall by cemented and un-cemented backfill materials are q_1 and q_2, respectively, the two walls of the same barrier pillar stope will have different deflections, as shown in Figure 3a. In this situation, the deformations of thin walls in adjacent barrier pillar stopes will affect each other, so the three-dimensional problem cannot be simplified into a two-dimensional plane problem.

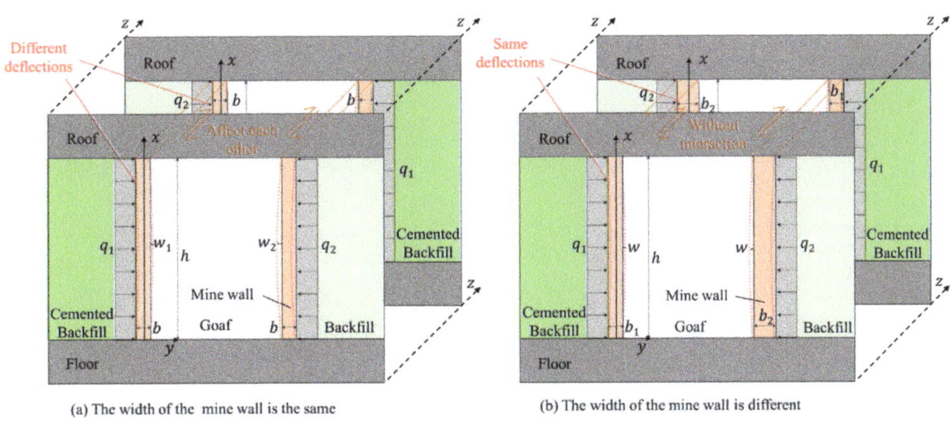

Figure 3. Mechanical model of the thin wall when using two backfill materials.

Since the thin wall deformation of adjacent barrier pillar stopes is different, the mutual influence is the root cause of the inability to reduce the three-dimensional problem to a two-dimensional one. Then, if it is guaranteed that the thin wall deformation of adjacent barrier pillar stopes is the same, the mutual influence between them will disappear. At this point, they can be regarded as two mutually independent planes, each calculated. As shown in Equation (8), the wall deflection is a function of the lateral load and the wall width. The width of the thin wall can be adjusted according to the lateral loads if it is desired to keep the deflection identical to ensure no interference of wall deformations between adjacent barrier pillar stopes. Based on Equation (8), this situation occurs when the values of wall width satisfy the following relationship:

$$\frac{q_1}{q_2} = \frac{b_1^3}{b_2^3} \tag{9}$$

As the backfill material bears the roof loads under lateral pressure support from the thin wall, the lateral pressure q and the roof pressure p satisfy the following relationship according to the Mohr–Coulomb criterion:

$$p = \sigma_{\text{cb}} + \frac{1 + \sin \varphi_b}{1 - \sin \varphi_b} q \tag{10}$$

where σ_{cb} is the uniaxial compressive strength and φ_b is the internal friction angle of the backfill material.

By combining Equations (9) and (10), to achieve the same wall deflection, the width of the thin wall should satisfy the following:

$$\frac{b_1}{b_2} = \sqrt[3]{\frac{(p - \sigma_{\text{cb1}})}{(p - \sigma_{\text{cb2}})}} \cdot \sqrt[3]{\frac{(1 - \sin \varphi_{\text{b1}})(1 + \sin \varphi_{\text{b2}})}{(1 + \sin \varphi_{\text{b1}})(1 - \sin \varphi_{\text{b2}})}} \tag{11}$$

where σ_{cb1} and σ_{cb2} are the uniaxial compressive strength values of the backfill materials in room stopes and pillar stopes, respectively; φ_{b1} and φ_{b2} are the internal friction angles of the backfill materials in room stopes and pillar stopes, respectively.

From Equation (11), one can observe that the thin wall width is mainly influenced by the roof load, uniaxial compressive strength, and the internal friction angle of the backfill materials. The wall width needs to increase as the roof load increases, the uniaxial compressive strength decreases, or the internal friction angle decreases. Without considering the influence of internal friction angles, when the roof load is much larger than the uniaxial compressive strength of both backfill materials and the uniaxial compressive strength of both backfill materials is close to each other, the width ratio of the two walls would be close to 1. When the roof load is close to the uniaxial compressive strength of one backfill material, but very different from the uniaxial compressive strengths of the other, the width ratio of the two walls would be far away from 1. From the perspective of internal friction angle, the smaller the difference in the internal friction angle between the two filling materials, the closer the width ratio of the two walls would be to 1. In summary, the different-width walls will be a better choice when the mine pressure and the backfill materials satisfy the following conditions: (a) there is a large difference between the uniaxial compressive strengths of the two backfill materials; (b) there is a large difference between the roof load and the uniaxial compressive strength of one of the backfill materials; and (c) there is a large difference between the internal friction angles of the two backfill materials.

3.3. Optimization of the Width for the Thin Wall

The roof of underground excavations will cave when it loses its support. Engineering practices have indicated [27,28] that the roof will reach a stable state after an arch structure is formed by caving. The arch structure can help re-distribute the ground stresses of the roof strata to unexcavated areas on both sides of the excavation. This stable arch structure formed by the phenomenon of stress redistribution is known as a natural arch.

For the mining operation described in this paper, backfill materials are used to replace the original rocks to support the roof strata. However, as the strength of the backfill materials is far lower than that of the original rock, they cannot completely replace the original rock to bear the stresses from the roof strata, so stress redistribution will inevitably occur. Based on the characteristics of a natural arch, the roof loads will gradually be re-distributed to rocks on both sides of the excavation if the bearing capacity of the substructure is insufficient. This is the case for the mining operation described, where the backfill will only bear the weight of caved loose rocks, as shown in Figure 4.

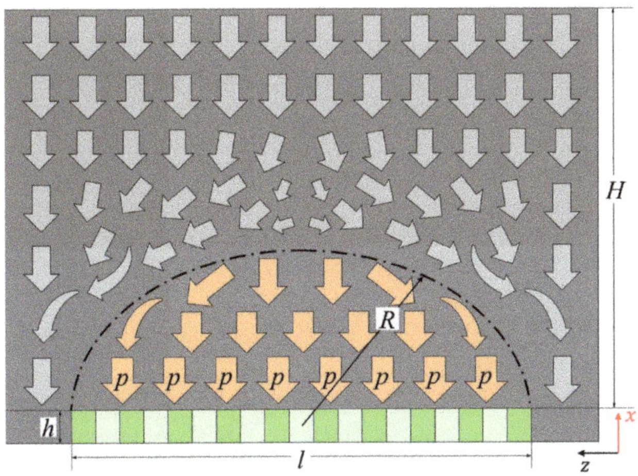

Figure 4. Schematic diagram of load distribution.

Based on the theory related to natural arches and the findings of our previous studies [29,30], the radius R of the rock mass in the disturbed region surrounding the mine with risk of falling can be determined as:

$$R = \left[\frac{(\rho_r g H + c_r \cot\varphi_r)(1-\sin\varphi_r)}{c_r \cot\varphi_r}\right]^{\frac{1-\sin\varphi_r}{2\sin\varphi_r}} \sqrt{\left(\frac{h}{2}\right)^2 + \left(\frac{na}{2}\right)^2} \tag{12}$$

where ρ_r is the density of the roof rock, g is the gravity acceleration (9.8 m/s^2), H is the depth of the stope from the ground, h is the height of the mine wall, c_r is the cohesion of the roof rock, φ_r is the internal friction angle of the roof rock and n is the number of stopes along the direction of the stope width.

The gravitational pressure of the caved rocks, namely, the roof load p, applied to the backfill is

$$p = \rho_r g\left(R - \frac{h}{2}\right) \tag{13}$$

According to the mine site test data, the mechanical parameters of different materials are shown in Table 1. Other required parameters are H = 720 m, n = 19, h = 25 m, and a = 18 m.

To ensure the stability of the thin wall, it is required that both the maximum compressive stress and the maximum tensile stress within the mine wall are less than their corresponding strength values. From Equation (5), $M(x)$ is a quadratic function whose maximum appears at $x = h/2$. When $x = h/2$, by combining Equations (2), (3) and (5), the stress inside the thin wall can be obtained:

$$\sigma = -\frac{qh^2}{2ab^3}y, -\frac{b}{2} \le y \le \frac{b}{2} \tag{14}$$

From this equation, the maximum compressive stress in the wall appears at $y = b/2$, and the maximum tensile stress at $y = -b/2$. From Equations (10), (12) and (13), when $b_1 < 1.23$ m or $b_2 < 3.94$ m, the tensile stress inside the wall is greater than the tensile strength, and therefore tensile damage may occur. From Equation (11), the deflections of the wall with different widths are the same when b_1:b_2 = 0.63. In this case, using b_1 = 2.48 m and b_2 = 3.94 m can ensure that the deflections of both walls are the same, and the wall will not be damaged as it has not exceeded its own ultimate stresses under the lateral loading of the backfill.

Table 1. Physical and mechanical parameters of different materials.

Name	Density (kg·m^{-3})	Elastic Modulus (GPa)	Cohesion (MPa)	Internal Friction Angle (°)	Poisson Ratio	Compressive Strength (MPa)	Tensile Strength (MPa)
Ores	3020	28.14	25.63	46.2	0.301	102.86	2.02
Roof rock	2600	6.13	15.20	35.1	0.327	51.32	1.16
Floor rock	2620	13.48	20.37	43.8	0.210	86.33	1.47
Cemented Backfill	1740	0.85	1.02	34.6	0.311	2.19	0.23
Backfill	1640	0.41	-	22.1	0.403	0	-

4. Numerical Simulation and Engineering Verification

4.1. Numerical Model and Initial Conditions

The geological conditions of the mine were simplified, and a 1000 m × 1200 m × 1200 m ($x \times y \times z$) numerical model was constructed using COMSOL Multiphysics 6.2 software, as shown in Figure 5a. The thickness of the roof strata in the upper part of the model is 720 m, the thickness of the ore body in the middle part of the model is 25 m, and the thickness of the floor rocks in the bottom part of the model is 255 m. Figure 5b,c illustrate the location of the ore body in the model and the layout of the stopes within the mine, respectively. The model was meshed using free tetrahedral units. Fine mesh was used in the mining area to ensure the accuracy of the model while coarse mesh was used in areas far away from the stopes to reduce the computational cost, resulting in a total of 640,207 elements. The minimum cell mass is 0.278, the average cell mass is 0.721, and the cell volume ratio is 2.43×10^{-4}, giving the mesh a high overall mass. In this case, the Drucker–Prager criterion was used. This criterion is a modified version of the Mohr–Coulomb criterion, retaining the ability to use the Mohr–Coulomb parameters while avoiding the problem of sharp corners that tend to occur in the principal stress space when the Mohr–Coulomb criterion is used. The parameters used in the simulations are listed in Table 1. As the model is large and the excavated area is located in the center of the model, far from the boundary, it can be assumed that there is no displacement at the bottom and only vertical displacement around the surrounding area. The following boundary conditions were used: fixed constraints on the bottom surface of the model, fixed displacement constraints in the normal directions of the four side surfaces of the model, and no constraint for the top surface of the model. Only gravitational loads were considered. The model is the result of a simplification of the actual mine, and in the simulation calculations, it is assumed that all material properties are homogeneous and isotropic, ignoring the anisotropy of materials in the actual geological environment. Moreover, the boundaries are set to be fixed or free, which may deviate from the actual situation. The current model is mainly applicable to laminated rock structures. For geological environments that are unstratified, highly fractured, or where there are a large number of faults, the predictive power of the model may be reduced.

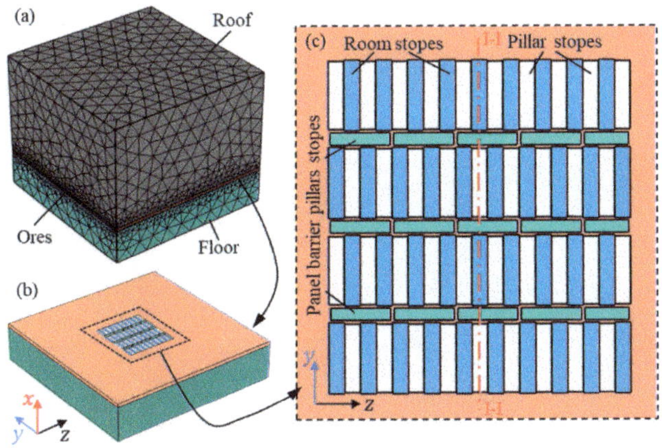

Figure 5. Numerical model: (**a**) meshing, (**b**) the location of the ore body, and (**c**) the layout of stopes.

4.2. Analysis of Simulation Results

The simulation was conducted according to the three-step mining processes. Figure 6 shows the contour plot of the vertical displacement in cross-section I-I (Figure 5) after the mining of the barrier pillar stopes is completed. The maximum displacement is found to be about 12 cm at the roof of the panel barrier pillar stope. The backfill materials of room stopes and pillar stopes play a partial bearing role and restrict the roof displacement, so the displacement at these locations is about 3 cm. The contour map of the vertical displacements shows a dome structure, which is in general agreement with the assumptions of a caved natural arch formed at the top of the excavation.

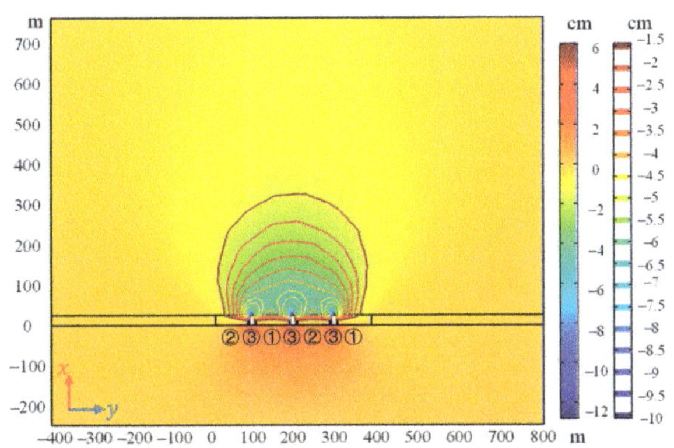

Figure 6. Contour plot of the vertical displacement on the I-I cross-section.

In this study, simulations were run for both equal-width and non-equal-width thin walls. For the first case, the room stopes and pillar stopes were set to the same size, and the width of the thin walls was set at 3.25 m, as shown in Figure 7a. As different backfill materials were used in the room stopes and pillar stopes, the horizontal displacement of the thin wall near the pillar stope is greater than that near the room stope, which is expected. For the case when the width of the thin walls is different, at 2.5 m and 4 m, respectively, the results are shown in Figure 7b. Compared with Figure 7a, the horizontal displacement of the thin wall close to the room stopes becomes much larger, while the horizontal displacement of the wall close to the pillar stope is slightly reduced.

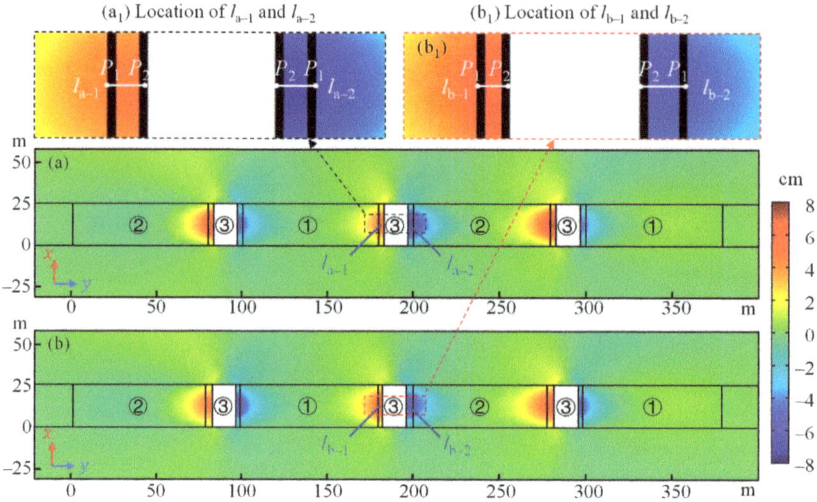

Figure 7. Horizontal displacement of the thin walls: (**a**) equal-width walls and (**b**) non-equal-width walls.

To have a better understanding of the horizontal displacement of the thin wall, the horizontal displacement in the middle of the thin walls from different stopes was extracted and is plotted in Figure 8. l_{a-1}, l_{a-2}, l_{b-1}, and l_{b-2} are the four segments, as shown in Figure 7, and their corresponding simulation results are listed in Table 2. When the width of the thin wall is 3.25 m, the average horizontal displacement of the wall close to the room stope is 4.92 cm, while the average horizontal displacement of the wall close to the pillar stope is 7.28 cm. The horizontal displacement of the wall close to the room stopes is only 67.6% of that close to the pillar stopes. In other words, the wall close to the pillar stopes may be damaged first, leading to stress concentration in the walls close to the room stopes which may be subsequently damaged. If the width of the wall was reduced to 2.5 m near the room stopes and increased to 4 m near the pillar stopes, it could be found that the average horizontal displacement of the wall increased to 5.96 cm for the 2.5 m wall and decreased to 5.72 cm for the 4 m wall. In this case, the horizontal displacements of both walls are approximately equal. In other words, the deformations of both walls were relatively more coordinated, and the risk of partial failure and overall collapse was reduced.

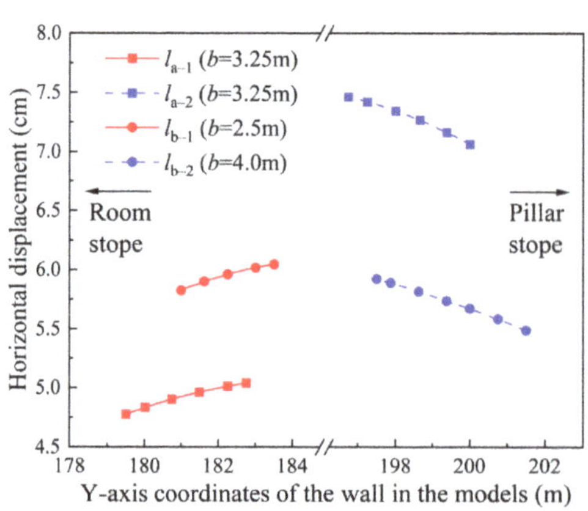

Figure 8. Horizontal displacement in the middle of the thin wall.

Table 2. Parameters of selected segments in the middle of the thin wall.

Name	Starting Point Coordinates (P_1)	Ending Point Coordinates (P_2)	Width	Average Horizontal Displacement
l_{a-1}	(12.5, 179.5)	(12.5, 182.75)	3.25	4.92
l_{a-2}	(12.5, 200)	(12.5, 196.75)	3.25	7.28
l_{b-1}	(12.5, 181)	(12.5, 183.5)	2.50	5.96
l_{b-2}	(12.5, 201.5)	(12.5, 197.5)	4.00	5.72

4.3. Verification Against Filed Data

The mine was mined in strict accordance with the design discussed above. From steps 1 to 3, the difficulty of mining gradually increases but the ore recovery rate increases. To understand the wall stability and the roof collapse within the barrier pillar stopes after mining, the Cavity Monitoring System (CMS) developed by Optech Company (Troy, MI, USA), was used to scan the goaf [30]. The collected data were further processed using the Qvol software, and then a 3D model of the goaf was built using the DIMINE modeling software. Examples of the typical cross-sectional profiles of the G6 goaf in the panel barrier pillar stope are shown in Figure 9.

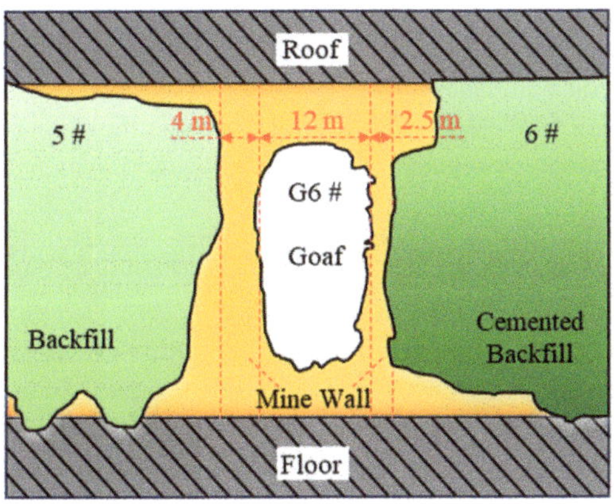

Figure 9. Typical profiles of the G6 goaf in the panel barrier pillar stope.

The scan results shown in Figure 9 indicate that the width of the retained walls is in good agreement with the design value without significant over-break or under-break. The scanning confirmed that both walls remain in good integrity with no significant peeling or cracking identified. No large quantities of backfill materials were found to have entered the stope during the mining process, and the walls performed well in isolating the backfill from the stopes. In addition, BGK-A3 multi-point displacement gauges were installed on the roof of the barrier pillar stopes to monitor the roof displacements. After 6 months of monitoring, the maximum displacements were 34 mm at the monitoring points in the G6 goaf. It is worth explaining that the maximum displacement monitored by numerical simulation is 12 cm, which is 3.5 times that of the on-site monitoring result, and there is some error. The main reason for this phenomenon is that the on-site monitoring results only represent the maximum displacement near the monitoring point, and do not represent the maximum displacement of the whole roof, so the maximum displacement of the roof on-site will be greater than 34 mm. In addition, the simulation parameters are the average of the rock tests at several sampling points, which may have some differences with the rock at the location of the monitoring point. Also, the simulation process simplifies the strata, and other reasons can lead to these errors. Although there was some error in the modeling and monitoring results, these displacements were within the safe range, proving that the backfill in this case had provided sufficient additional support to the roof strata.

The results demonstrated that the backfill, in this case, worked well, the thin wall did not appear to have large deformations or damages, the theoretical and numerical simulation results agreed well, and the layout of non-equal-width walls had a great application value in this type of mining operation.

5. Conclusions

This study modeled the performance of retained walls in panel barrier pillar stopes. By simplifying the three-dimensional problem based on the mining operation, a two-dimensional mechanical model of non-equal-width retained walls was established, and the stress and deflection were solved analytically. The following key conclusions were obtained:

(1) Retained barrier walls will be subjected to bending deformation under the lateral pressure created by the backfill materials. The degree of deformation is mainly influenced by the roof loads, as well as the uniaxial compressive strength and the internal friction angle of the backfill materials. A greater width of the retained walls will be required for higher roof loads, lower uniaxial compressive strength, or a lower internal friction angle of backfill materials.

(2) For the case when different backfill materials with different mechanical properties are used for room stopes and pillar stopes, non-equal width of the retained walls should be designed to ensure that the deformations of the walls are coordinated under the effect of different backfill materials. This will avoid the wall damage caused by over-deformation in part of the retained wall.

(3) The field project chose to use a non-equal-thickness spacing arrangement for the wall in the panel barrier pillar. The monitoring results indicated that the roof displacements of the G6 goaf at the third mining step were 34 mm and both sides of the backfill materials played a good role in supporting the roof. The backfill had sufficient stability in limiting its lateral displacement and the retained walls performed well in preventing the flow of the backfill materials into the stope during the mining operation. The theoretical model has been demonstrated to be able to provide answers to the problems discussed.

Author Contributions: Conceptualization, K.C.; methodology, K.C.; software, Z.Y.; validation, J.L.; resources, K.C.; writing—original draft preparation, K.C.; writing—review and editing, Z.Y.; visualization, J.L.; supervision, J.L.; funding acquisition, K.C. All authors have read and agreed to the published version of the manuscript.

Funding: This work was supported by the Key Scientific Research Project of Colleges and Universities of Henan Province (24A620003).

Institutional Review Board Statement: Not applicable.

Informed Consent Statement: Not applicable.

Data Availability Statement: Data is contained within the article.

Conflicts of Interest: The authors declare no conflicts of interest.

References

1. Yao, B.; Wei, J.; Wang, D.; Ma, D.; Chen, Z. Numerical study on seepage property of karst collapse columns under particle migration. *CMES Comput. Model. Eng. Sci.* **2013**, *91*, 81–100.
2. Yang, Z.; Tao, M.; Fei, W.; Yin, T.; Ranjith, P.G. Grain-based coupled thermo-mechanical modeling for stressed heterogeneous granite under thermal shock. *Undergr. Space* **2025**, *20*, 174–196. [CrossRef]
3. Yang, Z.; Tao, M.; Memon, M.B.; Zhuang, D.; Zhao, Y. Microwave irradiation-induced deterioration of rock mechanical properties and implications for mechanized hard rock excavation. *J. Rock Mech. Geotech. Eng.* **2025**, *17*, 275–290. [CrossRef]

4. Bai, J.; Dou, L.; Li, J.; Zhou, K.; Cao, J.; Kan, J. Mechanism of coal burst triggered by Mining-Induced fault slip under High-Stress conditions: A case study. *Front. Earth Sci.* **2022**, *10*, 884974. [CrossRef]
5. Zhao, Y.; Tao, M.; Du, K.; Wu, Y.; Wu, C. Development and application of gas adsorption model for coal based on particle flow code. *Gas Sci. Eng.* **2023**, *110*, 204858. [CrossRef]
6. Tuncay, D.; Tulu, I.; Klemetti, T. Re-analysis of abutment angle method for moderate and deep cover retreat room and pillar mines and investigation of loading mechanics using finite volume modeling. *Rock Mech. Rock Eng.* **2021**, *54*, 3447–3468. [CrossRef]
7. Rastiello, G.; Federico, F.; Screpanti, S. New soft rock pillar strength formula derived through parametric FEA using a critical state plasticity model. *Rock Mech. Rock Eng.* **2015**, *48*, 2077–2091. [CrossRef]
8. Naghadehi, M.; Mikaeil, R.; Ataei, M. The application of fuzzy analytic hierarchy process (FAHP) approach to selection of optimum underground mining method for Jajarm Bauxite Mine, Iran. *Expert Syst. Appl.* **2009**, *36*, 8218–8226. [CrossRef]
9. Guo, J.; Cheng, X.; Lu, J.; Zhao, Y.; Xie, X. Research on factors affecting mine wall stability in isolated pillar mining in deep mines. *Minerals* **2022**, *12*, 623. [CrossRef]
10. Sari, M.; Yilmaz, E.; Kasap, T. Long-term ageing characteristics of cemented paste backfill: Usability of sand as a partial substitute of hazardous tailings. *J. Clean. Prod.* **2023**, *401*, 136723. [CrossRef]
11. Huang, X.; Cai, X.; Bo, J.; Li, S.; Qi, W. Experimental study of the influence of gradation on the dynamic properties of centerline tailings sand. *Soil Dyn. Earthq. Eng.* **2021**, *151*, 106993. [CrossRef]
12. Gao, S.; Li, W.; Yuan, K.; Rong, C. Properties and application of thixotropic cement paste backfill with molybdenum tailings. *J. Clean. Prod.* **2023**, *391*, 136169. [CrossRef]
13. Li, J.; Huang, Y.; Qi, W.; Kong, G.; Song, T. Loose gangues backfill body's acoustic emissions rules during compaction test: Based on solid backfill mining. *Comput. Model. Eng. Sci.* **2018**, *115*, 85–103.
14. Hou, C.; Zhu, W.; Yan, B.; Yang, L.; Du, J.; Niu, L. Mechanical behavior of backfilled pillar under biaxial loading. *J. Cent. South Univ.* **2023**, *30*, 1191–1204. [CrossRef]
15. Wang, H.; Poulsen, B.; Shen, B.; Xue, S.; Jiang, Y. The influence of roadway backfill on the coal pillar strength by numerical investigation. *Int. J. Rock Mech. Min. Sci.* **2011**, *48*, 443–450. [CrossRef]
16. Hou, C.; Zhu, W.; Yan, B.; Guan, K.; Niu, L. Analytical and experimental study of cemented backfill and pillar interactions. *Int. J. Geomech.* **2019**, *19*, 04019080. [CrossRef]
17. Yan, B.; Ren, F.; Cai, M.; Qiao, C. Influence of new hydrophobic agent on the mechanical properties of modified cemented paste backfill. *J. Mater. Res. Technol.* **2019**, *8*, 5716–5727. [CrossRef]
18. Lu, B.; Li, Y.; Fang, S.; Lin, H.; Zhu, Y. Cemented backfilling mining technology for gently inclined coal seams using a continuous mining and continuous backfilling method. *Shock. Vib.* **2021**, *2021*, 6652309. [CrossRef]
19. Wang, X.; Zhang, J.; Li, M.; Gao, F.; Taheri, A.; Huo, B.; Jin, L. Expansion properties of cemented foam backfill utilizing coal gangue and fly ash. *Minerals* **2022**, *12*, 763. [CrossRef]
20. Zhu, W.; Yu, S.; Xuan, D.; Shan, Z.; Xu, J. Experimental study on excavating strip coal pillars using caving zone backfill technology. *Arab. J. Geosci.* **2018**, *11*, 554. [CrossRef]
21. Wang, J.; Qiao, D.; Han, R.; Li, G.; Xie, J. Strength model of cemented backfill in subsequent filling at the stage of open stope and its application. *Rock Soil Mech.* **2019**, *40*, 1105–1112. [CrossRef]
22. Du, X.; Feng, G.; Qi, T.; Guo, Y.; Zhang, Y.; Wang, Z. Failure characteristics of large unconfined cemented gangue backfill structure in partial backfill mining. *Constr. Build. Mater.* **2019**, *194*, 257–265. [CrossRef]
23. Le, Z.; Yu, Q.; Yang, T.; Cao, Y.; Zheng, H.; Deng, W. Compressive bearing and force transmission characteristics of granular backfill under lateral confinement. *Rock Soil Mech.* **2023**, *43*, 362–372. [CrossRef]
24. Wu, D.; Guo, W.; Luo, F.; Li, M.; Wen, P. Stiffness of gangue backfilling body in goaf and its influence mechanism on rock strata control and stress evolution in gangue backfill mining. *Environ. Sci. Pollut. Res.* **2023**, *30*, 61789–61807. [CrossRef]
25. Wang, J.; Fu, J.; Song, W.; Zhang, Y. Mechanical properties, damage evolution, and constitutive model of rock-encased backfill under uniaxial compression. *Constr. Build. Mater.* **2021**, *285*, 122898. [CrossRef]
26. Zhu, X.; Guo, G.; Liu, H.; Peng, X.; Yang, X. Stability analysis of the composite support pillar in backfill-strip mining using particle flow simulation method. *Environ. Earth Sci.* **2022**, *81*, 124. [CrossRef]
27. He, M.; Wang, Q. Excavation compensation method and key technology for surrounding rock control. *Eng. Geol.* **2022**, *307*, 106784. [CrossRef]
28. Yang, Y.; Zhao, Y.; Ma, J.; Han, P. Study on stability and bearing characteristics of macroscopic pressure arch of surrounding rock in western deep buried stope of China. *Front. Earth Sci.* **2023**, *11*, 1125689. [CrossRef]

29. Guo, J.; Zhao, Y.; Zhang, W.; Xie, X. Stress analysis of mine wall in panel barrier pillar-stope under multi-directional loads. *J. Cent. South Univ. (Sci. Technol.)* **2018**, *49*, 3020–3028. [CrossRef]
30. Tao, M.; Zhao, Y.; Guo, J. Arch model of roof and optimization of roof-contacted filling rate in two-step mining. *Trans. Nonferrous Met. Soc. China* **2023**, *33*, 1893–1905. [CrossRef]

applied sciences

MDPI

Article

Parameter Study and Engineering Verification of the Hardening Soil Model with Small-Strain Stiffness for Loess in the Xi'an Area

Jiayuan Hu and Qinwen Du *

School of Highway, Chang'an University, Xi'an 710054, China; chdhujy@163.com
* Correspondence: duqinwen@chd.edu.cn

Abstract: With the advancement of the construction of urban underground spaces, it is inevitable that new tunnels will pass through existing pipelines. To ensure the safety and stability of these pipelines, it is essential to strictly control the impact of shield tunneling. The hardening soil model with small-strain stiffness (HSS) comprehensively accounts for the small-strain behavior of soil, and the calculated results are closer to the values measured in engineering compared to those of other models. Consequently, it has been widely adopted in the development and utilization of underground spaces. The selection of parameters for the HSS model is particularly critical when performing numerical simulations. This article establishes the proportional relationships between the small-strain moduli of the HSS model in the loess region of Xi'an through standard consolidation tests, triaxial consolidation drained shear tests, and triaxial consolidation drained loading–unloading shear tests. Additionally, an empirical formula for the static lateral pressure coefficient applicable to loess was derived and validated through engineering examples.

Keywords: HSS model; indoor testing; tunnel engineering; numerical simulation; loess

1. Introduction

With the rapid development of China's economy and the acceleration of urbanization, traffic congestion has become an increasingly severe problem, causing significant inconvenience to residents during travel. The construction of urban underground railways has emerged as an important measure to address these issues. However, the layout of urban underground pipelines is intricate and complex, and tunnel excavation or foundation pit excavation must often occur close to existing pipelines [1,2]. This increases the complexity of underground engineering and places higher demands on deformation control during the project.

Numerical simulation has been widely used to predict the deformation and disturbance caused by underground engineering projects such as tunnels and foundation pits [3,4]. When finite element software is used for modeling, the choice of soil constitutive model has a significant impact on the prediction results.

The traditional Mohr–Coulomb constitutive model is an ideal elastoplastic model that does not account for the correlation between soil stiffness, stress, and stress path. Generally, it is considered to effectively reflect the failure stress state of soil and is widely used in engineering. However, this model does not account for soil unloading and rebound behaviors, resulting in calculation outcomes that are often overly conservative and unsuitable for engineering projects with stringent deformation-control requirements.

Atkinson and Smallfors [5] defined the small-strain range for soil as being between 0.0001% and 0.1%. Numerous engineering practices have shown [6–9] that a considerable portion of the soil surrounding underground structures is in a small-strain state under working loads. For instance, Burland [7] analyzed the excavation deformation of a deep foundation pit in the UK and found that the deformation of the soil was within 0.03%. Figure 1 [9] shows the strain range of the soil under various engineering conditions. The stiffness of soil under small-strain conditions exhibits a high degree of nonlinearity, and the true stiffness value of soil under small-strain conditions is significantly higher than the nominal elastic stiffness value obtained from conventional experiments. The HSS (hardening soil model with small-strain stiffness) model is a high-order elastoplastic hyperbolic model that considers the increased stiffness of soil during unloading, comprehensively considers the small-strain characteristics of soil, and can more accurately describe the plastic characteristics and small-strain conditions of soil. After years of research, this constitutive model has matured and is widely applied in engineering practice [10–14]. After practical engineering verification [11,15–17], compared with simple models such as M−C, the prediction results of the HSS model were found to be closer to the measured values in engineering applications.

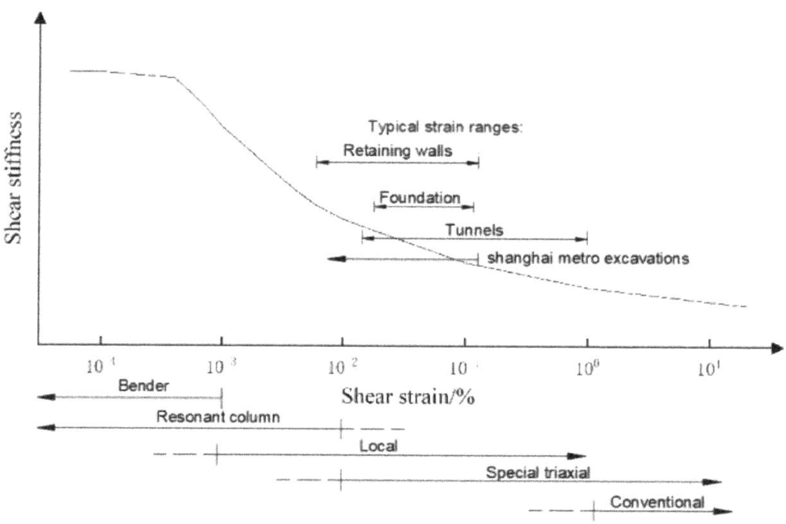

Figure 1. Strain limits for soil under conditions associated with various geotechnical engineering projects.

The difficulties and key point in using the HSS model for analysis and calculation lie in determining soil parameters. Currently, extensive relevant research has been conducted in only a few areas, such as Shanghai [18,19], while there are few instances of using the HSS constitutive model to study underground engineering in the loess region of Xi'an. In practical engineering, the HSS model parameters for loess are often based on empirical data from other regions, which significantly affects the accuracy of model predictions. Against this backdrop, this paper focuses on the newly constructed subway tunnel in Xi'an, and the research combines indoor tests and back-analysis methods to recommend HSS parameter values suitable for the loess region of Xi'an.

2. The Parameters and Introduction of the Hardening Soil Model with Small-Strain Stiffness Model

The HSS model was proposed by Benz [20] based on the hardening soil (HS) model, taking into account the nonlinear relationship between the shear stiffness and strain of soil in the small-strain range. Compared to the HS model, it can more comprehensively reflect the mechanical properties of soil such as small-strain behavior, compressive hardening, and dilatancy. The characteristics of common soil constitutive models are shown in Table 1.

Table 1. Characteristics of common constitutive models.

Model Category	Elasticity or Plasticity	Shear Hardening	Compression Hardening	Dilatancy	Small-Strain Characteristic	Stress Path Dependence
Linear Elastic	Elasticity	×	×	×	×	×
D–C	Elasticity	√	√	×	×	√
M––C	Perfect Elastoplasticity	×	×	√	×	×
DP	Elastoplasticity	√	×	√	×	×
HS	Plasticity	√	√	√	×	√
HSS	Plasticity	√	√	√	√	√

The HS model contains a total of 11 parameters, while the HSS model adds two additional small-strain parameters based on this, as shown in Table 2. In the Tables 1 and 2, "√" indicates that the indicator can be reflected, while "×" and "-" indicate that the indicator can not be reflected.

Table 2. The parameters and names of parameter in the HSS and HS models.

Parameter	Parameter Name	HS	HSS
c'	Effective cohesion of soil	√	√
φ'	Effective internal friction angle of soil	√	√
ψ	Dilation angle of soil	√	√
R_f	Failure ratio as determined by triaxial drainage shear test	√	√
v_{ur}	Loading–unloading Poisson's ratio	√	√
K_0	Initial coefficient of earth pressure at rest	√	√
p^{ref}	Reference stress	√	√
m	Power exponent related to the modulus stress	√	√
E_{50}^{ref}	Reference secant modulus as determined by triaxial consolidation drainage shear test	√	√
E_{oed}^{ref}	Tangent modulus under reference stress as determined by standard consolidation test	√	√
E_{ur}^{ref}	Loading and unloading modulus as determined by triaxial drainage shear loading–unloading test	√	√
G_0^{ref}	Initial dynamic shear modulus	-	√
$\gamma_{0.7}$	Shear strain corresponding to the attenuation of shear modulus to 70% of the initial shear modulus	-	√

Although these parameters can be obtained through indoor testing, they are quite complex, and it is difficult to provide all parameters in geological exploration reports.

Two of the parameters, c' and φ', can be taken from geological exploration reports. This article provides recommended values for E_{50}^{ref}, E_{oed}^{ref}, E_{ur}^{ref}, and R_f based on indoor testing, while the estimation of other parameters relies on relevant research experience or parameter inversion methods to obtain values.

3. The Indoor Tests on Loess Materials

3.1. Test Purpose

The parameters E_{50}^{ref}, E_{oed}^{ref}, and E_{ur}^{ref} in the HSS model, as well as the proportional relationship between these parameters and the compression modulus E_{s1-2}, can be obtained through standard consolidation tests and triaxial consolidation drained tests. The soil sample was taken from the construction site of Xi'an East Chang'an Street Subway Line 15, and the basic parameters of the soil sample are shown in Table 3.

Table 3. The basic properties of the soil used in the indoor test.

Soil Sample	Dry Density $\rho_s/(\mathrm{kg \cdot m^{-3}})$	Water Content $w/\%$	Void Ratio /m
Loess	1.562	21	0.75

The instrument used for the standard consolidation test was a single-lever consolidation apparatus, as shown in Figure 2a; the instrument used for the triaxial test was a stress-strain controlled triaxial shear permeameter, as shown in Figure 2b.

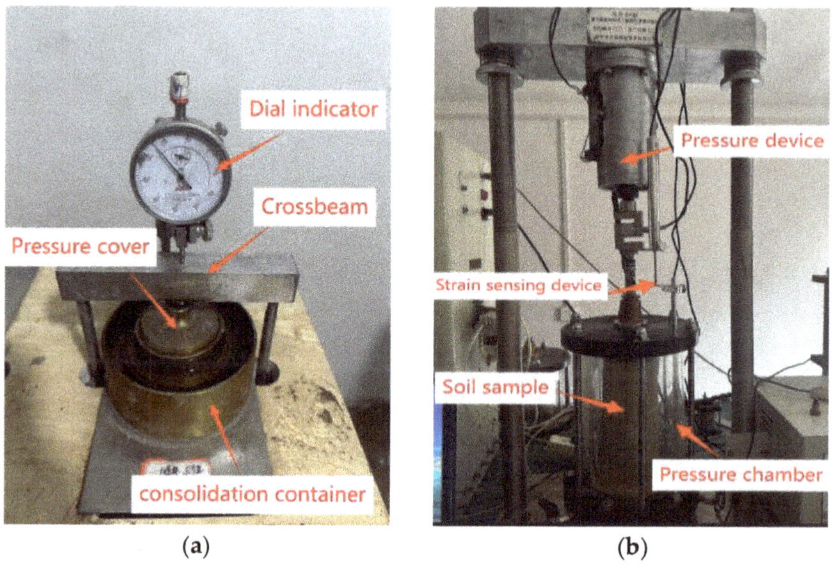

Figure 2. Photos of instruments used in the tests.

3.2. Test Procedure

(1) Standard consolidation test

(a) In the consolidation container, the following were placed in order from bottom to top: large permeable stones, filter paper, a ring-knife containing the saturated soil sample (with the ring-knife facing downward), filter paper, and small permeable stones. Then, the guide ring was placed and covered with the pressure cap. The permeable stones and filter paper had been pre-soaked to moisten them.

(b) Water was injected into the grooves around the ring-knife to ensure that the saturated sample remained unaffected by water evaporation during consolidation. During the test, the water in the grooves was replenished as needed.

(c) The crossbeam used to apply vertical pressure was lowered, ensuring that the bolts at the bottom of the crossbeam were securely fitted into the concave holes on the upper part of the pressure cover. Then, the dial indicator was installed and adjusted to zero.

(d) The vertical load was applied in five sequential stages: 50 kPa, 100 kPa, 200 kPa, 400 kPa, and 800 kPa. For each load level, the dial indicator reading was allowed to stabilize before the next load level was applied.

(e) The readings of the dial indicator were recorded after each level of load had stabilized.

(2) Triaxial consolidation drained shear test

(a) Water-head saturation: The water head outside the pressure chamber was set to approximately 1 m higher than the top of the sample. Under the influence of gravitational potential energy, water gradually seeped upward from the bottom of the sample, while gas was expelled through the exhaust pipe at the top of the sample.

(b) Back-pressure saturation: Water-head saturation often fails to fully saturate the sample, making back-pressure saturation necessary. The back pressure was set to 100 kPa, the confining pressure to 110 kPa, and the test duration to 3 h.

(c) B-value detection: The back pressure and undrained conditions were kept unchanged, and the confining pressure was increased by 30 kPa. Then, the pore water pressure coefficient B was measured. When $B = \Delta u/\Delta\sigma > 95\%$, the sample was considered saturated.

(d) Consolidation: The back pressure was maintained at the level of the saturated back pressure, and the effective confining pressure was set to 100 kPa, meaning that the confining pressure was 100 kPa greater than the back pressure. The consolidation time was 1 to 2 days.

(e) Drained shear: The confining pressure was kept constant, and the strain rate was controlled with an equal strain rate of 0.008 mm/min to ensure that the pore water in the sample was fully drained. The test was stopped when the strain of the sample reached 20% or when the sample exhibited significant strain softening.

(3) Triaxial consolidation drained loading–unloading shear test

The only difference between the triaxial consolidation drained loading–unloading shear test and the triaxial consolidation drained shear test was in step (e).

Step (e): Loading–unloading shear: During the drained shear process, when the axial strain reached approximately 40% of the axial strain corresponding to the peak strength of the sample in the triaxial consolidated drained test, shearing was stopped and axial unloading was carried out to zero. After the unloading was complete, the drained shear process was repeated until the end of the test.

3.3. Tests Result

The reference tangent modulus $E_{\text{oed}}^{\text{ref}}$ and compression modulus $E_{\text{s1}-2}$ of the soil sample were obtained through standard consolidation tests. Three parallel tests were conducted, and the relationship between the axial load and axial strain of the soil sample is presented in Figure 3. The fitted curve function R^2 for all three tests exceeded 0.99, resulting in $E_{\text{oed}}^{\text{ref}}$ values of 6.78, 6.86, and 6.41 MPa for the three soil samples, respectively, with an average value of 6.68 MPa.

Figure 4 illustrates the relationship between the void ratio and axial load in the standard consolidation test. Through parallel testing of samples a, b, and c, the compression moduli $E_{\text{s1}-2}$ of the soil samples were determined to be 7.08 MPa, 6.93 MPa, and 6.84 MPa, respectively, with an average value of 6.95 MPa.

In Figure 5, the stress–strain relationship curves for soil samples a, b, and c obtained from the triaxial consolidation-drained shear tests are presented. From these curves, the failure values q_{f} and the reference secant modulus E_{50}^{ref} at the reference stress can be determined. The q_{f} values were set to the peak stress values, which were 268.0 kPa, 264.3 kPa, and 266.6 kPa, respectively. The value of E_{50}^{ref} is equal to the slope of the line connecting the origin of the coordinate system with the point on the curve corresponding to $0.5q_{\text{f}}$. These values were 8.75 MPa, 8.19 MPa, and 9.22 MPa, respectively, yielding an average E_{50}^{ref} value of 8.72 MPa.

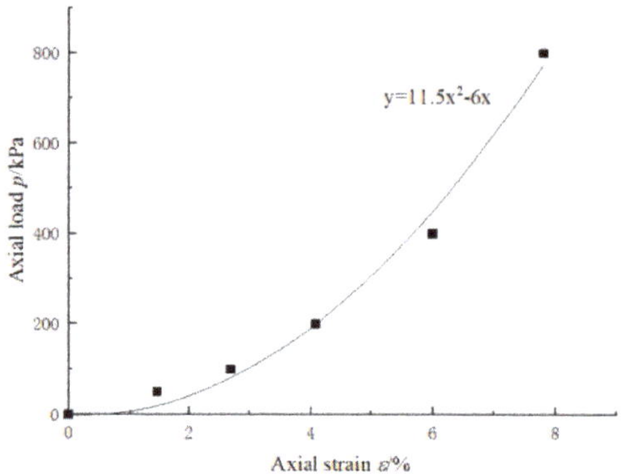

(**a**) The test curve of sample a

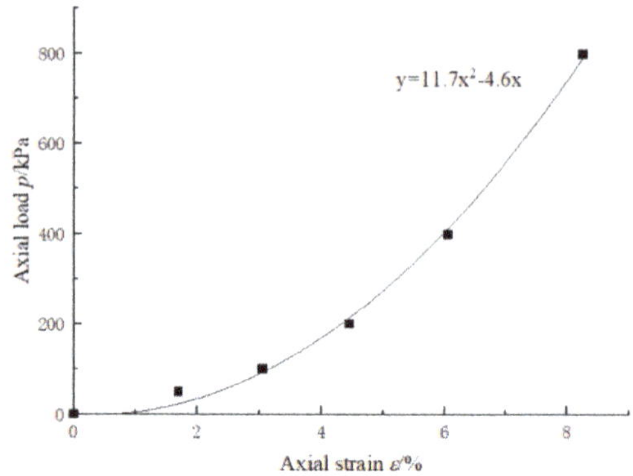

(**b**) The test curve of sample b

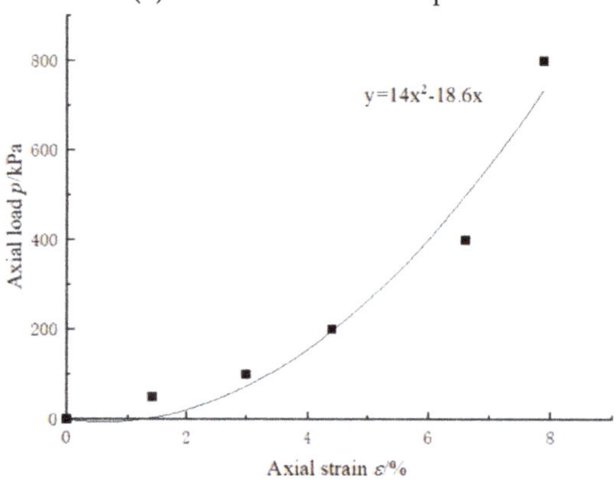

(**c**) The test curve of sample c

Figure 3. Relationships between axial load and axial strain, as determined by the standard consolidation test.

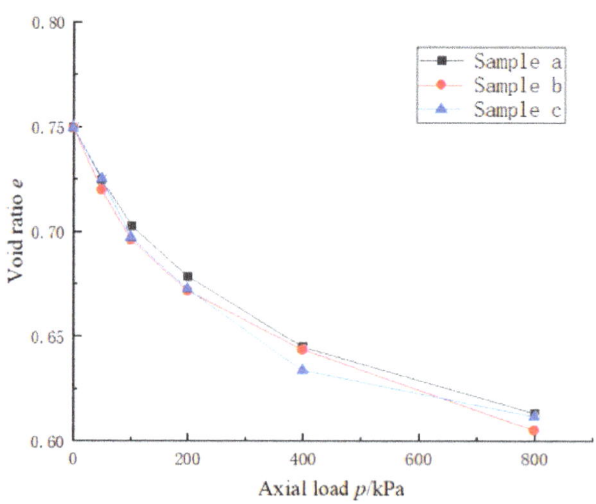

Figure 4. Relationship between void ratio and axial load diagram, as determined by the standard consolidation test.

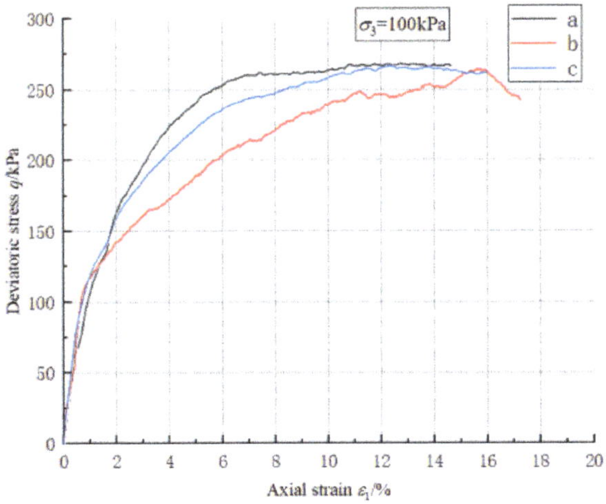

Figure 5. Stress–strain relationship based on the triaxial test.

In the definition of the HSS model, the axial strain ε_1 has a hyperbolic relationship with the deviatoric stress q, as shown in Formula (1):

$$\varepsilon_1 = \frac{1}{2E_{50}} \cdot \frac{q}{1 - q/q_a}(q < q_f) \tag{1}$$

After transformation, a linear relationship between ε_1/q and ε_1 can be obtained, as shown in Formula (2):

$$\frac{\varepsilon_1}{q} = \frac{\varepsilon_1}{q_a} + \frac{1}{2E_{50}}(q < q_f) \tag{2}$$

Due to the difficulty of perfectly aligning the actual stress–strain relationship with a hyperbolic curve, the transformed $\varepsilon_1/q - \varepsilon_1$ relationship curve will also deviate from linearity. To reduce human factors and perform a better linear fitting, the fitting line was made to pass through the points where the stress levels q/q_f were 70% and 95%, and the $\varepsilon_1/q - \varepsilon_1$ relationship line was plotted [11] (as shown in Figure 6); by taking the reciprocal of the slope of this line, the asymptotic value q_a could be obtained, and the failure ratio $R_f = q_f/q_a$, resulting in R_f values of 0.75, 0.84, and 0.86, with R_f set to 0.85.

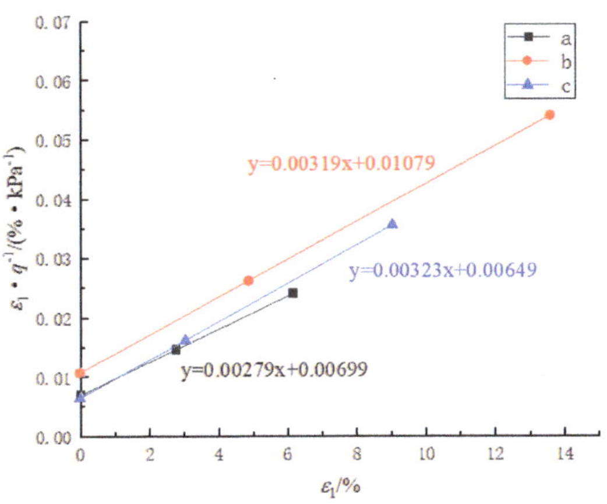

Figure 6. Curve graph of the relationship between ε_1/q and ε_1.

To obtain the reference loading and unloading modulus E_{ur}^{ref}, triaxial consolidation drained loading and unloading tests were required. The stress−strain curve obtained during the loading and unloading test at the reference confining pressure $p^{ref} = 100$ kPa is shown in Figure 7. During the loading and unloading process, the test curve of the soil sample exhibited a hysteresis loop. Connecting the two endpoints of the hysteresis curve, the slope of the resulting straight line is the reference loading and unloading modulus of the soil sample. The E_{ur}^{ref} values obtained from three sets of tests were 44.1 MPa, 41.7 MPa, and 43.4 MPa, with an average value of 43.1 MPa.

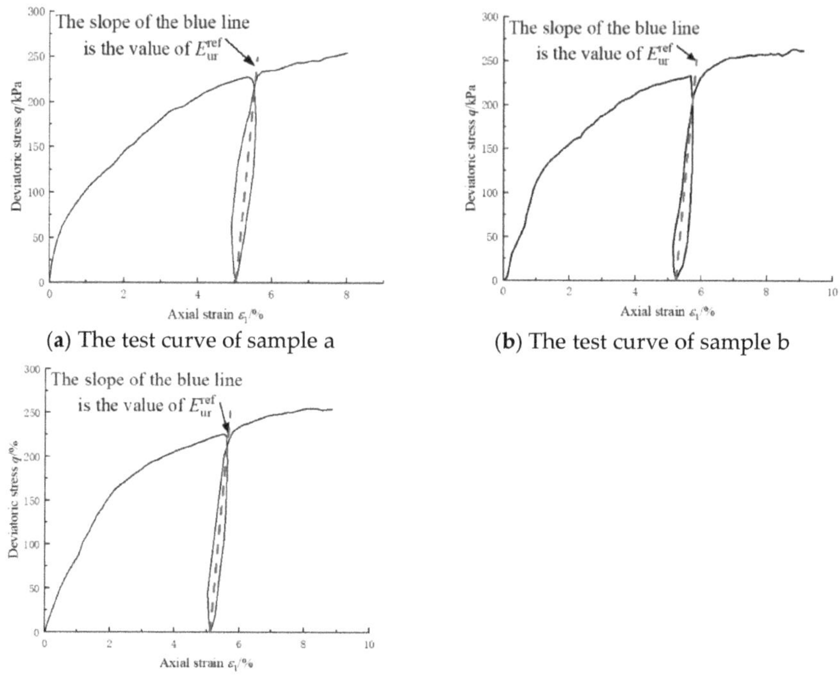

(**a**) The test curve of sample a (**b**) The test curve of sample b

(**c**) The test curve of sample c

Figure 7. Stress−strain curves of the triaxial loading and unloading test.

4. Selection of HSS Model Parameters

Through the above three parts of the test, the compression modulus E_{s1-2} of the soil sample can be obtained, as well as the small-strain moduli E_{oed}^{ref}, E_{50}^{ref}, E_{ur}^{ref} at the reference stress $p^{ref} = 100$ kPa, and the failure ratio R_f. This allows us to determine the ratio of the soil sample's compression modulus to its small-strain moduli, which is $E_{s1-2} : E_{oed}^{ref} : E_{50}^{ref} : E_{ur}^{ref} = 1 : 0.96 : 1.25 : 6.2$.

The value of G_0^{ref} needs to be determined through resonant column tests; it can also be calculated using empirical formulas. The initial dynamic shear modulus of soil G_0 is also known as the maximum shear modulus G_{max}; Hardin [21] provided an expression for G_{max} based on the soil's initial void ratio e_0 through extensive tests, and this is Equation (3); after summarizing the effects of e_0 and confining pressure on G_{max}, Hardin [22] improved this formula, thus obtaining Formula (4).

$$G_{max} = 33 \times \frac{(2.97 - e_0)^2}{1 + e_0} \quad (3)$$

$$G_{\max} = \frac{A(2.973-e)^2}{1+e}(\frac{p'}{p_a})^n \quad (4)$$

In the formula, p' represents the average effective confining pressure and p_a is the reference confining pressure, typically equal to standard atmospheric pressure (100 kPa).

After considering the effect of moisture content on the initial dynamic shear modulus of loess, Jian Tao et al. [23] improved Hardin's empirical Formula (4), and fitted the functions of A and n with respect to water content, making it more suitable for loess regions, as shown in Formulas (5) and (6).

$$A = 129.094 \times (100w)^{-0.364} \quad (5)$$

$$n = 0.301(100w)^{0.060} \quad (6)$$

Gu Xiaoqiang [17] derived the fitting functions (7) and (8) between the void ratio e and G_0^{ref} based on field wave-speed tests.

$$G_0^{ref} = 67.5e^{-1.57} \quad (7)$$

$$G_0^{ref} = 98.9e^{-0.45} \quad (8)$$

In this context, Formula (7) is more suitable for clay, while Formula (8) is more suitable for sand.

In actual engineering practice, the initial dynamic shear modulus G_0^{ref} can also be determined based on empirical values. As can be seen from Table 4, the value of G_0^{ref} in various regions generally ranges from 1.5 to 5 times that of the $E_{\text{ur}}^{\text{ref}}$. In loess regions, the ratio $G_0^{\text{ref}}/E_{\text{ur}}^{\text{ref}}$ typically falls between 1.5 and 3.

Table 4. The modulus ratios of HSS models in different regions.

Region	Soil Sample	$E_{\text{oed}}^{\text{ref}}/E_s$	$E_{50}^{\text{ref}}/E_{\text{oed}}^{\text{ref}}$	$E_{\text{ur}}^{\text{ref}}/E_{50}^{\text{ref}}$	$G_0^{\text{ref}}/E_{\text{ur}}^{\text{ref}}$
Gaoxiong [17]	Sand	-	1	3	2.42
Tongchuan [24]	Loess	0.93	12.46	7.6	2.23
Xi'an [25]	Loess	1.77	1.61	5.29	1.73
Yan'an [26]	Loess	-	1.2	3	3
Jinan [27]	Clay	1	1	8	1.5
Jinan [28]	Silty clay	-	0.91	6.97	1.74
Shanghai [18]	Soft clay	0.63~1.06	1.08~1.2	6.7~11.6	0.8~1.2
Shanghai [29]	Clay	0.9	1.2	6~8	2.5~4.9
Wuhan [30]	Silty clay	0.7~0.8	0.7~1.5	3.7~5.8	2.1~3.0
Hangzhou [31]	Clay	0.92	1.18	6.53	2
Hangzhou [32]	Silty clay	0.73~1	1~2.22	5~8.97	1.3~3
Suzhou [33]	Silty clay, Silt	-	0.8	4	3.8~4.55
Yancheng [34]	Silty clay	0.84	1.5	5.56	1.57~1.73
Changzhou [35]	Clay, Silty clay	-	1.33~1.50	3.6~4.3	1.6~2.0
Paris, France [36]	Sand	-	1	2.5	1.89
London, Britain [37]	Clay	-	2	5.3	4.16
Lisbon, Portugal [38]	Stiff clay	-	1	3	2.3
Jurmala, Latvia [39]	Subglacial till	-	1	3	2.96
Incheon, South Korea [40]	Sediment	-	1.25	3	2.65
Ledsgård, Sweden [41]	Clay	-	1	3	2.53~3.03

However, since the value of G_0^{ref} has a significant impact on the prediction generated by the numerical simulation, this paper will further analyze the value of G_0^{ref} in the loess region.

The coefficient of earth pressure at rest K_0 can be obtained through stress-path triaxial tests, and some geological survey reports may also provide it. However, in practical applications, it is usually calculated using empirical formulas, such as those for clay ($K_0 = 0.95 - \sin\varphi'$) and for sand ($K_0 = 1 - \sin\varphi'$). According to the theory of elasticity, it can also be calculated using Poisson's ratio ν (see Formula (9)):

$$K_0 = \nu/(1-\nu) \tag{9}$$

Gu Xiaoqiang et al. [19] established the relationship between K_0 and the void ratio e based on indoor tests. For cohesive soils, the relationship is as expressed in Formula (10), while for sandy soils, it is as shown in Formula (11):

$$K_0 = 0.049e + 0.02I_p + 0.139 \tag{10}$$

$$K_0 = 0.154e + 0.224 \tag{11}$$

In the formulas mentioned, I_p represents the plasticity index.

Since Formulas (10) and (11) are derived from the K_0 test values of typical soils in the Shanghai area, they are subject to regional limitations. Shanghai is characterized by soft clay soil, the mechanical properties of which may differ significantly from those of loess in the Xi'an area.

Wang Yu-chuan [42] obtained the fitting formula for K_0 of loess by fitting it with the over-consolidation ratio (OCR), as in Formula (12):

$$K_{OCR} = 0.42 \times OCR^{0.36} \tag{12}$$

Due to the significant influence of water content and stress history on the K_0 value of loess, Formulas (10) and (11), provided by Gu Xiaoqiang, do not account for the impact of stress history. Although Formula (12) is a fitting formula specifically for loess, it also fails to consider the effects of water content w and void ratio e. Consequently, using these formulas to calculate the K_0 value of loess may lead to some deviation.

This article fits the water content w, void ratio e, over-consolidation ratio, and the coefficient of earth pressure at rest K_0 for loess, resulting in Formula (13):

$$K_0 = 1.4w + 0.033e + 0.11OCR^{-0.28} \tag{13}$$

The deviation between the K_0 calculated using this formula and the experimental results does not exceed 15%, as shown in Figure 8. This indicates that Equation (13) is more suitable than the previous equation for calculating the K_0 of loess.

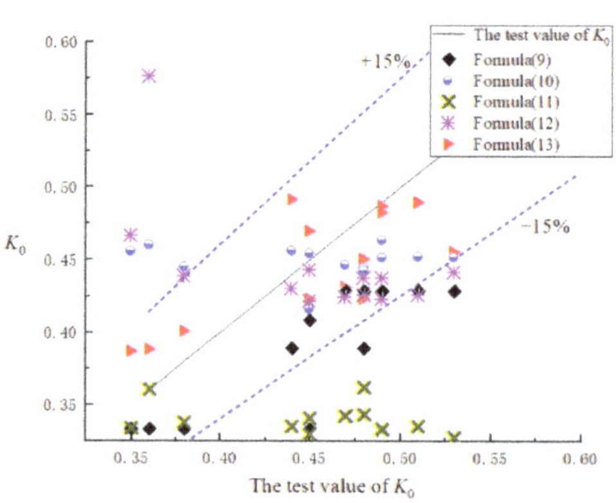

Figure 8. Comparison chart between K_0 calculated by various formulas and test results.

Other parameters can be determined based on the geological survey report and relevant research experience.

m was set to 0.75.

The value of ψ can be set according to Bolton's [43] recommendation: for sandy soil, $\psi = \varphi' - 30°$, and when $\varphi' < 30°$, ψ is set to 0; for clayey soil, ψ is set to 0.

Based on soil parameters in Shanghai, when the void ratio e is less than 1.0, R_f is set to 0.95. However, Zhong Peng [24] suggests that for loess, R_f should be between 0.8 and 0.9. This article, based on the results of tests, set R_f to 0.85.

When there are no experimental data available, v_{ur} can be set according to the recommended value in the PLAXIS (version, 20.1.0.98) manual, which is 0.2.

$\gamma_{0.7}$ can be set in the range 1×10^{-4} to 4×10^{-4} [44,45]. This article used the value 2×10^{-4}.

5. Engineering Verification

5.1. Project Overview

The proposed construction of the Xi'an Metro Line 15, from Shenzhou 2nd Road Station to East Chang'an Street Station, involves tunneling beneath the Heihe River water-supply pipeline (Phase II, single pipe). The right line is 1018.923 m long, and the left line is 1018.952 m long (including a short chain of 0.271 m). Both the left and right lines feature a single slope, with maximum gradients of 30.147‰ and 30.138‰, respectively. The burial depth of the tunnel arch top ranges from 19.94 m to 43.55 m, and the stable water-level burial depth is between 21 m and 35.8 m. The surrounding rock and soil are mainly classified as Class IV and Class V surrounding rocks. The minimum net distance between the shield-tunnel structure and the Heihe pipeline is only 3.867 m. The Heihe River water-supply pipeline has been in service for more than 20 years; it has low structural strength and is easily damaged due to long-term neglect. During the tunneling process, it is necessary to strictly control the disturbance to the water pipeline.

Figure 9 is a plan view of the relationship between the tunnel and the water pipeline. Figure 10 shows the internal view after tunnel construction.

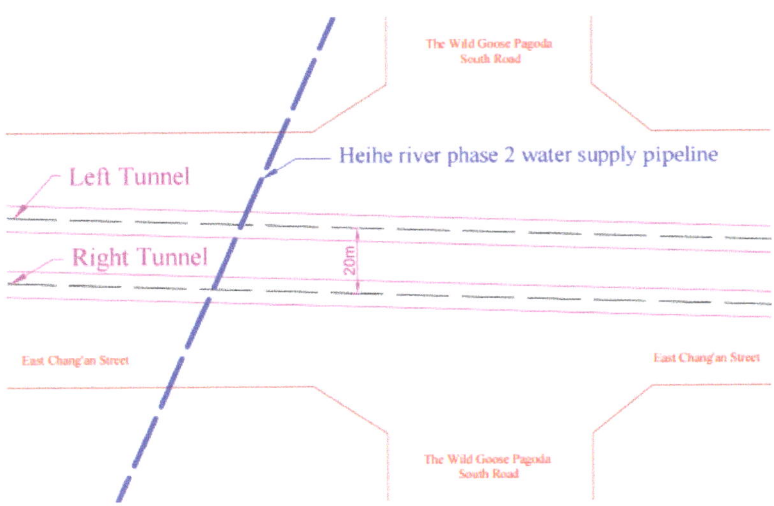

Figure 9. Plane diagram of the relationship between the tunnel and the Heihe River phase 2 water-supply pipeline.

Figure 10. Internal photo of tunnel.

The geological survey report provided soil-layer parameters, as shown in Table 5.

Table 5. Partial physical and mechanical parameters of soil.

Soil Layer Name	Unit Weight	Natural Water Content	Void Ratio	Over-Consolidation Ratio	Compression Modulus	Consolidated Quick Shear Test	
						Cohesion	Internal Friction Angle
	γ kN·m^{-3}	w %	e -	OCR -	E_s MPa	c kPa	φ °
I Plain Fill	15.7	-	-	-	-	0	5.0
II 3-1 New Loess	16.97	19.5	0.887	2.41	10.1	34.0	23.5
III 3-2 Paleosol	18.52	18.7	0.713	1.34	11.2	34.0	24.0
IV 4-1-1 Old Loess	16.87	21.2	0.925	1.30	9.7	34.0	24.0
V 4-1-2 Old Loess	19.52	21.2	0.768	1.03	6.6	33.5	22.5
VI 4-2-2 Old Loess	19.72	24.1	0.688	1.01	7.5	34.0	24.0

5.2. Modeling

Using the PLAXIS 3D finite element software for modeling, this study analyzed the left line tunnel passing beneath the Heihe River water-supply pipeline. The geometric dimensions of the model are 70 m $\times$ 90.5 m $\times$ 70 m. The tunnel has a circular cross-section with an inner diameter of 5.5 m and an outer diameter of 6.2 m. The cross-section of the Heihe River phase 2 water-supply pipeline is horseshoe-shaped, with a width of 2.5 m, a height of 2.5 m, and a wall thickness of 0.3 m. The model structure is shown in Figure 11. Considering the properties of various materials, the plain fill is modeled using the Mohr-Coulomb model, while the concrete lining of the tunnel and the water-supply pipeline are modeled using the linear elastic model. All other soil types are modeled with the HSS model.

The partial parameters of the HSS model for each soil layer are shown in Table 6.

The basic parameters of the concrete lining for tunnels and the Heihe Water Pipeline are shown in Table 7.

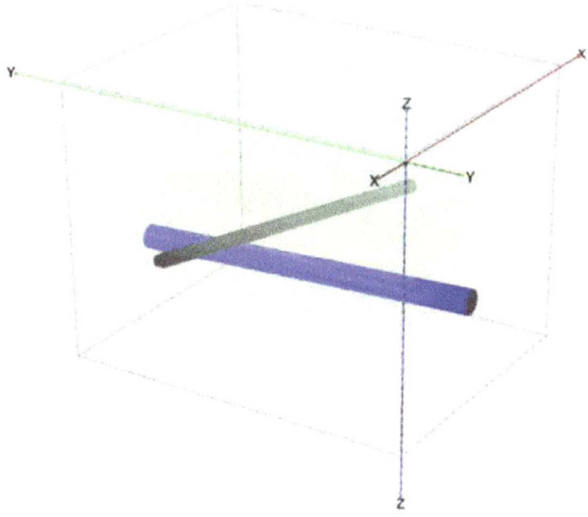

Figure 11. Model structure diagram.

Table 6. Partial parameters of HSS model.

Soil Layer Name	E_{oed}^{ref}/MPa	E_{50}^{ref}/MPa	E_{ur}^{ref}/MPa	K_0
II 3-1 New Loess	9.7	12.6	62.6	0.40
III 3-2 Paleosol	10.8	14	69.4	0.39
IV 4-1-1 Old Loess	9.3	12.1	60.1	0.43
V 4-1-2 Old Loess	6.3	8.3	40.9	0.43
VI 4-2-2 Old Loess	7.2	9.4	46.5	0.47

Table 7. The table of concrete lining.

Structure	Thickness	Unit Weight	Elastic Modulus	Poisson's Ratio
	m	kN·m^{-3}	GPa	-
Lining of tunnel	0.35	24.1	34.5	0.2
Lining of Heihe Water Pipeline	0.3	23.5	25.5	0.2

5.3. Analysis of Simulated Result

Numerical simulations were conducted using the HSS model, HS model, and M−C model for the shield tunnel passing beneath the Heihe River phase 2 water-supply pipeline. For the simulation using the HSS model, calculations were performed based on the improved Formulas (4), (7) and (8), as well as on results with $1.5E_{ur}^{ref}$, $2E_{ur}^{ref}$, and $3E_{ur}^{ref}$ multipliers. The simulation results were compared and verified against on-site monitoring data, as shown in Figure 12a. The comparison between the calculation results of the M−C and HS models and the field monitoring data is presented in Figure 12b.

As the shield tunnel advances, the settlement at the center of the bottom of the Heihe River water-supply pipeline gradually increases. The maximum value monitored on-site is 2.5 mm. The maximum settlement value simulated by the M−C model is 13.6 mm, while the maximum settlement simulated by the HSS model when $G_0^{ref} = 1.5E_{ur}^{ref}$ is 4.1 mm. It can be seen that the HSS model fits the on-site monitoring data better than the M−C and HS models. The engineering requirement is that the cumulative settlement of the Heihe River water pipeline must not exceed 8 mm. It is evident that the calculation results of the M−C and HS models are overly conservative, and the accuracy of these results no longer meets the needs of this project.

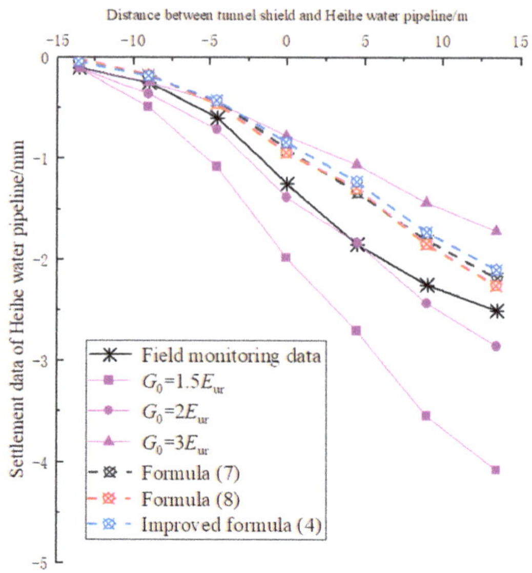

(**a**)

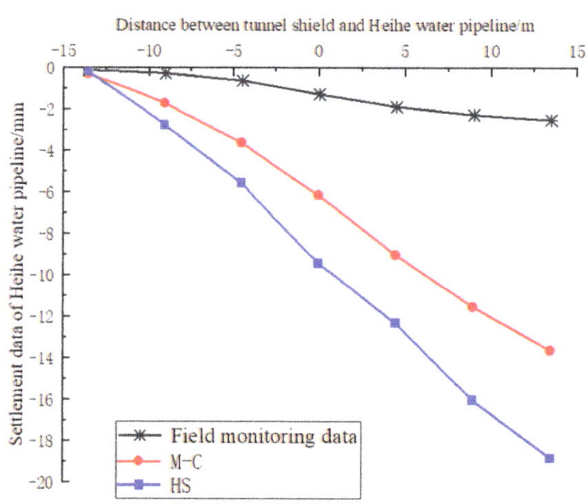

(**b**)

Figure 12. Variations in the bottom centerline settlement of the Heihe water pipeline due to shield tunneling. (**a**) The settlement prediction results of the HSS model under different values of G_0. (**b**) The prediction results of the HS model and M-C model.

Among the various methods of determining the value of G_0^{ref}, the simulation results from the three empirical formulas are similar, but all are lower than the on-site monitoring data. When G_0^{ref} is set to $2E_{ur}^{ref}$, the simulation results from the HSS model are the closest to the on-site monitoring data, and the settlement values obtained from the simulation are slightly greater than the monitoring data, which could be more conducive to construction safety.

6. Conclusions

This article obtained several parameters of the HSS model for Xi'an loess through indoor tests, established proportional relationships between the parameters, and vali-dated them using engineering examples. The conclusions are as follows:

(1) The proportional relationship between the small-strain moduli in the HSS model for Xi'an loess was obtained through indoor tests. Numerical simulations were performed for various values of the initial dynamic shear modulus G_0^{ref}. Through a comparison of the simulation results with the field monitoring data, it was found that when the ratio of E_{s1-2}, $E_{\text{oed}}^{\text{ref}}$, E_{50}^{ref}, $E_{\text{ur}}^{\text{ref}}$ and G_0^{ref} in the HSS model is 1:0.96:1.25:6.2:12.4, and $R_f = 0.85$, the simulation results are closest to the field monitoring data. This can provide a reference for similar engineering analyses in the Xi'an loess region.

(2) By fitting the data, a functional relationship between the coefficient of earth pressure at rest K_0 and the water content w, void ratio e, and over-consolidation ratio OCR for loess was established. The deviation between the values calculated using this fitting formula and the test data is within 15%, providing an empirical method for determining the K value of Xi'an loess.

(3) Through numerical analysis of the section between Shen Zhou 2nd Road Station and Dong Chang'an Street Station on Xi'an Metro Line 15, it was found that compared to the M−C and HS models, the HSS model results are closer to the field monitoring data. The HSS model is more suitable for projects with stricter settlement requirements, and it also validates the feasibility of applying the HSS parameters proposed in this paper.

(4) The above conclusions are derived from the study of loess in the Xi'an area, and further research is needed to determine whether they are applicable to loess in other regions. Future studies will conduct experiments on loess from more regions and at different depths to analyze and summarize the patterns of HSS model parameters for loess in various areas.

Author Contributions: Conceptualization, J.H. and Q.D.; methodology, J.H. and Q.D.; software, J.H. and Q.D.; validation, J.H. and Q.D.; formal analysis, J.H.; investigation, J.H.; resources, Q.D.; data curation, J.H.; writing—original draft preparation, J.H.; writing—review and editing, J.H. and Q.D.; visualization, J.H.; supervision, Q.D.; project administration, Q.D.; funding acquisition, Q.D. All authors have read and agreed to the published version of the manuscript.

Funding: This research received no external funding.

Institutional Review Board Statement: Not applicable.

Informed Consent Statement: Not applicable.

Data Availability Statement: The raw data supporting the findings of this study are available from the corresponding author upon reasonable request.

Acknowledgments: The authors thank Yanrui Chang (Y.C.) for his guidance on indoor tests and thank Xu Li (X.L.) for providing us with resources.

Conflicts of Interest: The authors declare no conflicts of interest.

References

1. Jafari, M. System identification of a soil tunnel based on a hybrid artificial neural network–numerical model approach. *Iran. J. Sci. Technol.-Trans. Civ. Eng.* **2020**, *44*, 560–572. [CrossRef]
2. Zhao, X.; Li, Z.; Dai, G.; Wang, H.; Yin, Z.; Cao, S. Numerical Study on the Effect of Large Deep Foundation Excavation on Underlying Complex Intersecting Tunnels. *Appl. Sci.* **2022**, *12*, 4530. [CrossRef]
3. Han, Y.; Ye, W.; Wei, Q. Research on the Influence of New Shield Tunnel to Adjacent Existing Tunnel. *Appl. Mech. Mater.* **2013**, *295–298*, 2985–2989. [CrossRef]

4. Zhang, Z.; Huang, M. Geotechnical influence on existing subway tunnels induced by multiline tunneling in Shanghai soft soil. *Comput. Geotech.* **2014**, *56*, 121–132. [CrossRef]
5. Atkinson, J.; Sallfórs, G. Experimental determination of stress-strain-time characteristics in laboratory and in-situ tests. General Report. In Proceedings of the 10th European Conference on Soil Mechanics and Foundation Engineering, Florence, Italy, 26–30 May 1991; pp. 915–956.
6. Wang, Z.; Wang, H.; Qi, K.; Zai, J. Summary and evaluation of experimental investigation on small strain of soil. *Rock Soil Mech.* **2007**, *28*, 1518–1524.
7. Burland, J. Ninth Laurits Bjerrum Memorial Lecture: "Small is beautiful"—The stiffness of soils at small strains. *Can. Geotech. J.* **1989**, *26*, 499–516. [CrossRef]
8. Attewell, P.; Farmer, I. Ground deformations resulting from shield tunnelling in London Clay. *Can. Geotech. J.* **1974**, *11*, 380–395. [CrossRef]
9. Mair, R. Developments in geotechnical engineering research: Application to tunnels and deep excavations. In Proceedings of the Institution of Civil Engineers, London, UK, 20 January 1993; pp. 27–41.
10. Zheng, G.; Yang, X.; Zhou, H.; Du, Y.; Sun, J.; Yu, X. A simplified prediction method for evaluating tunnel displacement induced by laterally adjacent excavations. *Comput. Geotech.* **2018**, *95*, 119–128. [CrossRef]
11. Ardakani, A.; Bayat, M.; Javanmard, M. Numerical modeling of soil nail walls considering Mohr Coulomb, hardening soil and hardening soil with small-strain stiffness effect models. *Geomech. Eng.* **2014**, *6*, 391–401. [CrossRef]
12. Mu, L.; Huang, M. Small strain based method for predicting three-dimensional soil displacements induced by braced excavation. *Tunn. Undergr. Space Technol. Inc. Trenchless Technol. Res.* **2016**, *52*, 12–22. [CrossRef]
13. Zhang, R.; Zhang, W.; Goh, A. Numerical investigation of pile responses caused by adjacent braced excavation in soft clays. *Int. J. Geotech. Eng.* **2018**, *15*, 783–797. [CrossRef]
14. Zhang, J.; Zhao, G.; Zhang, L.; Jiang, H. Application of HSS Model in Shield Simulation and Parameter Sensitivity Research. *Chin. J. Undergr. Space Eng.* **2020**, *16*, 618–625.
15. Xiao, S.; Xu, M.; Lan, R. Choice of Soil Constitutive Models in Numerical Analysis of Foundation Pit Excavation Based on FLAC3D. In Proceedings of the 2023 International Conference on Green Building, Civil Engineering and Smart City (GBCESC 2023); Lecture Notes in Civil Engineering. Springer: Singapore, 2023; Volume 328. [CrossRef]
16. Wang, T.; Deng, T.; Deng, Y.; Yu, X.; Zou, P.; Deng, Z. Numerical Simulation of Deep Excavation Considering Strain-Dependent Behavior of Soil: A Case Study of Tangluo Street Station of Nanjing Metro. *Int. J. Civil. Eng.* **2023**, *21*, 541–550. [CrossRef]
17. Hsiung, B.; Dao, S. Evaluation of Constitutive Soil Models for Predicting Movements Caused by a Deep Excavation in Sands. *Electron. J. Geotech. Eng.* **2014**, *19*, 17325–17345.
18. Liang, F.; Jia, Y.; Ding, Y.; Huang, M. Experimental study on parameters of HSS model for soft soils in Shanghai. *Chin. J. Geotech. Eng.* **2017**, *39*, 269–278.
19. Gu, X.; Wu, R.; Liang, F.; Gao, G. On HSS model parameters for Shanghai soils with engineering verification. *Rock Soil Mech.* **2021**, *42*, 833–845.
20. Benz, T. Small Strain Stiffness of Soils and Its Numerical Consequences. Ph.D. Thesis, University of Stuttgart, Stuttgart, Germany, 2006.
21. Hardin, B.O.; Black, W.L. Closure to "Vibration Modulus of Normally Consolidated Clay". *J. Soil. Mech. Found. Div.* **1969**, *95*, 1531–1537. [CrossRef]
22. Hardin, B.O.; Drnevich, V.P. Shear Modulus and Damping in Soils: Design Equations and Curves. *J. Soil. Mech. Found. Division.* **1972**, *98*, 667–692. [CrossRef]
23. Jian, T.; Kong, L.; Bo, W.; Wang, J.; Liu, B. Experimental study on the influence of water content on the small strain shear modulus of undisturbed loess. *Chin. J. Geotech. Eng.* **2022**, *44*, 160–165.
24. Zhong, P. Research on Deformation Characteristics of Loess Foundation Pit Excavation Based on HSS Model. Master's Thesis, Chang'an University, Xi'an, China, 2023.
25. Liu, D. Parameter Inversion of Small Strain Hardened Soil Constitutive Model Based on GA-BP and Its Application in Foundation Pit Engineering. Master's Thesis, Chang'an University, Xi'an, China, 2022.
26. Liao, J.; Zhu, Q.; Wang, Z.; Dai, Z. Comparison of Simulating Loess Deformation by Two Advanced and Conventional Constitutive Models. *Chin. J. Undergr. Space Eng.* **2023**, *19*, 784–792.
27. Li, L.; Liu, J.; Li, K.; Huang, H.; Ji, X. Study of parameters selection and applicability of HSS model in typical stratum of Jinan. *Rock Soil Mech.* **2019**, *40*, 4021–4029. [CrossRef]
28. Lu, Y. *Research on the Impact and Control Measures of Shield Tunnel Construction on Adjacent Bridges Based on HSS Model*; University of Jinan: Jinan, China, 2018.
29. *DG/TJ 508–61–2018*; Technical Code for Excavation Engineering. Tongji University Press: Shanghai, China, 2018.
30. Xu, W.; Liu, C.; Hu, K.; Fan, Y.; Tai, J.; Ren, X.; Feng, X. Experimental study on parameters of hardening soil small-strain model for soft soil in Wuhan. *Saf. Environ. Eng.* **2024**, *31*, 88–99. [CrossRef]

31. Lin, D. Experimental Study and Engineering Application of Small Strain Soil Hardening Model Parameters. Master's Thesis, Zhejiang University, Zhejiang, China, 2022. [CrossRef]
32. Meng, R. Parameter tests and engineering applications of small-strain hardening soil model of mucky silty clay in Hangzhou. *J. Chengdu Univ. Technol. Sci. Technol. Ed.* **2024**, *51*, 303–315.
33. Lu, T.; Liu, S.; Cai, G.; Wu, K.; Li, Z. Method and application of parameter determination for small strain hardened soil model based on SCPTU testing. *J. Ground Improv.* **2023**, *5*, 451–459.
34. Wang, X.; Gu, X.; Liu, H.; Lai, H.; Wu, C. Experimental Study on Main Parameters of HSS Model for Marine Soil in Dafeng Sea Area, Jiangsu Province. *Mech. Eng.* **2024**, *46*, 316–322.
35. Dong, X.; Zhou, F.; Zhu, R.; Wang, X.; Zeng, L. Sensitivity analysis and determination method of parameters for small strain soil hardening model. *Sci. Technol. Eng.* **2024**, *24*, 6366–6376.
36. Rafa, S. Cyclic lateral response of piles in dry sand: Effect of pile slenderness. In Proceedings of the Seventh International Jordanian Civil Engineering Conference (JICEC07) Reconstruction of Damaged Zones "The Role of Civil Engineering", Chengdu, China, 9–11 May 2017.
37. Brinkgreve, R.B.J.; Kappert, M.H.; Bonnier, P.G. Hysteretic damping in a small-strain stiffness model. *Environ. Sci.* **2007**, 737–742. Available online: https://www.researchgate.net/profile/R-Brinkgreve/publication/267411874_Hysteretic_damping_in_a_small-strain_stiffness_model/links/54b4edbc0cf26833efd0453b/Hysteretic-damping-in-a-small-strain-stiffness-model.pdf (accessed on 2 November 2024).
38. Fernando, S. Importance of small-strain stiffness on the prediction of the displacements of a flexible retaining wall on stiff clay. In Proceedings of the 17th European Conference on Soil Mechanics and Geotechnical Engineering, ECSMGE 2019—Proceedings, Reykjavik, Iceland, 1–6 September 2019.
39. Peteris, S.; Kaspars, B. Applicability of Small Strain Stiffness Parameters for Pile Settlement Calculation. *Procedia Eng.* **2017**, *172*, 999–1006. [CrossRef]
40. Kima, T.; Jung, Y. Optimizing Material Parameters to Best Capture Deformation Responses in Supported Bottom-up Excavation: Field Monitoring and Inverse Analysis. *KSCE J. Civ. Eng.* **2022**, *26*, 3384–3401. [CrossRef]
41. Jesús, F.; Cláudio, P. Non-linear critical speed in high-speed ballasted railways with ground reinforcement. *Transp. Geotech.* **2024**, *48*, 101325.
42. Wang, Y. Study on Triaxial Shear Characteristics of Loess Under K_0 Consolidation Conditions. Master's Thesis, Chang'an University, Xi'an, China, 2023. [CrossRef]
43. Bolton, M. The strength and dilatancy of sands. *Géotechnique* **1986**, *36*, 65–78. [CrossRef]
44. Zhang, W.; Goh, A.; Feng, X. A simple prediction model for wall deflection caused by braced excavation in clays. *Comput. Geotech.* **2015**, *63*, 67–72. [CrossRef]
45. Brinkgreve, R.; Kumarswamy, S.; Swolfs, W. PLAXIS Version 2015 Manual. The Netherlands. 2015. Available online: https://pdfcoffee.com/plaxis-10-pdf-free.html (accessed on 2 November 2024).

applied sciences

MDPI

Article

Open-Pit Bench Blasting Fragmentation Prediction Based on Stacking Integrated Strategy

Yikun Sui [1], Zhiyong Zhou [1,*], Rui Zhao [1], Zheng Yang [1] and Yang Zou [2,*]

[1] School of Resource and Safety Engineering, Central South University, Changsha 410083, China
[2] Division of Mining and Geotechnical Engineering, Luleå University of Technology, 97187 Luleå, Sweden
* Correspondence: csuzzy@csu.edu.cn (Z.Z.); yang.zou@ltu.se (Y.Z.)

Abstract: The size distribution of rock fragments significantly influences subsequent operations in geotechnical and mining engineering projects. Thus, accurate prediction of this distribution according to the relevant blasting design parameters is essential. This study employs artificial intelligence methods to predict the fragmentation of open-pit bench blasting. The study employed a dataset comprising 97 blast fragment samples. Random forest and XGBoost models were utilized as base learners. A prediction model was developed using the stacking integrated strategy to enhance predictive performance. The model's performance was evaluated using the coefficient of determination (R^2), the mean square error (MSE), the root mean square error (RMSE), and the mean absolute error (MAE). The results indicated that the model achieved the highest prediction accuracy, with an R^2 of 0.943. In the training set, the model achieved MSE, RMSE, and MAE values of 0.00269, 0.05187, and 0.03320, while in the testing set, these values were 0.00197, 0.04435, and 0.03687, respectively. The model was validated using five sets of actual blasting block data from a northeastern mining area, which yielded more accurate prediction results. These findings demonstrate that the stacking strategy effectively enhances the prediction performance of a single model and offers innovative approaches to predicting blasting block size.

Keywords: blasting fragmentation prediction; stacking integrated strategy; random forest; XGBoost

1. Introduction

The proper fragmentation of blasting has always been the goal of rock mining production. Many subsequent operations after blasting, including loading, transport, and grinding, are affected by the degree of fragmentation [1]. High-quality breakage can avoid the need for the secondary blasting of large rock pieces, reduce the energy consumed by crushers and grinders, increase digging capacity and loading productivity, and reduce production costs [2,3]. The breakage of rock is a complex and dynamic process, and rock fragmentation can be affected by a number of variables. Numerous previous studies have shown that the parameters that determine fragmentation can be categorized into controllable and uncontrollable. The design parameters of blasting and explosive-related parameters are controllable, whereas uncontrollable parameters include the physical and mechanical properties of the rock and its structure [4,5]. When establishing the prediction model for the blasting situation, the controllable and uncontrollable parameters should be considered comprehensively. However, due to the difficulty in obtaining parameters in actual projects, ensuring the validity of collected parameters is challenging. Therefore, it is urgent to establish a feasible model considering a variety of factors to predict the size of fragmentation.

As early as the last century, attempts have been made to establish empirical formulas for predicting blasting fragmentation by summarizing and analyzing data. Rosin and Rammler [6] proposed the Rosin–Rammler semi-empirical formula, which provides a rough judgment of block size through the relationship between characteristic bulkiness and the cumulative passing percentage. The Kuznetsov formula combines factors such as explosive quality and rock strength for block size prediction [7]. Cunningham further developed the Kuz–Ram model based on the Rosin–Rammler and Kuznetsov formulations [8]. Due to the shortcomings of the Kuz–Ram model in fine fragment prediction, many scholars have subsequently continued to modify and improve it. Ouchterlony et al. [9] developed the Kuznetsov–Cunningham–Ouchterlony (KCO) model based on the Kuz–Ram model, which compensates for its shortcomings in predicting fine fragments and its limitations regarding the upper limit of block size.

During blasting, uncontrolled parameters affect rock fragmentation to a greater extent compared to controlled parameters [10], making it difficult to propose a fitting equation for fragmentation that accounts for all blasting influences. With the extensive development and application of AI technology in many fields, it has gradually become an effective method for solving complex prediction problems [11–18]. Kulatilake et al. [19] used a Back Propagation Neural Network (BPNN) model to predict the average particle size of blast fragmentation, training it with four learning algorithms. Its predictive ability was confirmed to be superior to multiple regression models used by previous scholars. Monjezi et al. [20] constructed an artificial neural network (ANN) to predict the degree of rock fragmentation by comparing several methods. He pointed out through sensitivity analysis that the most influential parameters in the fragmentation process were powder factor, burden, and bench height. In addition to neural networks, there are multiple other methods for prediction. For example, Philip et al. [21] established the XGBoost model to predict rock fragmentation and compared it with the prediction results of BPNN, confirming that XGBoost has better prediction ability and faster prediction speed. Esmaeili et al. [22] constructed a Support Vector Machine (SVM) model and an Adaptive Network-based Fuzzy Inference System (ANFIS) based on Principal Component Analysis (PCA) and compared the prediction results with the Kuz–Ram method. Their findings further proved that the accuracy of AI methods is better compared to previous empirical formulas. It is evident that random forest, XGBoost, SVM, and neural network algorithms are common choices among researchers for prediction.

To ensure and improve prediction accuracy, many scholars choose to optimize the prediction process using optimization algorithms. For example, Asl et al. [23] optimized the ANN model prediction process using the firefly algorithm, resulting in coefficients of determination R^2 of 0.94 and 0.93 for the prediction of rock fragments and fly-rock. Jia et al. [24] established an extreme learning machine model using Grey Wolf Optimization (GWO) and compared the prediction results of the average size of blast fragmentation before and after optimization, proving that the optimization algorithm can effectively improve the model's prediction accuracy. Some scholars have further improved optimization algorithms. Rong et al. [25] conducted a study on the parameters of Convolutional Neural Network (CNN) and Multilayer Perceptron (MLP) for predicting the mean fragment size, and utilized the GWO algorithm to optimize the training process, further improving model performance. Rather than using a single algorithm to optimize the model, using multiple optimization algorithms and comparing the prediction quality can make the prediction results more convincing. Li et al. [26] used five optimization algorithms, including the Salp Swarm Algorithm, GWO, and Particle Swarm Optimization (PSO), to optimize the forecasting performance of the Support Vector Regression (SVR) model. They combined these algorithms with three mathematical indicators to compare and evaluate the best prediction model.

Although the prediction method for optimizing the basic algorithm has gradually developed and matured, enhancing the model's accuracy and prediction ability, there are still some limitations of the single prediction method due to the relatively complex factors affecting the fragmentation produced by rock blasting. To improve the generalization ability of a single algorithm, some scholars have integrated algorithms to predict the blasting effect. Barkhordari et al. [27] established various integrated models, including the simple averaging ensemble, weighted averaging ensemble, separate stacking model, and Bayesian–eXtreme Gradient Boosting, filling the gap in predicting blasting fly-rock using integrated models. Kaffayatullah et al. [28] also introduced the AdaBoost algorithm in their study on predicting compressive strength. However, AdaBoost and XGBoost are essentially integrated models of homogeneous algorithms, with poor integration of the advantages of non-homogeneous algorithms. Koopialipoor et al. [29] combined the different characteristics of four models and achieved the highest prediction accuracy for rock deformation using the stacking-tree-RF-KNN-MLP structure. Kadingdi et al. [30] utilized random forest, Gaussian Process (GP), and Gradient Boosting Machine (GBM) as base learners to establish a stacking ensemble model for predicting ground vibration, significantly improving the model's predictive performance. Wu et al. [31] employed a stacking ensemble learning method that combined LightGBM, random forest, and XGBoost as base learners, and Ridge, Lasso, and Linear Regression as meta-learners to predict rock compressive strength (RCS). The model demonstrated excellent prediction accuracy, achieving an R^2 of 0.946. It can be seen that the stacking integrated model gives better prediction results than a single model in all cases. It also offers better performance in fusing different qualitative single models to combine their respective strengths, as well as in model generalization. Currently, the effectiveness of applying the stacking ensemble model to blast fragmentation prediction remains unknown, and this area requires further exploration.

This study aims to improve prediction accuracy by applying a stacking ensemble model to blast fragmentation prediction. In this study, 97 samples of data collected from blasts conducted around the world by T. Hudaverdi et al. [4] are used as datasets. The model performance of random forest, XGBoost, and SVM was compared, and the base learner was selected. An integrated model was developed using the stacking integration strategy to predict the size of rock fragments produced by blasting. The predictive performance of the model was evaluated using four mathematical indicators: R^2, MSE, RMSE, and MAE. The importance of features using random forest and XGBoost was analyzed, and the impact of various input parameters on the size of blasting blocks was compared. Finally, the predictive performance of this model was validated using actual engineering blasting data.

2. Ensemble Machine Learning (ML) Models

Unlike the traditional approach of developing a single learner, ensemble learning aims to construct a model by combining multiple weak learners, seeking an approach to create a strong learner that surpasses the performance of a single one, thereby improving the efficiency, stability, and resilience of the model. Common ensemble learning methods include bagging, boosting, and stacking.

2.1. Bagging and Boosting

The bagging algorithm, also known as 'bootstrap aggregating', was proposed by Breiman in 1996 [32]. The random forest algorithm, also proposed by Breiman, is based on the bagging concept, combining multiple single decision trees. The principle of the bagging algorithm involves randomly sampling multiple subsets from the original dataset with replacement, ensuring that the size of each subset is the same as the original dataset. Each subset is used to independently train a learner, and during prediction, the bagging algorithm combines the predictions of all learners to derive the final prediction result. Bagging can significantly reduce model variance and improve generalization performance, but it is sensitive to noisy data and prone to consistency issues.

The boosting algorithm, on the other hand, combines weak learners into a strong learner by chaining them together. It enhances the accuracy of the algorithm and is more generalizable [33]. Boosting trains its base learners using the initial training set and, based on their performance, constructs the next base learner. If the previous round's prediction was incorrect, higher weights are assigned to the misclassified instances in the next round of prediction. Ultimately, the outputs of the weighted combination of multiple learners are used to derive the final result. Compared to bagging, boosting is more effective in reducing bias and improving training accuracy, thus addressing the problem of underfitting. However, both bagging and boosting are essentially ensemble algorithms of homogeneous learners, and their schematic diagrams are shown in Figure 1.

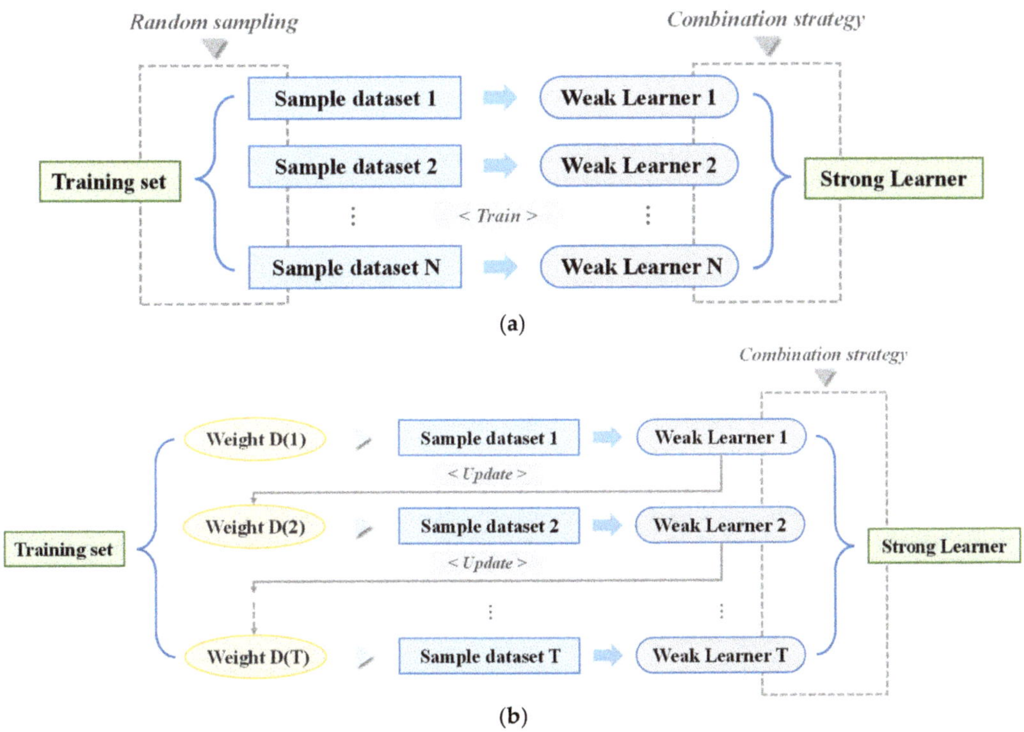

Figure 1. Schematic diagram of the bagging and boosting algorithms: (**a**) the bagging algorithm; (**b**) the boosting algorithm.

2.2. Stacking Strategy

The stacking strategy was first proposed by David H. Wolpert and further developed by Leo Breiman, who introduced 'Stacked Regressions' using generalized linear models [34]. Unlike bagging and boosting algorithms, the stacking algorithm combines multiple base learners using a nonlinear integration method to form a multilayered strong learner. These base learners can be different types of heterogeneous learners. For example, in a two-layer learner structure, the stacking algorithm first trains the base learners in the first layer. The results from the first layer are then used as a new training and testing set for the second layer (meta-learners) to train and predict. The basic principle of the stacking strategy is shown in Figure 2.

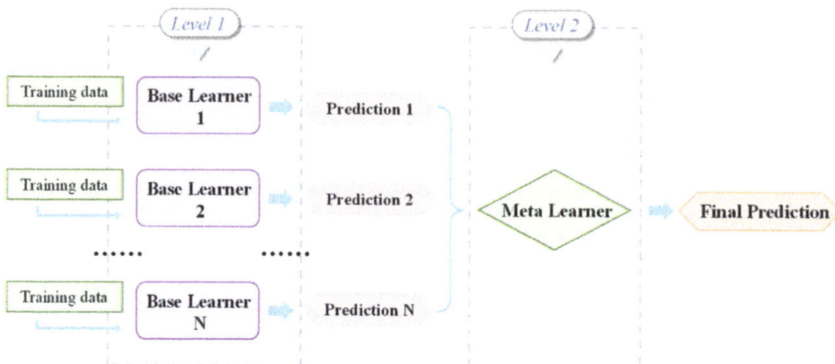

Figure 2. Stacking strategy schematic diagram.

The blending algorithm is very similar to the stacking algorithm, as both of them are effective ensemble learning methods designed to enhance prediction performance. Compared to the stacking algorithm, the blending algorithm directly uses non-overlapping data to train different base learners. Therefore, when sample data are insufficient, the blending method uses fewer data in the second layer, making it prone to overfitting. In contrast, the stacking method uses all data at each training layer, resulting in a more robust forecasting ability. In this study, with only 97 sets of prediction data, the sample data are relatively insufficient. Thus, the stacking method was chosen for model fusion.

2.3. Base Learner

2.3.1. Random Forest

Random forest is an ensemble algorithm based on classification and regression trees, proposed by Breiman by combining the bagging method and the Random Subspace method [35]. By combining multiple individual trees to form a forest, random forest can achieve more reliable prediction results. The randomness of the random forest algorithm is mainly reflected in two aspects: first, samples are drawn from the original dataset with replacement using bootstrap sampling, forming training subsets to train decision trees. Next, the Random Subspace method is used to randomly select the feature variables required for each node split in the decision trees, generating multiple decision trees. Finally, the predictions of all decision trees are aggregated to produce the final prediction result. The random forest method requires few parameters to be tuned during implementation, and its randomness effectively prevents overfitting issues, making it widely used in the prediction of blasting effects [36–38]. The principle formula for generating random forests is as follows:

$$\mathrm{RF(x)} = \frac{1}{\mathrm{N_{trees}}} \sum_{\mathrm{i=1}}^{\mathrm{N_{trees}}} \mathrm{f_i(x)} \tag{1}$$

where $RF(x)$ is the prediction result of the random forest method on sample x; N_{trees} represents the number of decision trees in a random forest; $f_i(x)$ indicates the prediction result of the ith decision tree on sample x.

2.3.2. XGBoost

Extreme Gradient Boosting (XGBoost) was proposed by Chen et al. based on improvements to the Gradient Boosting Decision Tree (GBDT) algorithm [39]. Based on decision trees, the XGBoost algorithm leverages advantages such as parallel tree boosting, regularization, and efficient tree pruning to handle regression and classification problems more effectively [40]. The decision trees in XGBoost are sequentially constructed, with each new tree considering the prediction errors of the previous tree. Adjustments are made based on the modified sample distribution to train the next tree. The core idea is to optimize the objective function. XGBoost can be seen as a soft computing library that combines new algorithms with gradient-boosting decision tree algorithms to improve prediction accuracy [41]. Figure 3 shows the pseudocode for the XGBoost algorithm.

```
Exact Greedy Algorithm for Split Finding
Input: I, instance set of current node
Input: d, feature dimension
gain ← 0
G ← Σ_{i∈I} g_i , H ← Σ_{i∈I} h_i
for k = 1 to m do
    G_L ← 0, H_L ← 0
    for j in sorted(I, by x_jk) do
        G_L ← G_L + g_i , H_L ← H_L + h_i
        G_R ← G − G_L , H_R ← H − H_L
        score ← max(score, G_L^2/(H_L+λ) + G_R^2/(H_R+λ) − G^2/(H+λ))
    end
end
Output: Split with max score
```

Figure 3. Splitting pseudocode of a single leaf node in XGBoost algorithm.

2.3.3. SVM

Support Vector Machine (SVM) can be regarded as a supervised learning technique suitable for classification and regression problems, and it is widely used to solve various engineering problems [42,43]. Support Vector Regression (SVR) is a non-probabilistic algorithm based on VC dimension theory and a branch of SVM. It fits the regression function by constructing an optimal hyperplane, optimizing the model by minimizing total loss, and maximizing the margin. Its innovation lies in introducing an artificially set 'margin band', which transforms the regression problem into an optimization problem and allows for certain errors. In SVR, input features are mapped into a high-dimensional feature space, giving it a strong capability to handle high-dimensional data [44]. The computational principles of SVR can be represented by the following formulas [45]:

$$f(x) = [\omega \cdot \varphi(x)] + b \tag{2}$$

where $f(x)$ is the constructed linear regression function; ω is the weight coefficient; $\varphi(x)$ is a nonlinear mapping function; $b \in R$ is the deviation term.

3. Database

3.1. Data Description

The database used in this study was established by Hudaverdi et al. [4] based on the collection and evaluation of several blasting tests in different mines and rock formations around the world, all of which involved open-pit bench blasting. This database contains a total of 97 samples of blasting data covering various open-pit mines in the world, and takes into account five blasting design parameters, including burden (B), spacing (S), bench height (H), stemming (T), and hole diameter (D). The inputs for the blast design parameters are in the form of ratios to avoid the influence of different parameter units on predictions. Additionally, the explosive parameters, structural parameters, and physical and mechanical properties of the rock were used as input parameters. Specifically, this study used bench height to drilled burden (H/B), spacing to burden (S/B), burden to hole diameter (B/D), stemming to burden (T/B), powder factor (Pf), in situ block size (X_B), and modulus of elasticity (E) as input variables, with mean fragment size (X_{50}) as the output variable.

In order to better characterize the data used, Figure 4 shows the detailed distribution of the seven input parameters through a combination of bar charts and violin plots, with the mean and median of the data indicated by solid and hollow dots in the half violin plots. It can be observed that the sample point distribution of most input parameters is relatively concentrated, particularly for input parameter E, which is mostly concentrated in the 0–20 and 40–70 intervals. Table 1 details the four statistical characteristics of all input parameters, including minimum, maximum, mean, and standard deviation. Figure 5 illustrates the degree of correlation between the input parameters using a heat map. It can be observed that there is a certain correlation between the input parameters, which makes them reasonable as input variables for the model. The dataset was randomly divided into two groups: 80% was used as the training set to construct the prediction model, and the remaining 20% was used as the testing set to evaluate the model's performance. To reduce bias among the input variables and enhance the model's accuracy, the data were normalized using the StandardScaler function before prediction. The standardization formula is as follows:

$$x^* = \frac{x - \mu}{\sigma} \tag{3}$$

where μ is the mean of the data set, σ is the standard deviation of the dataset, and x and x^* are the data values before and after standardization, respectively.

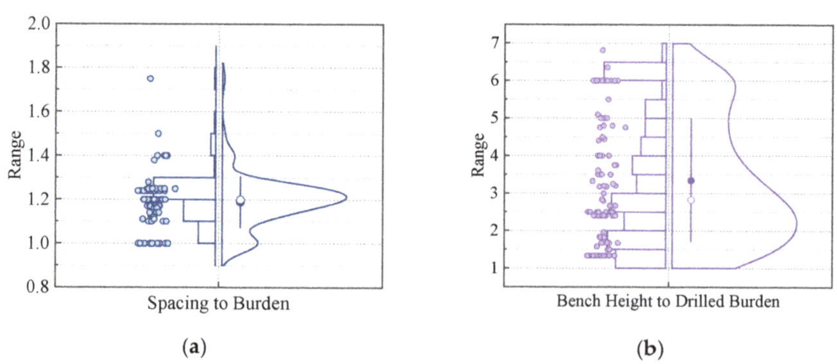

Figure 4. *Cont.*

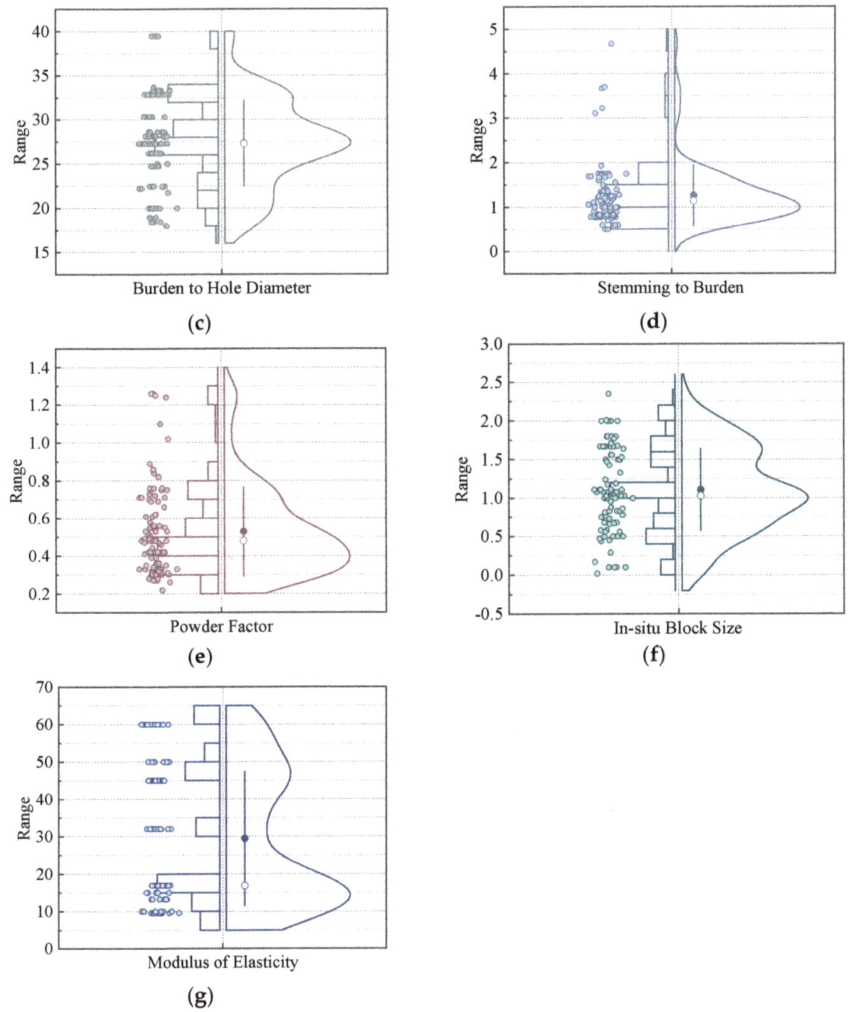

Figure 4. Distribution of input parameters: (**a**) spacing to burden; (**b**) bench height to drilled burden; (**c**) burden to hole diameter; (**d**) stemming to burden; (**e**) powder factor; (**f**) insitu block size; (**g**) modulus of slasticity.

Table 1. Statistical characterization of model input parameters.

Parameters	Minimum	Maximum	Mean	Standard Deviation
S/B	1.00	1.75	1.20	0.109
H/B	1.33	6.82	3.44	1.64
B/D	17.98	39.47	27.21	4.77
T/B	0.50	4.67	1.27	0.688
Pf(kg/m^3)	0.22	1.26	0.53	0.238
XB(m)	0.02	2.35	1.17	0.479
E(GPa)	9.57	60.00	30.74	17.72

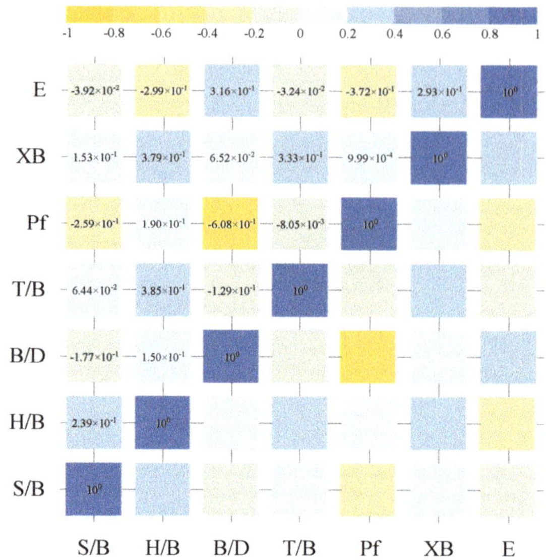

Figure 5. Heat map of input parameter correlation distribution.

3.2. Model Validation and Evaluation

To ensure the accuracy and predictive effectiveness of the established model, the validation and evaluation process of model performance is essential. This study compared and screened the predictive performance of base learners and utilized the stacking strategy to establish a prediction model. Figure 6 illustrates the overall analysis and modelling process of the prediction model. To assess the effectiveness of the model, the mean square error (MSE), mean absolute error (MAE), determination coefficient (R^2), and root mean squared error (RMSE) were utilized as the evaluation metrics. When the value of R^2 is closer to 1, and the values of MSE, MAE, and RMSE are closer to 0, it indicates a better predictive performance of the model. The evaluation indicators are calculated using the following formulas [26,46–48]:

$$\mathrm{MSE} = \frac{1}{n}\sum_{i=1}^{n} (\hat{y}_i - y_i)^2 \tag{4}$$

$$\mathrm{MAE} = \frac{1}{n}\sum_{i=1}^{n} |\hat{y}_i - y_i| \tag{5}$$

$$\mathrm{RMSE} = \sqrt{\frac{1}{n}\sum_{i=1}^{n} (\hat{y}_i - y_i)^2} \tag{6}$$

$$R^2 = 1 - \frac{\sum_{i=1}^{n} (\hat{y}_i - y_i)^2}{\sum_{i=1}^{n} (\overline{y_i} - y_i)^2} \tag{7}$$

where n is the number of input–output data in the algorithm, $\hat{y}$ is the predicted value of the model, y is the true value of the model, and $\overline{y}$ is the average value of the model predictions.

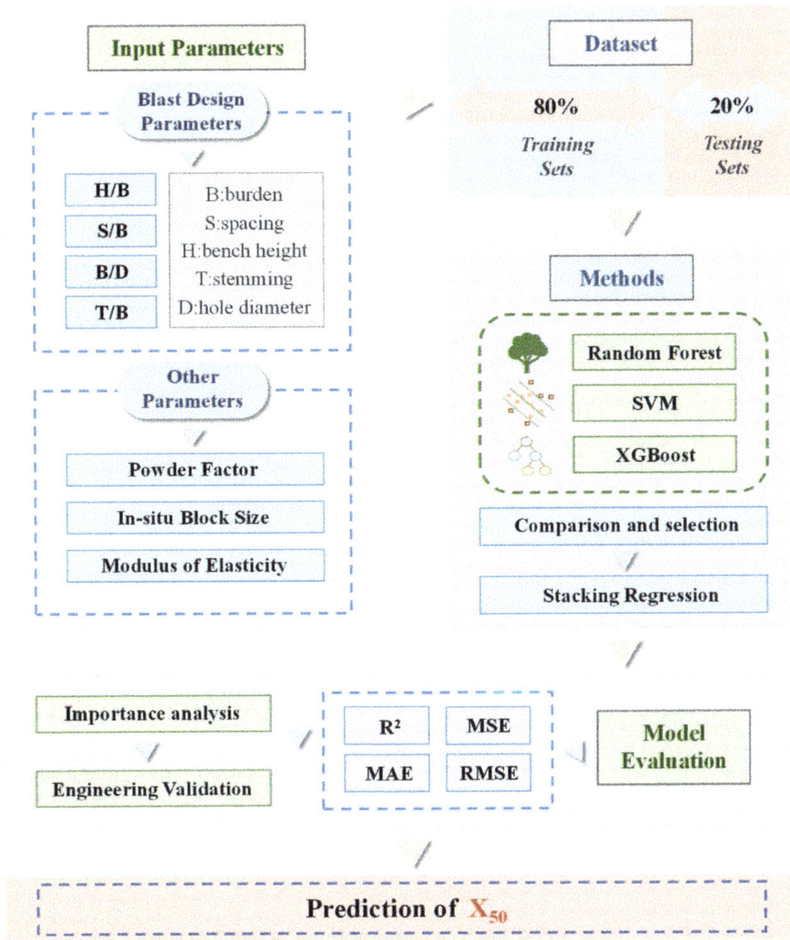

Figure 6. Stacking integrated model prediction flowchart.

4. Results and Discussion

4.1. Model Construction and Prediction Results

To construct the prediction model, this study followed three steps: determining evaluation indexes, selecting the best-performing single model (random forest, SVM, or XGBoost) as the base learner, and training the meta-learner to predict the average fragment size (X_{50}) of the blasted rock. This study used Python 3.9 as the programming language, Pycharm 2023.3.6 as the program's running software, and the packages in the configured environment, including Pandas, Numpy, and Scikit-learn.

As mentioned earlier, after constructing the models, their parameters were selected and tuned to improve the prediction of the three single learners. Specifically, the parameters for the random forest model were set to random_state = 1 and n_estimators = 50; the parameters of the SVM model were set to kernel = poly, degree = 5, and C = 1; and the parameter of the XGBoost model was set to learning_rate = 1.9. Some of the prediction results of the three models are presented in Table 2.

Table 2. Partial prediction results of the testing sets.

Model	True Value	Predicted Value
Random Forest	0.44	0.4154
	0.1	0.1259
	0.35	0.4042
	0.25	0.2578
	0.74	0.6014
Support Vector Machine	0.44	0.2504
	0.1	0.1826
	0.35	0.2503
	0.25	0.3287
	0.74	0.4342
XGBoost	0.44	0.3273
	0.1	0.2011
	0.35	0.2740
	0.25	0.2437
	0.74	0.7453

By drawing predictive regression diagrams, the predicted values of the four models are compared with the actual values, facilitating a clearer judgment of the models' prediction performance. To compare the prediction accuracy of each model in the training and the testing sets, the results for each set were plotted in the same regression plot, as shown in Figure 7. In these plots, the vertical axis represents the predicted values, and the horizontal axis represents the true values. Diagonal dashed lines are added to each regression plot to indicate a perfect fit. When the predicted value is larger than the true value, the point is located above the diagonal; when the predicted value is smaller, the point is below the diagonal. As the predicted values converge closer to the true values, their distance to the diagonal decreases until they appear on the line when they are equal. The more target points located on the diagonal, the better the predictive performance of the model.

The results in Figure 7 show that the XGBoost model has the best prediction accuracy in the training set, where the predicted values are essentially the same as the true values, and only a few sample points fall outside the perfect fit diagonal. The prediction accuracy of the random forest model is slightly weaker, but most of the sample points are distributed near the diagonal. The model is more accurate in predicting samples with an X_{50} range of 0–0.6. The SVR model's predictions have a larger bias, and the sample points are very dispersed, making the model less accurate. However, excellent model performance in the training set does not represent the overall performance of the model. In the testing set, the predictions of all three models show certain deviations, and the prediction accuracy for the minimum and maximum sample points is lower than for those with intermediate values. Among them, the prediction bias of the XGBoost model in the testing set is more pronounced, and its prediction accuracy is much worse than in the training set, indicating that the model is overfitting.

Overall, the random forest and XGBoost models provide better predictions, while the SVR algorithm provides poorer predictions. Thus, random forest and XGBoost were chosen as the base learners for the integrated model. The parameters of the random forest and XGBoost models were continuously adjusted during the model construction process, and the final parameters were random_state = 27, n_estimators = 76 for the random forest model; random_state = 42, learning_rate = 0.5 for the XGBoost model.

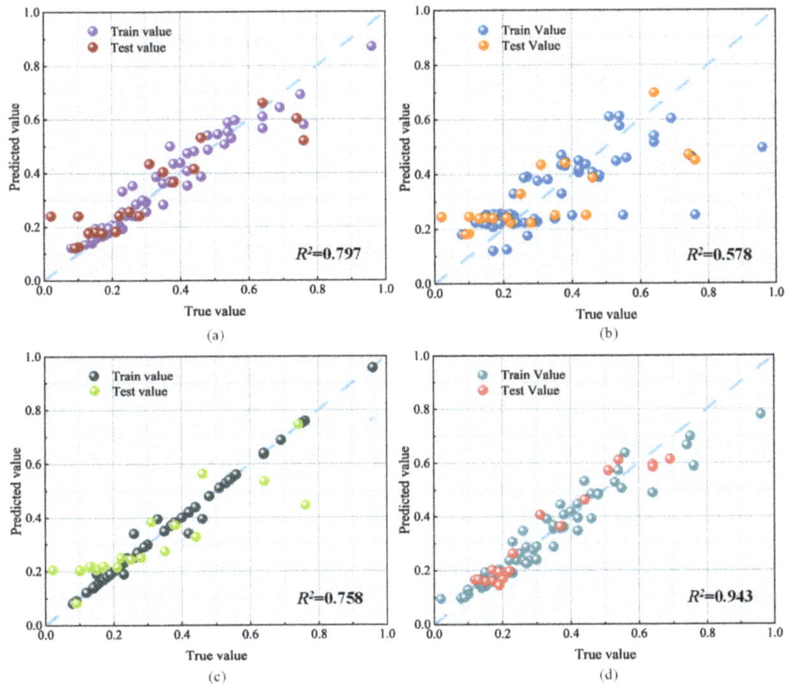

Figure 7. Comparison of the prediction results between the training and testing sets of the four models: (**a**) random forest; (**b**) SVR; (**c**) XGBoost; (**d**) stacking integrated model.

When constructing the stacking integrated model, considering the size of the dataset, the first attempt was to use cross-validation to increase the generalization ability of the model, but the cross-validated model had a poor prediction effect on the test set, and the stacking fusion model itself could effectively improve the residuals, thus canceling the cross-validation in order to improve the prediction effect. At the same time, the XGBoost algorithm has the phenomenon of overfitting, so when building the meta-learner, linear regression, a relatively simple model, was chosen to avoid overfitting caused by excessive complexity.

The meta-model was trained, and the prediction results of the stacking integrated model were obtained, as shown in Figure 7d. For the training set, the prediction effect of the stacking integrated model is relatively close to that of the random forest, but the accuracy is not as good as that of XGBoost; for the testing set, the prediction effect of the stacking integrated model is the best, with the sample points closer to the diagonal. In summary, the stacking integrated model demonstrates better prediction capability in the training and testing sets, particularly in the testing set, indicating that the integrated model is more stable and can avoid overfitting.

4.2. Model Performance Analysis and Discussion

During the research process, four evaluation indicators were used to assess the predictive performance of the model. Table 3 presents the results of the metrics for the training and testing sets of the stacking integrated model. From the results in the table, it is evident that the MSE and RMSE values of the integrated model on the testing set are lower than those on the training set, while the MAE values on the testing set are slightly higher than those on the training set, though they are very close to each other. A comparison of the indicators of the four models is shown in Figure 8. It is evident that the R^2 value of the stacking integrated model is the highest, at 0.943. The R^2 value of the random forest model is relatively close to that of the XGBoost model, with values of 0.797 and 0.758, respectively. The R^2 value of the SVR model is the lowest, at 0.578. The stacking ensemble algorithm demonstrates significant improvements in R^2, achieving increases of 0.146, 0.365, and 0.185 compared to the respective base algorithms.

Table 3. Results of stacking integrated model evaluation indicators.

	MSE	RMSE	MAE
Training set	0.00269	0.05187	0.03320
Testing set	0.00197	0.04435	0.03687

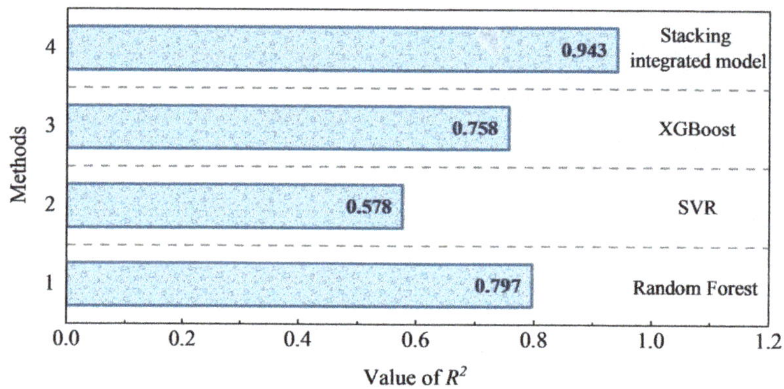

Figure 8. Comparison of R^2 values of four models.

The comparison of the MSE, RMSE, and MAE results of the four prediction models, shown in Figure 9, reveals that the MSE values of the four models are relatively close, with the stacking integrated model having the lowest value of 0.00197. The MSE values of the random forest and XGBoost models are 0.00903 and 0.01081, respectively. The SVR model has the highest MSE value, at 0.01883. The distribution of RMSE and MAE values is very similar to MSE, with values of 0.04436 and 0.03688 for the stacking integrated model, 0.09503 and 0.0668 for the random forest model, 0.10397 and 0.05264 for the XGBoost model, and 0.13722 and 0.11388 for the SVM model. Figure 9 clearly shows that the MSE, RMSE, and MAE results of the stacking integrated model are the lowest among the models, proving that the integrated model has the best predictive performance. By integrating models, the issues of insufficient prediction accuracy in base algorithms like random forest and SVR, as well as the overfitting problem in XGBoost on the training set, can be effectively addressed and avoided. In summary, using the stacking strategy to construct an integrated model for predicting X_{50} can effectively improve predictive performance and better avoid overfitting.

Feature importance analysis can effectively assess the degree of influence of input variables on X_{50} prediction. In this study, the importance assessment tools in random forest and XGBoost were used to extract the importance of input variables of the stacking integrated model and compare the assessment results of the two. The importance assessment of input features in the random forest algorithm is based on impurity calculation, and the features are sorted before taking the average value. The final result also requires normalizing the scores of all features, ensuring that the sum of the importance values of the features equals 1. The XGBoost algorithm determines the importance of features based on their ability to split data at nodes. The stronger the splitting ability, the greater the contribution to the final result.

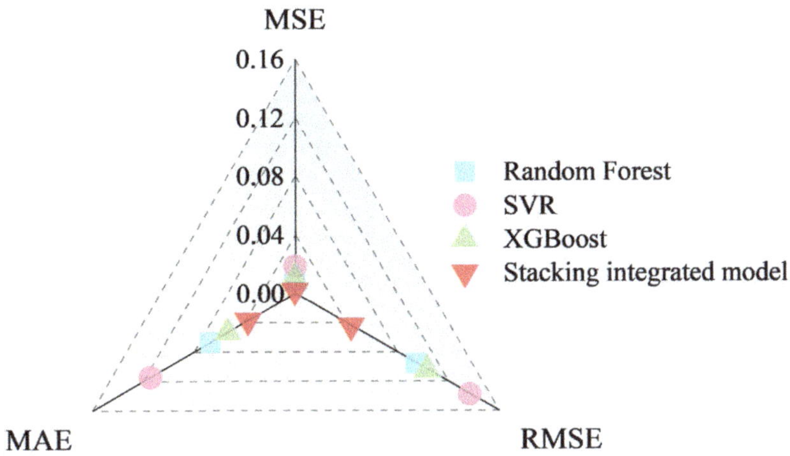

Figure 9. Comparison of assessment indicators for the four models.

Figure 10 illustrates the extraction results of the two methods. The input parameter E is identified as the most important feature variable for predicting X_{50} in both methods, with importance values of 0.7129 and 0.4608, respectively. X_B and T/B follow, exhibiting similar importance values. In the random forest method, the importance value of the input variable T/B is slightly higher, at 0.1183, while in the XGBoost method, X_B has a relatively higher importance value of 0.1503. The importance of the input variables S/B, H/B, and B/D is low, with B/D having the lowest importance, at 0.0023 in the random forest method and S/B having the lowest importance, at 0.0335 in the XGBoost method. Among the seven input parameters in this study, it can be concluded that the input parameter E has the greatest effect on the X_{50} prediction results, followed by X_B and T/B, while the three parameters S/B, H/B, and B/D have the least effect on the X_{50} prediction results.

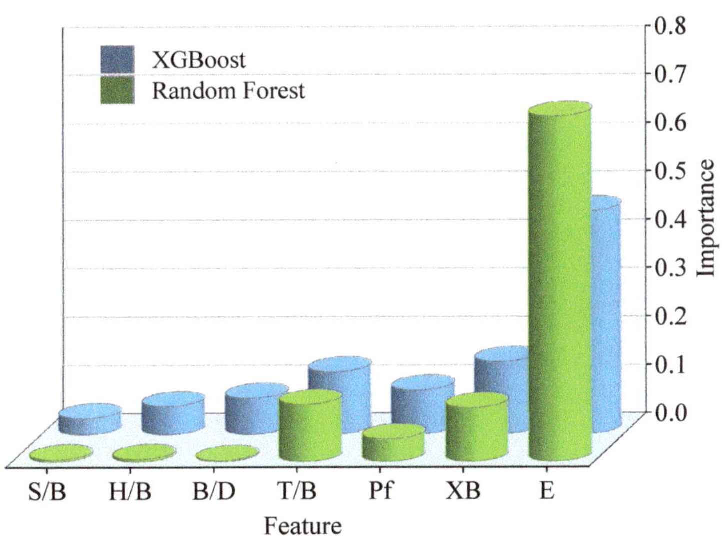

Figure 10. Comparison of importance values results for input parameters.

In the field of rock engineering, the elastic modulus (*E*) reflects the stiffness of rock and serves as an indicator of the material's resistance to deformation. The elastic modulus influences wave velocity, which in turn affects the extent of energy dissipation, resulting in variations in fragment size during rock blasting. The in situ block size, representing the fundamental block size of rock mass in its natural state, is determined by the characteristics of joints and fractures. Rock masses with well-developed fractures are more susceptible to fragmentation. Stemming length and burden distance are critical parameters in blast design, influencing the utilization and release of energy by constraining gas escape and controlling the distance between the blast hole and the free face, respectively. Consequently, they affect fragmentation efficiency [49–51]. Scholars, in related studies, validated that E, XB, and T/B are the primary influencing factors in predicting outcomes [52], which further supports the correctness of the aforementioned analytical results.

All of the above results demonstrate that high prediction accuracy can be obtained by constructing a stacking integrated model for the prediction of blasting fragmentation. Previous research on the prediction of blasting effects mainly compared some single prediction models or optimized a single model [53–55]. Although these methods can achieve good prediction results, they lack the combination of various prediction methods. Additionally, the tuning process of these models is usually quite complex and time-consuming. One major advantage of utilizing the stacking integrated strategy is that it can combine the strengths of multiple single models and allows for the combination of non-homogeneous models. Moreover, the construction process of the model is relatively simple, and the training speed is faster.

4.3. Engineering Validation Results of the Model

In this study, field data from five blasting operations in a northeastern mining area were collected to validate the performance of the constructed model. The rocks in this mining area are mainly granite, with moderate hardness, making it a relatively explosive region. Based on the model's required input parameters, the specific values of the collected relevant parameters are shown in Table 4.

Table 4. Blasting parameter statistics.

No.	B (m)	S (m)	H (m)	D (mm)	T (m)	Pf (kg/m^3)	X_B (m)	E (GPa)	X_{50} (m)
1	7	8.5	15	250	7.5	0.62	1.11	5.6	0.1456
2	7	9	15	250	7.5	0.64	1.11	5.6	0.163
3	7	10	15	250	7.5	0.61	1.11	5.6	0.1962
4	7	9.5	15	250	7.5	0.57	1.11	5.6	0.1997
5	7	9.5	15	250	7.5	0.62	1.11	5.6	0.1786

The fragmentation size X_{50} of the blasted rock was obtained using the fragmentation recognition software Split Desktop v4.0.0.42 [56]. Figure 11 illustrates the software's recognition of the blasting results on-site. The software effectively maps the boundaries of rock fragmentation blocks and calculates the distribution of block sizes for this blasting, using a 1-meter-long ruler placed on-site as a reference.

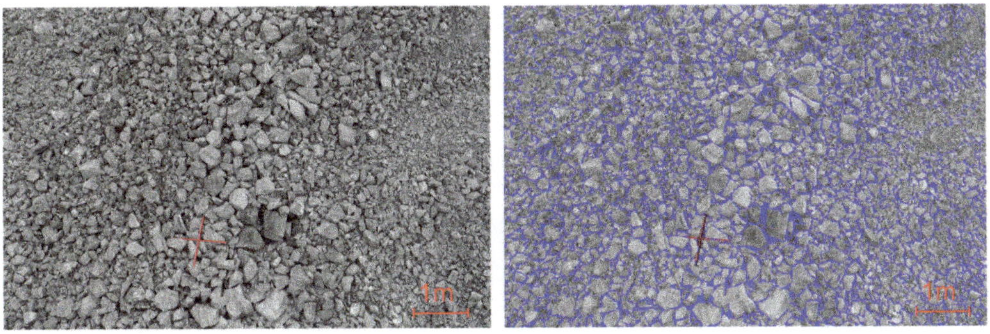

Figure 11. Block size recognition results from Split-Desktop software.

To maintain consistency with the input parameters used during model training, the validation still uses the value of the ratio between the parameters as the model's input parameters. Table 5 shows the input parameters and their values for the validation process. Figure 12 shows the comparison between the model-predicted results and the actual results of rock fragmentation. To make it easier to compare the difference between the true and predicted values, a purple color legend representing the true values has been added to the figure. The model's predicted values closely match the actual values, with prediction errors within 0.03 m. This indicates that the prediction model developed in this study can accurately predict rock fragmentation in open blast projects, which is of practical significance. However, the cross-validation method commonly adopted by most scholars was not employed in constructing the stacking model. This may be due to the small dataset size in this study, which resulted in lower prediction accuracy when using cross-validation methods. Additionally, during the engineering verification process, Split-Desktop software v4.0.0.42 inevitably has some errors in recognizing the block size. Therefore, in future research, it is advisable to consider more diverse datasets and increase the number of base learners. In practical engineering applications, blasting outcomes can be predicted in advance based on various design parameters. This enables iterative adjustments to parameter values, ensuring that the desired fragmentation index is achieved for the project.

Table 5. Validation process input parameters and values.

No.	S/B	H/B	B/D	T/B	Pf	X_B	E
1	1.2143	2.1429	28	1.0714	0.62	1.11	5.6
2	1.2857	2.1429	28	1.0714	0.64	1.11	5.6
3	1.4286	2.1429	28	1.0714	0.61	1.11	5.6
4	1.3571	2.1429	28	1.0714	0.57	1.11	5.6
5	1.3571	2.1429	28	1.0714	0.62	1.11	5.6

Based on the results of the engineering verification mentioned above, the model can be used to predict the fragmentation size of the blast in advance, according to the blasting parameters, before production blasting. The parameters of on-site blasting operations can be further optimized based on the predicted results to meet the required fragmentation size for production. Similarly, under the guidance of a large amount of actual on-site parameters, the predictive accuracy of the stacking model established in this study can be further improved, providing the potential to develop an intelligent prediction system for on-site blasting.

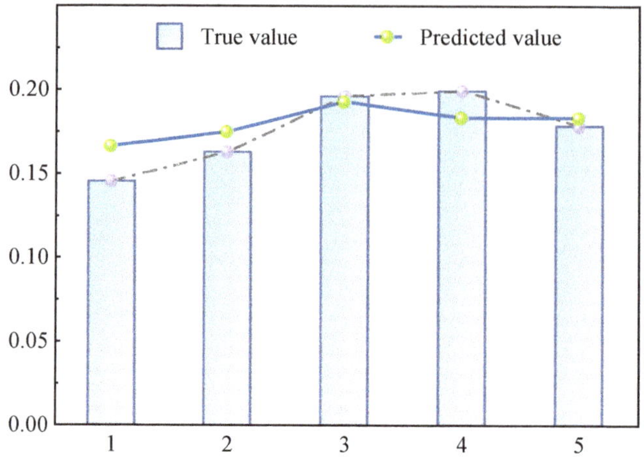

Figure 12. Comparison of results from engineering validation.

5. Conclusion and Summary

During the blasting process, numerous factors, including controllable and uncontrollable ones, affect the blasting results. Thus, it is challenging to use a single-fitting equation to comprehensively describe the relationship between various influencing factors and fragmentation size. Compared to traditional empirical equation prediction methods, artificial intelligence methods can effectively simulate the relationship between various input and output parameters. This study utilized a dataset containing 97 data samples and set seven input parameters to predict the fragmentation size X_{50}. To improve prediction performance, the prediction effects of three single models—random forest, SVR, and XGBoost—were compared and selected. A stacking integrated strategy was employed, using random forest and XGBoost models as base learners to construct the fusion model. This method has been applied in engineering fields, but its application in predicting blasting fragmentation is still relatively limited. The predictive performance of the established models was evaluated using four mathematical indicators. The following conclusions can be drawn from this study:

- The stacking integrated model outperformed the single models and had higher predictive accuracy. The model evaluation yielded an R^2 value of 0.943, with MSE, RMSE, and MAE values of 0.00269, 0.05187, and 0.03320, respectively, on the training set; and 0.00197, 0.04435, and 0.03687, respectively, on the testing set;
- From the feature importance evaluation results of the two methods, it can be seen that in the process of constructing the model, the input feature E has the highest influence on the predicted fragmentation size, followed by T/B and X_B;
- Compared with the prediction methods of other researchers, the prediction method established in this study better integrates the advantages of individual algorithms. The engineering verification results also demonstrated that the constructed algorithm has good predictive accuracy, and its prediction results can provide references for blasting design.

Author Contributions: Validation, Y.S.; formal analysis, Y.S., R.Z. and Z.Y.; investigation, Y.S.; methodology, Y.S.; writing—original draft preparation, Y.S.; writing—review and editing, Z.Z., R.Z., Z.Y. and Y.Z. All authors have read and agreed to the published version of the manuscript.

Funding: The research presented in this paper was funded by the National Natural Science Foundation of China (Grant Nos. 12072376 and 52274105).

Institutional Review Board Statement: Not applicable.

Informed Consent Statement: Not applicable.

Data Availability Statement: The raw data supporting the conclusions of this article will be made available by the authors on request.

Conflicts of Interest: The authors declare no conflicts of interest.

Abbreviations

The following abbreviations are used in this manuscript:

R^2	Coefficient of Determination
MSE	Mean Square Error
RMSE	Root Mean Square Error
MAE	Mean Absolute Error
KCO	Kuznetsov-Cunningham-Ouchterlony
BPNN	Back Propagation Neural Network
ANN	Artificial Neural Network
SVM	Support Vector Machine
ANFIS	Adaptive Network-based Fuzzy Inference System
PCA	Principal Component Analysis
GWO	Grey Wolf Optimization
CNN	Convolutional Neural Network
MLP	Multilayer Perceptron
PSO	Particle Swarm Optimization
SVR	Support Vector Regression
KNN	K-Nearest Neighbor
GP	Gaussian Process
RCS	Rock Compressive Strength

References

1. Abbaspour, H.; Drebenstedt, C.; Badroddin, M.; Maghaminik, A. Optimized design of drilling and blasting operations in open pit mines under technical and economic uncertainties by system dynamic modelling. *Int. J. Min. Sci. Technol.* **2018**, *28*, 839–848. [CrossRef]
2. Kanchibotla, S.S.; Valery, W.; Morrell, S. Modeling fines in blast fragmentation and its impact on crushing and grinding. In Proceedings of the Explo'99: A Conference on Rock Breaking, Kalgoorlie, WA, Australia, 7–11 November 1999; AusIMM: Melbourne, Australia, 1999.
3. Dimitraki, L.; Christaras, B.; Marinos, V.; Vlahavas, I.; Arampelos, N. Predicting the average size of blasted rocks in aggregate quarries using artificial neural networks. *Bull. Eng. Geol. Environ.* **2019**, *78*, 2717–2729. [CrossRef]
4. Hudaverdi, T.; Kulatilake, P.; Kuzu, C. Prediction of blast fragmentation using multivariate analysis procedures. *Int. J. Numer. Anal. Methods Geomech.* **2011**, *35*, 1318–1333. [CrossRef]
5. Chandrahas, N.S.; Choudhary, B.S.; Teja, M.V.; Venkataramayya, M.S.; Prasad, N.S.R.K. XGBoost Algorithm to Simultaneous Prediction of Rock Fragmentation and Induced Ground Vibration Using Unique Blast Data. *Appl. Sci.* **2022**, *12*, 5269. [CrossRef]
6. Peleg, M. Determination of the parameters of the Rosin-Rammler and beta distributions from their mode and variance using equation-solving software. *Powder Technol.* **1996**, *87*, 181–184. [CrossRef]
7. Kuznetsov, V.M. The mean diameter of the fragments formed by blasting rock. *Sov. Min. Sci.* **1973**, *9*, 144–148. [CrossRef]
8. Cunningham, C.V.B. The kuz-ram fragmentation model—20 years on. In *Brighton Conference Proceedings*; European Federation of Explosives Engineer: Brighton, UK, 2005.
9. Ouchterlony, F. The Swebrec function: Linking fragmentation by blasting and crushing. *Min. Technol.* **2005**, *114*, 29–44. [CrossRef]
10. Mehrdanesh, A.; Monjezi, M.; Sayadi, A.R. Evaluation of effect of rock mass properties on fragmentation using robust techniques. *Eng. Comput.* **2017**, *34*, 253–260. [CrossRef]
11. Ebrahimi, E.; Monjezi, M.; Khalesi, M.R.; Armaghani, D.J. Prediction and optimization of back-break and rock fragmentation using an artificial neural network and a bee colony algorithm. *Bull. Eng. Geol. Environ.* **2016**, *75*, 27–36. [CrossRef]

12. Hasanipanah, M.; Amnieh, H.B.; Arab, H.; Zamzam, M.S. Feasibility of PSO–ANFIS model to estimate rock fragmentation produced by mine blasting. *Neural Comput. Appl.* **2018**, *30*, 1015–1024. [CrossRef]
13. Armaghani, D.J. Rock Fragmentation Prediction through a New Hybrid Model Based on Imperial Competitive Algorithm and Neural Network. *Smart Constr. Res.* **2018**, *2*, 1–12.
14. Cui, X.J.; Li, Q.Y.; Tao, M.; Hong, Z.X.; Zhao, M.S.; Li, J.; Zhou, J.M.; Yu, H.B. Research on Explosive-Rock Matching System based on XGBoost. *Blasting* **2023**, *40*, 31–38+58.
15. Li, C.Q.; Zhou, J.; Du, K. Towards lightweight excavation: Machine learning exploration of rock size distribution prediction after tunnel blasting. *J. Comput. Sci.* **2024**, *78*, 20. [CrossRef]
16. Rui, Y.C.; Chen, J.; Chen, J.K.; Qiu, J.D.; Zhou, Z.L.; Wang, W.Z.; Fan, J.Y. A robust triaxial localization method of AE source using refraction path. *Int. J. Min. Sci. Technol.* **2024**, *34*, 521–530. [CrossRef]
17. Chen, J.; Chen, J.K.; Rui, Y.C.; Pu, Y.Y. Joint Inversion of AE/MS Sources and Velocity with Full Measurements and Residual Estimation. *Rock Mech. Rock Eng.* **2024**, *57*, 7371–7386. [CrossRef]
18. Chen, J.; Tong, J.; Rui, Y.C.; Cui, Y.; Pu, Y.Y.; Du, J.S.; Apel, D.B. Step-path failure mechanism and stability analysis of water-bearing rock slopes based on particle flow simulation. *Theor. Appl. Fract. Mech.* **2024**, *131*, 14. [CrossRef]
19. Kulatilake, P.; Qiong, W.; Hudaverdi, T.; Kuzu, C. Mean particle size prediction in rock blast fragmentation using neural networks. *Eng. Geol.* **2010**, *114*, 298–311. [CrossRef]
20. Monjezi, M.; Mohamadi, H.A.; Barati, B.; Khandelwal, M. Application of soft computing in predicting rock fragmentation to reduce environmental blasting side effects. *Arab. J. Geosci.* **2014**, *7*, 505–511. [CrossRef]
21. Philip, D.; Abbaspour, H.; Kansake, B.A.; Drebenstedt, C. Unraveling the Capability of Artificial Intelligence for Prediction of Rock Fragmentation. In Proceedings of the REAL TIME MINING—Conference on Innovation on Raw Material Extraction, Freiberg, Germany, 27 March 2019.
22. Esmaeili, M.; Salimi, A.; Drebenstedt, C.; Abbaszadeh, M.; Bazzazi, A.A. Application of PCA, SVR, and ANFIS for modeling of rock fragmentation. *Arab. J. Geosci.* **2015**, *8*, 6881–6893. [CrossRef]
23. Asl, P.F.; Monjezi, M.; Hamidi, J.K.; Armaghani, D.J. Optimization of flyrock and rock fragmentation in the Tajareh limestone mine using metaheuristics method of firefly algorithm. *Eng. Comput.* **2018**, *34*, 241–251. [CrossRef]
24. Jia, Z.Z.; Song, Z.L.; Fan, J.F.; Jiang, J.Y. Prediction of Blasting Fragmentation Based on GWO-ELM. *Shock. Vib.* **2022**, *2022*, 7385456. [CrossRef]
25. Rong, K.; Xu, X.; Wang, H.B.; Yang, J. Prediction of the mean fragment size in mine blasting operations by deep learning and grey wolf optimization algorithm. *Earth Sci. Inform.* **2024**, *17*, 2903–2919. [CrossRef]
26. Li, E.M.; Yang, F.H.; Ren, M.H.; Zhang, X.L.; Zhou, J.; Khandelwal, M. Prediction of blasting mean fragment size using support vector regression combined with five optimization algorithms. *J. Rock Mech. Geotech. Eng.* **2021**, *13*, 1380–1397. [CrossRef]
27. Barkhordari, M.S.; Armaghani, D.J.; Fakharians, P. Ensemble machine learning models for prediction of flyrock due to quarry blasting. *Int. J. Environ. Sci. Technol.* **2022**, *19*, 8661–8676. [CrossRef]
28. Khan, K.; Ahmad, W.; Amin, M.N.; Ahmad, A.; Nazar, S.; Alabdullah, A.A.; Arab, A.M.A. Exploring the Use of Waste Marble Powder in Concrete and Predicting Its Strength with Different Advanced Algorithms. *Materials* **2022**, *15*, 4108. [CrossRef]
29. Koopialipoor, M.; Asteris, P.G.; Mohammed, A.S.; Alexakis, D.E.; Mamou, A.; Armaghani, D.J. Introducing stacking machine learning approaches for the prediction of rock deformation. *Transp. Geotech.* **2022**, *34*, 100756. [CrossRef]
30. Kadingdi, F.A.; Ayawah, P.E.A.; Azure, J.W.A.; Bruno, K.A.; Kaba, A.G.A.; Frimpong, S. Stacked Generalization for Improved Prediction of Ground Vibration from Blasting in Open-Pit Mine Operations. *Min. Metall. Explor.* **2022**, *39*, 2351–2363. [CrossRef]
31. Wu, L.Y.; Li, J.H.; Zhang, J.W.; Wang, A.F.; Tong, J.B.; Ding, F.; Li, M.; Feng, Y.; Li, H. Prediction model for the compressive strength of rock based on stacking ensemble learning and shapley additive explanations. *Bull. Eng. Geol. Environ.* **2024**, *83*, 439. [CrossRef]
32. Breiman, L. Bagging Predictors. *Mach. Learn.* **1996**, *24*, 123–140. [CrossRef]
33. Yu, L.; Wu, T.J. Assemble learning: A survey of boosting algrithms. *Pattern Recognit. Artif. Intell.* **2004**, *17*, 52–59.
34. Breiman, L. Stacked regressions. *Mach. Learn.* **1996**, *24*, 49–64. [CrossRef]
35. Breiman, L. Random forests. *Mach. Learn.* **2001**, *45*, 5–32. [CrossRef]
36. Han, H.; Armaghani, D.J.; Tarinejad, R.; Zhou, J.; Tahir, M.M. Random Forest and Bayesian Network Techniques for Probabilistic Prediction of Flyrock Induced by Blasting in Quarry Sites. *Nat. Resour. Res.* **2020**, *29*, 655–667. [CrossRef]
37. Yu, Z.; Shi, X.Z.; Qiu, X.Y.; Zhou, J.; Chen, X.; Gou, Y.G. Optimization of postblast ore boundary determination using a novel sine cosine algorithm-based random forest technique and Monte Carlo simulation. *Eng. Optim.* **2021**, *53*, 1467–1482. [CrossRef]
38. Zhou, J.; Asteris, P.G.; Armaghani, D.J.; Pham, B. Prediction of ground vibration induced by blasting operations through the use of the Bayesian Network and random forest models. *Soil. Dyn. Earthq. Eng.* **2020**, *139*, 106390. [CrossRef]
39. Chen, T.; He, T.; Benesty, M. XGBoost: Extreme Gradient Boosting. *arXiv* **2016**, arXiv:1603.02754.
40. Nabavi, Z.; Mirzehi, M.; Dehghani, H.; Ashtari, P. A Hybrid Model for Back-Break Prediction using XGBoost Machine learning and Metaheuristic Algorithms in Chadormalu Iron Mine. *J. Min. Environ.* **2023**, *14*, 689–712.

41. Qiu, Y.G.; Zhou, J.; Khandelwal, M.; Yang, H.T.; Yang, P.X.; Li, C.Q. Performance evaluation of hybrid WOA-XGBoost, GWO-XGBoost and BO-XGBoost models to predict blast-induced ground vibration. *Eng. Comput.* **2022**, *38* (Suppl. S5), 4145–4162. [CrossRef]
42. Yu, Q.; Monjezi, M.; Mohammed, A.S.; Dehghani, H.; Armaghani, D.J.; Ulrikh, D.V. Optimized Support Vector Machines Combined with Evolutionary Random Forest for Prediction of Back-Break Caused by Blasting Operation. *Sustainability* **2021**, *13*, 12797. [CrossRef]
43. Mehrdanesh, A.; Monjezi, M.; Khandelwal, M.; Bayat, P. Application of various robust techniques to study and evaluate the role of effective parameters on rock fragmentation. *Eng. Comput.* **2023**, *39*, 1317–1327. [CrossRef]
44. Rad, H.N.; Hasanipanah, M.; Rezaei, M.; Eghlim, A.L. Developing a least squares support vector machine for estimating the blast-induced flyrock. *Eng. Comput.* **2018**, *34*, 709–717. [CrossRef]
45. Xu, G.Q.; Wang, X.Y. Support vector regression optimized by black widow optimization algorithm combining with feature selection by MARS for mining blast vibration prediction. *Measurement* **2023**, *218*, 113106. [CrossRef]
46. Amoako, R.; Jha, A.; Zhong, S. Rock Fragmentation Prediction Using an Artificial Neural Network and Support Vector Regression Hybrid Approach. *Mining* **2022**, *2*, 233–247. [CrossRef]
47. Zhou, J.; Chen, Y.X.; Chen, H.; Khandelwal, M.; Monjezi, M.; Peng, K. Hybridizing five neural-metaheuristic paradigms to predict the pillar stress in bord and pillar method. *Front. Public Health* **2023**, *11*, 1119580. [CrossRef]
48. Zhou, J.; Zhang, R.; Qiu, Y.G.; Khandelwal, M. A true triaxial strength criterion for rocks by gene expression programming. *J. Rock Mech. Geotech. Eng.* **2023**, *15*, 2508–2520. [CrossRef]
49. Jern, M. Determination of the in situ block size distribution in fractured rock, an approach for comparing in-situ rock with rock sieve analysis. *Rock Mech. Rock Eng.* **2004**, *37*, 391–401. [CrossRef]
50. Tao, M.; Xu, Y.Q.; Zhao, R.; Liu, Y.L.; Wu, C.Q. Energy control and block performance optimization of bench blasting. *Int. J. Rock. Mech. Min. Sci.* **2024**, *180*, 105830. [CrossRef]
51. Mpofu, M.; Ngobese, S.; Maphalala, B.; Roberts, D.; Khan, S. The influence of stemming practice on ground vibration and air blast. *J. S. Afr. Inst. Min. Metall.* **2021**, *121*, 1–10. [CrossRef]
52. Zhao, J.; Li, D.; Zhou, J.; Armaghani, D.J.; Zhou, A.H. Performance evaluation of rock fragmentation prediction based on RF-BOA, AdaBoost-BOA, GBoost-BOA, and ERT-BOA hybrid models. *Deep. Undergr. Sci. Eng.* 2024, *early view*.
53. Gao, W.; Karbasi, M.; Hasanipanah, M.; Zhang, X.; Guo, J. Developing GPR model for forecasting the rock fragmentation in surface mines. *Eng. Comput.* **2018**, *34*, 339–345. [CrossRef]
54. Shams, S.; Monjezi, M.; Majd, V.J.; Armaghani, D.J. Application of fuzzy inference system for prediction of rock fragmentation induced by blasting. *Arab. J. Geosci.* **2015**, *8*, 10819–10832. [CrossRef]
55. Mojtahedi, S.F.F.; Ebtehaj, I.; Hasanipanah, M.; Bonakdari, H.; Amnieh, H.B. Proposing a novel hybrid intelligent model for the simulation of particle size distribution resulting from blasting. *Eng. Comput.* **2019**, *35*, 47–56. [CrossRef]
56. Xie, C.Y.; Nguyen, H.; Bui, X.N.; Choi, Y.S.; Zhou, J.; Thao, N.T. Predicting rock size distribution in mine blasting using various novel soft computing models based on meta-heuristics and machine learning algorithms. *Geosci. Front.* **2021**, *12*, 101108. [CrossRef]

applied sciences

MDPI

Article

Hypoplastic Modeling of Soil–Structure Contact Surface Considering Initial Anisotropy and Roughness

Jingtao Yu [1], Junwang Cao [1], Zixuan Chen [2], Jintao Zhu [2], Yulong Zhang [2] and Pengqiang Yu [2,*]

[1] China Road & Bridge Corporation, Beijing 100011, China; yu_crabc@163.com (J.Y.); caojw@crbc.com (J.C.)
[2] Department of Civil Engineering, University of Science and Technology Beijing, Beijing 100083, China; sdczx1998@163.com (Z.C.); m202210064@xs.ustb.edu.cn (J.Z.); m202210061@xs.ustb.edu.cn (Y.Z.)
* Correspondence: pq_yu@ustb.edu.cn

Abstract: The development of a constitutive model for soil–structure contact surfaces remains a pivotal area of research within the field of soil–structure interaction. Drawing from the Gudehus–Bauer sand hypoplasticity model, this paper employs a technique that reduces the stress tensor and strain rate tensor components to formulate a hypoplastic model tailored for sand–structure interfaces. To capture the influence of initial anisotropy, a deposition direction peak stress coefficient is incorporated; meanwhile, a friction parameter is introduced to address the surface roughness of the contact. Consequently, a comprehensive hypoplastic constitutive model is developed that takes into account both initial anisotropy and roughness. Comparative analysis with experimental data from soils on contact surfaces with diverse boundary conditions and levels of roughness indicates that the proposed model accurately forecasts shear test outcomes across various contact surfaces. Utilizing the finite element software ABAQUS 2021, an FRIC subroutine was developed, which, through simulating direct shear tests on sand–structure contact surfaces, has proven its efficacy in predicting the shear behavior of these interfaces.

Keywords: hypoplasticity constitutive model; sand–structure contact surface; initial anisotropy; roughness

1. Introduction

The interaction at the soil–structure contact surface is commonly found in engineering, occurring in scenarios such as the interface between soil and retaining walls, and between piles and the encircling soil. Serving as a critical conduit for stress and deformation transfer between soil and structures, the contact surface is often the weak link in the failure of soil–structure systems, with its mechanical properties exerting a substantial impact on the overall stress and deformation of both the soil and the structure. Consequently, the mechanical characteristics and the development of a constitutive model for soil–structure contact surfaces have consistently been a significant area of focus in the field of soil–structure interaction research.

The constitutive model for soil–structure contact surfaces is a fundamental method for describing their stress–strain relationships. Currently, the constitutive models used in practical applications include elastic constitutive models, elastoplastic constitutive models, hypoplastic constitutive models, and damage constitutive models primarily developed for easily breakable particles [1–7]. However, the equations and theoretical underpinnings of these models are quite intricate, presenting numerous challenges for research endeavors. In contrast, hypoplasticity theory is more straightforward in form, with fewer assumptions, and has proven to be highly amenable to numerical implementation. This has led to its broad application in characterizing the mechanical properties of geotechnical engineering materials.

The basic framework of hypoplasticity theory was first proposed by Kolymbas [8] in 1977, and Dafalias [9] was the first to introduce the term "hypoplasticity". Based on the hypoplasticity framework, many scholars proposed constitutive models focusing on different soil types and aspects. Kolymbas [10] first proposed an isotropic nonlinear tensor function to describe the relationship between the stress rate and strain rate of elastoplastic materials. Based on this equation, Wu and Bauer [11] introduced a simple four-parameter model known as the Wu–Bauer hypoplasticity model. This model is capable of describing the nonlinear stress–strain relationships of sandy soil under both drained and undrained conditions, as well as the contraction of loose sand and the dilation of dense sand. Gudehus [12] and Bauer [13] introduced the concept of critical state into the Wu–Bauer model and proposed the well-known eight-parameter hypoplasticity model, i.e., the Gudehus–Bauer eight-parameter model. The model captures the strain softening behavior of dense sand and the mechanical responses of sandy soils under a variety of stress paths effectively, including triaxial extension, triaxial compression, laterally confined compression, and consolidated undrained conditions. Wolffersdorff et al. [14] addressed the issue of excessive lateral contraction of the Wu–Bauer model under large strains by establishing the von Wolffersdorff hypoplasticity basic model, where two nonlinear material parameters were assumed to be equal. Wu et al. [15] introduced a new item that disappears at the critical state in the von Wolffersdorff hypoplasticity model, resulting in a wildly used hypoplasticity constitutive model.

As hypoplasticity theory has evolved, researchers in recent years have concentrated on delineating the mechanical behavior of soil–structure contact surfaces, leading to the establishment of diverse forms of hypoplasticity constitutive models tailored for these interfaces [16–21]. Herle [22] pioneered the adaptation of the sandy soil hypoplasticity model to a sandy soil–structure contact surface hypoplasticity model. This was achieved by positing that the stress tensor and strain rate tensor at the contact surface are composed solely of normal and tangential components. Based on that, Gutjahr [23] introduced the reduced component stress tensor and strain rate tensor into the von Wolffersdorff eight-parameter hypoplasticity model, establishing a one-dimensional contact surface hypoplasticity model with a predefined limit state surface.

Herle and Arnold [24] proposed a new transformation method, converting the von Wolffersdorff eight-parameter model into a two-dimensional contact surface hypoplasticity model. This model can be used to model line contact in a two-dimensional state and surface contact under three-dimensional conditions. Based on this model, Stutz et al. [25] introduced an in-plane stress and established an enhanced three-dimensional sandy soil–structure contact surface hypoplasticity constitutive model. Following that, Stutz and Mašín [26] applied the reduced component stress and strain rate tensors to the hypoplasticity Cambridge clay model by introducing a variable that reduces the strength and stiffness of the contact surface, thereby establishing a three-dimensional Cambridge clay contact surface hypoplasticity model. As a relatively recent constitutive theory, hypoplasticity has garnered significant applications in the development of contact surface constitutive models [27–30].

Initial anisotropy, as an inherent property of soil, has a significant impact on soil–structure contact surfaces [31–34]. Wu [31] proposed a constitutive model grounded in a nonlinear tensor function to depict the three-dimensional deformation–strength behavior of naturally anisotropic sandy soils, achieving a comprehensive treatment of both deformation and strength anisotropy. Yang et al. [32] introduced a fabric tensor into the Gudehus–Bauer hypoplasticity model and proposed a hypoplasticity constitutive model for granular soils based on anisotropic critical state theory. Bauer et al. [33] analyzed the formation of shear bands in cohesionless and initially transversely isotropic granular materials, presenting a constitutive equation for stress evolution.

Surface roughness of structures is also an important factor influencing the mechanical properties of contact surfaces, which has been considered by researchers from different perspectives [24,26,35]. Herle and Arnold [24] proposed a hypoplasticity model that uses a parameter, K_r, to describe contact surface roughness. Stutz and Mašín [26] argued that parameter K_r depends on contact surface roughness, establishing an enhanced hypoplasticity contact surface model. Stutz et al. [35] conducted axisymmetric contact surface shear simulations using hypoplasticity contact and soil models, comparing the results with shear tests on contact surfaces of varying roughness.

In many existing intrinsic models for hypoplasticity contact surfaces, the stress and strain rate components within the contact surface are often overlooked, and initial anisotropy is not considered. This paper introduces a hypoplasticity constitutive model for sand–structure contact surfaces, derived from the widely adopted Gudehus–Bauer low-plasticity model by simplifying the stress tensor and strain rate tensor components. The model incorporates the effects of initial anisotropy and surface roughness. The developed model is then applied to predict and compare with experimental data from sand–structure contact surfaces of varying materials and boundary conditions, thereby validating the model's rationality.

2. Fundament of the Hypoplastic Constitutive Model for the Soil–Structure Contact Surface

2.1. Hypoplastic Model for Soil

Traditional geotechnical plasticity mechanics theories are primarily constrained by two significant limitations [8]: (1) the deformation of the soil is continuous and cannot objectively be decomposed into elastic and plastic components; and (2) the theory does not adequately describe the verified incremental nonlinear phenomena. Therefore, Wu and Kolymbas [36] proposed a tensor function to describe the relationship between stress rate and strain rate by considering the stress rate as a function of stress and strain rate. The fundamental equation is as follows:

$$\overset{\circ}{\sigma} = \mathrm{H}(\sigma, \dot{\varepsilon}) \tag{1}$$

where $\overset{\circ}{\sigma}$ is the Jaumann stress rate tensor, defined as follows:

$$\overset{\circ}{\sigma} = \dot{\sigma} + \sigma\dot{\omega} - \dot{\omega}\sigma \tag{2}$$

Gudehus [12] and Bauer [13] introduced the critical state (critical stress surface) into the hypoplastic model and proposed a widely used eight-parameter hypoplastic model (referred to as the Gudehus–Bauer model), with the basic formula as follows:

$$\overset{\circ}{\sigma} = \mathrm{L}(\sigma, \dot{\varepsilon}) - \mathrm{N}(\sigma, \dot{\varepsilon}) \tag{3}$$

In Equation (3), the expressions for tensor functions $\mathrm{L}(\hat{\sigma}, \dot{\varepsilon})$ and $\mathrm{N}(\hat{\sigma})$ are as follows:

$$\mathrm{L}(\hat{\sigma}, \dot{\varepsilon}) = {a_1}^2\dot{\varepsilon} + \hat{\sigma}\mathrm{tr}(\hat{\sigma}\dot{\varepsilon}) \tag{4}$$

$$\mathrm{N}(\hat{\sigma}) = a_1(\hat{\sigma} + \hat{\sigma}^*) \tag{5}$$

where a_1 is a dimensionless scalar factor representing the radius of the critical stress point, and its expression is as follows:

$$a_1 = \left[c_1 + c_2\|\hat{\sigma}^*\|(1+\cos 3\theta)\right]^{-1} \tag{6}$$

$$c_1 = \sqrt{\frac{3}{8}} \times \frac{3-\sin\varphi_c}{\sin\varphi_c} \tag{7}$$

$$c_2 = \frac{3}{8} \times \frac{3+\sin\varphi_c}{\sin\varphi_c} \tag{8}$$

where φ_c is the critical friction angle.

In addition, the parameters f_s and, in Equation (3), f_d are the stiffness factor and the density factor, respectively. The stiffness factor, f_s, is introduced to reflect the influence of stress level and void ratio:

$$f_s = \frac{h_s}{nh_i}\left(\frac{e_i}{e}\right)^{\beta}\frac{1+e_i}{e_i}\left(\frac{3p_s}{h_s}\right)^{1-n} \tag{9}$$

where

$$h_i = \frac{1}{c_1^2} + \frac{1}{3} - \left(\frac{e_{i0}-e_{d0}}{e_{c0}-e_{d0}}\right)^{\alpha}\frac{1}{\sqrt{3}c_1} \tag{10}$$

The density factor, f_d, is related to the material density:

$$f_d = \left(\frac{e-e_d}{e_c-e_d}\right)^{\alpha} \tag{11}$$

In Equations (9)–(11), the exponents α and β are dimensionless constants, while h_s and n are two compression constants (with h_s defined as the particle hardness). Moreover, e_i, e_c, and e_d represent the upper void ratio, critical void ratio, and lower void ratio under the current stress state, respectively. They are related to the stress level and can be determined by the following equation:

$$\frac{e_i}{e_{i0}} = \frac{e_c}{e_{c0}} = \frac{e_d}{e_{d0}} = \exp\left[-\left(\frac{3p_s}{h_s}\right)^{n}\right] \tag{12}$$

where e_{i0}, e_{c0}, and e_{d0} represent the upper, critical, and lower void ratios of the initial state, respectively.

Totally, the Gudehus–Bauer hypoplastic model includes 8 parameters: α, β, e_{i0}, e_{c0}, e_{d0}, h_s, n, and φ_c. This model effectively simulates the nonlinear constitutive relationships of granular materials while considering the effects of stress level, void ratio, and density.

2.2. Establishment for the Hypoplastic Constitutive Model of the Soil–Structure Contact Surface

The shear behavior and control mechanisms of the contact surface are fundamentally in line with those of soil. The contact surface can be regarded as a miniature continuous soil mass, and its deformation pattern is generally akin to that of soil. Consequently, the contact surface model can be extrapolated from soil constitutive models.

Herle and Arnold [24] established the contact surface model using the von Wolffersdorff eight-parameter hypoplastic model for sand by assuming an isotropic stress state at the contact surface. They also assumed that only in-plane shear stress exists at the contact surface, with all shear stresses outside the contact surface being zero, i.e., $\sigma_{23} = \sigma_{32} = 0$. The coordinate system for the contact surface is shown in Figure 1.

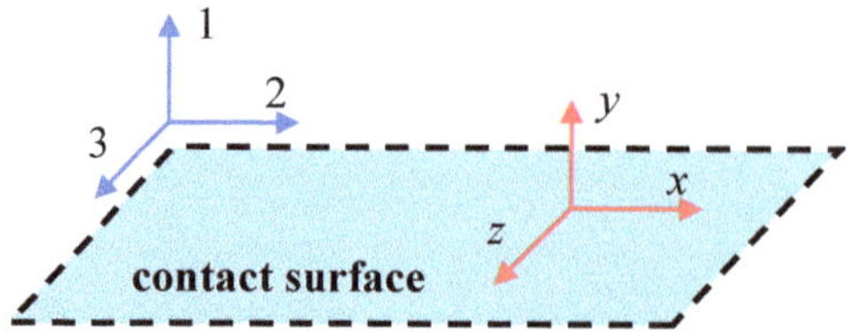

Figure 1. Schematic diagram of the contact surface coordinate system.

The stress tensor for the Arnold–Herle contact surface model is given by the following:

$$\sigma = \begin{bmatrix} \sigma_{11} & \sigma_{12} & \sigma_{13} \\ \sigma_{21} & \sigma_{22} & \sigma_{23} \\ \sigma_{31} & \sigma_{32} & \sigma_{33} \end{bmatrix} = \begin{bmatrix} \sigma_n & \tau_x & \tau_z \\ \tau_x & \sigma_n & 0 \\ \tau_z & 0 & \sigma_n \end{bmatrix} \tag{13}$$

Similarly, the strain rate tensor is given by the following:

$$\dot{\varepsilon} = \begin{bmatrix} \dot{\varepsilon}_{11} & \dot{\varepsilon}_{12} & \dot{\varepsilon}_{13} \\ \dot{\varepsilon}_{21} & \dot{\varepsilon}_{22} & \dot{\varepsilon}_{23} \\ \dot{\varepsilon}_{31} & \dot{\varepsilon}_{32} & \dot{\varepsilon}_{33} \end{bmatrix} = \begin{bmatrix} \dot{\varepsilon}_n & \frac{\dot{\gamma}_x}{2} & \frac{\dot{\gamma}_z}{2} \\ \frac{\dot{\gamma}_x}{2} & \dot{\varepsilon}_n & 0 \\ \frac{\dot{\gamma}_z}{2} & 0 & \dot{\varepsilon}_n \end{bmatrix} \tag{14}$$

where $\dot{\varepsilon}_n$ is the normal strain rate at the contact surface; and $\dot{\gamma}_x/2$ and $\dot{\gamma}_z/2$ are the shear strain rates in the x and z directions, respectively.

However, the stresses within the contact surface are autonomous and do not equate to the normal stress, as depicted in Figure 2. To address this, Mašín [25] revised the assumptions of the Arnold–Herle contact surface model, proposing that an isotropic stress state exists within the contact surface, i.e., $\sigma_{22} = \sigma_{33} \neq \sigma_{11}$, which develops independently of the normal stress during shear. The in-plane stress is denoted by σ_p. The modified stress tensor and strain rate tensor can be expressed in vector form as follows:

$$\sigma = \begin{bmatrix} \sigma_n & \sigma_p & \tau_x & \tau_z \end{bmatrix}^T \tag{15}$$

$$\dot{\varepsilon} = \begin{bmatrix} \dot{\varepsilon}_n & 0 & \frac{\dot{\gamma}_x}{2} & \frac{\dot{\gamma}_z}{2} \end{bmatrix}^T \tag{16}$$

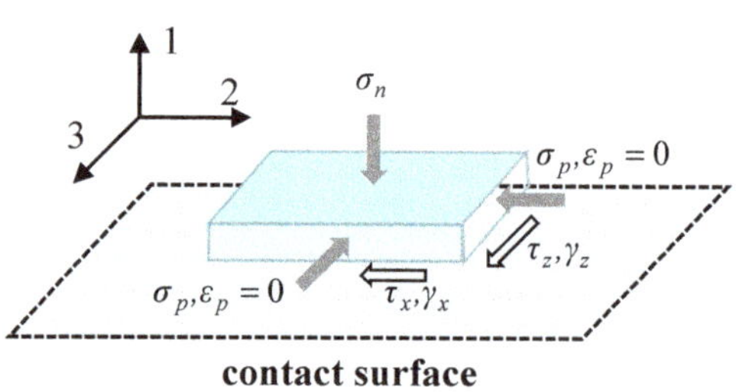

Figure 2. Schematic diagram of contact surface stress and strain components.

The three-dimensional hypoplastic constitutive model for the soil–structure contact surface developed by Mašín [25] can better predict the behavior of the contact surface and offers higher computational efficiency in numerical simulations. Therefore, this paper will adopt the assumptions of Mašín [25] to transform the Gudehus–Bauer hypoplastic model into a hypoplastic constitutive model for the soil contact surface.

2.3. Hypoplasticity Constitutive Expression of Soil–Structure Contact Surfaces

According to the stress tensor and strain rate tensor in Equations (15) and (16), the trace of the stress tensor can be obtained as follows:

$$\mathrm{tr}\sigma = \sigma_n + 2\sigma_p \tag{17}$$

Then, the stress ratio tensor can be given as follows:

$$\hat{\sigma} = \frac{\sigma}{\mathrm{tr}\sigma} = \begin{bmatrix} \frac{\sigma_n}{\sigma_n+2\sigma_p} \\ \frac{\sigma_p}{\sigma_n+2\sigma_p} \\ \frac{\tau_x}{\sigma_n+2\sigma_p} \\ \frac{\tau_z}{\sigma_n+2\sigma_p} \end{bmatrix} \tag{18}$$

And the deviatoric stress ratio is expressed as follows:

$$\hat{\sigma}^* = \begin{bmatrix} \frac{2}{3}\cdot\frac{\sigma_n-\sigma_p}{\sigma_n+2\sigma_p} \\ \frac{1}{3}\cdot\frac{\sigma_p-\sigma_n}{\sigma_n+2\sigma_p} \\ \frac{\tau_x}{\sigma_n+2\sigma_p} \\ \frac{\tau_z}{\sigma_n+2\sigma_p} \end{bmatrix} \tag{19}$$

By substituting the stress rate tensor into the formula of norm, we can obtain the following:

$$\|\dot{\varepsilon}\| = \sqrt{\dot{\varepsilon}\cdot\dot{\varepsilon}} = \sqrt{\dot{\varepsilon}_n{}^2 + 2\left(\frac{\dot{\gamma}_x}{2}\right)^2 + 2\left(\frac{\dot{\gamma}_z}{2}\right)^2} = \sqrt{\dot{\varepsilon}_n{}^2 + \frac{1}{2}\dot{\gamma}_x{}^2 + \frac{1}{2}\dot{\gamma}_z{}^2} \tag{20}$$

Similarly, substituting the deviatoric stress ratio tensor into the norm formula yields the following:

$$\|\hat{\sigma}^*\| = \sqrt{\left(\frac{2}{3}\cdot\frac{\sigma_n-\sigma_p}{\sigma_n+2\sigma_p}\right)^2 + 2\left(\frac{1}{3}\cdot\frac{\sigma_p-\sigma_n}{\sigma_n+2\sigma_p}\right)^2 + 2\left(\frac{\tau_x}{\sigma_n+2\sigma_p}\right)^2 + 2\left(\frac{\tau_z}{\sigma_n+2\sigma_p}\right)^2} \tag{21}$$

Then, by substituting Equations (15)–(21) into the Gudehus–Bauer eight-parameter equation, the linear and nonlinear terms of this equation can be calculated as follows:

$$\mathrm{L}(\hat{\sigma},\dot{\varepsilon}) = a_1{}^2\dot{\varepsilon} + \hat{\sigma}\mathrm{tr}(\hat{\sigma}\dot{\varepsilon}) = a_1{}^2\dot{\varepsilon} + \frac{\sigma_n\dot{\varepsilon}_n + \frac{3}{2}\tau_x\dot{\gamma}_x + \frac{1}{2}\tau_z\dot{\gamma}_z}{(\sigma_n+2\sigma_p)^2}\sigma \tag{22}$$

$$\mathrm{N}(\hat{\sigma}) = a_1(\hat{\sigma}+\hat{\sigma}^*) = a_1\begin{bmatrix} \frac{2\sigma_n}{\sigma_n+2\sigma_p} - \frac{1}{3} \\ \frac{2\sigma_p}{\sigma_n+2\sigma_p} - \frac{1}{3} \\ \frac{2\tau_x}{\sigma_n+2\sigma_p} \\ \frac{2\tau_z}{\sigma_n+2\sigma_p} \end{bmatrix} \tag{23}$$

The final constitutive equation of contact surface can be obtained in vector form as follows:

$$\begin{bmatrix}\dot{\sigma}_n\\ \dot{\sigma}_p\\ \dot{\tau}_x\\ \dot{\tau}_z\end{bmatrix} = f_s\left[\begin{array}{c} a_1^2\begin{bmatrix}\dot{\varepsilon}_n\\ 0\\ \frac{\dot{\gamma}_x}{2}\\ \frac{\dot{\gamma}_z}{2}\end{bmatrix} + \frac{\sigma_n\dot{\varepsilon}_n+\frac{3}{2}\tau_x\dot{\gamma}_x+\frac{1}{2}\tau_z\dot{\gamma}_z}{(\sigma_n+2\sigma_p)^2}\begin{bmatrix}\sigma_n\\ \sigma_p\\ \tau_x\\ \tau_z\end{bmatrix}\\ +f_d a_1\frac{\sqrt{\dot{\varepsilon}_n^2+\frac{1}{2}\dot{\gamma}_x^2+\frac{1}{2}\dot{\gamma}_z^2}}{\sigma_n+2\sigma_p}\begin{bmatrix}\frac{2\sigma_n}{\sigma_n+2\sigma_p}-\frac{1}{3}\\ \frac{2\sigma_p}{\sigma_n+2\sigma_p}-\frac{1}{3}\\ \frac{2\tau_x}{\sigma_n+2\sigma_p}\\ \frac{2\tau_z}{\sigma_n+2\sigma_p}\end{bmatrix}\end{array}\right] \tag{24}$$

The basic parameters of the above model are the same as those of the Gudehus–Bauer model, but with modified stress and strain rate tensors, which should be calculated using the simplified vector expressions. The upper, critical, and lower void ratios are as follows:

$$\frac{e_i}{e_{i0}} = \frac{e_c}{e_{c0}} = \frac{e_d}{e_{d0}} = exp\left[-\left(\frac{\sigma_n+2\sigma_p}{h_s}\right)^n\right] \tag{25}$$

Given that the mechanical response of the material at the contact surface significantly diverges from that of the soil at a distance under both static and dynamic loading, the critical state line of the proposed constitutive model is affected by these transformations. In this regard, the computational equations for the parameters linked to the critical state within the model are also adjusted accordingly. In addition, these parameters corresponding to the equations of stiffness factor and density factor need to be replaced, i.e., by substituting the modified expressions of e_i, e_c, and e_d into Equation (25) into Equations (9) and (10).

2.4. Contact Surface Model Considering Initial Anisotropy and Roughness

To enhance the model's predictive accuracy, crucial factors such as structural surface roughness and initial anisotropy, which significantly impact the mechanical response of the contact surface, are integrated into the model derived from Equation (24).

2.4.1. Quantitative Description of Initial Anisotropy

Based on the simulation results of direct shear test of sand–structure contact band in Li [37] using DEM (discrete element method), the peak shear stress of the initial anisotropic sand–structure contact band under various simulation conditions is listed in Table 1.

Table 1. Peak shear stress of the contact bands with different initial deposition angles under different normal stresses.

Deposition Angle/°	Peak Stress/kPa (100 kPa)	Peak Stress/kPa (200 kPa)	Peak Stress/kPa (400 kPa)	Peak Stress/kPa (800 kPa)
0	80	150	230	435
30	62	110	215	379
60	75	145	235	420
90	87	145	260	481

The peak stresses presented in Table 1 are normalized against the peak stress under the horizontal deposition direction, which serves as the isotropic reference state, to analyze the relationship with initial anisotropy. This normalized peak stress is referred to as the peak stress coefficient. Figure 3 illustrates the variation in the peak stress coefficient with respect to the deposition direction. A fitting approach is utilized to delineate this relationship, which can be expressed as follows:

$$\chi = b_1 + b_2 \exp\left[-\frac{(\delta-\delta_c)^2}{b_3}\right] \tag{26}$$

where χ is the deposition direction peak stress coefficient characterizing the influence of deposition direction on the peak stress; δ is the deposition direction; δ_c is approximated by taking the angle between the failure surface and the horizontal direction when the deposition direction is horizontal, i.e., $45^\circ - \varphi_c/2$; and parameters b_1, b_2, and b_3 are constant.

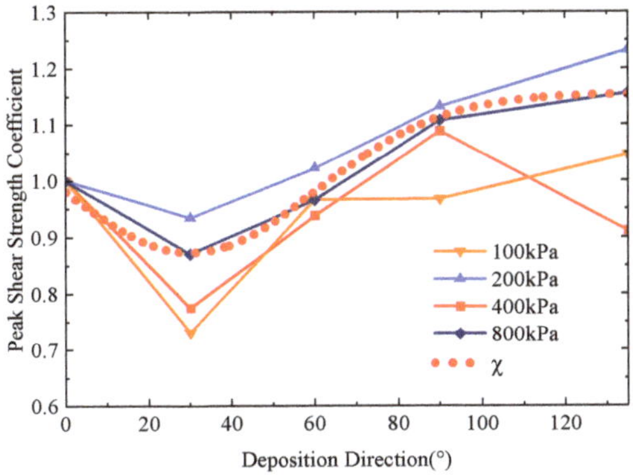

Figure 3. Variations in peak stress coefficient with the deposition direction.

Then, the deposition direction peak stress coefficient is introduced into the constitutive model proposed in Equation (24). By considering the influence of initial anisotropy, the index a_1 in Equation (6) is modified as follows:

$$a_{1i} = \chi^m [c_1 + c_2 \|\hat{\sigma}^*\| (1+\cos\theta)]^{-1} \tag{27}$$

where a_{1i} denotes the radius of the critical stress surface considering the effect of initial anisotropy, and the exponent m is a dimensionless positive constant.

2.4.2. Quantitative Description of Roughness Effect

The mechanical behavior of the contact surface when the structural surface is in a completely rough condition can be described by the hypoplasticity model in Equation (24). In order to extend the model to an arbitrary roughness case, the friction parameter (λ_R) and the relative roughness (R_n), as shown in Equation (28), referring to Gutjahr [23], are introduced, as they are widely used in various types of contact surface constitutive model such as elastic–plastic and hypoplasticity models:

$$\lambda_R = 0.25 \lg R_n + 1.05 \tag{28}$$

where $R_n = R/d_{50}$, d_{50} is the average particle size, and R is the peak-to-valley distance of the structural surface, as shown in Figure 4.

Figure 4. Diagram of peak-to-valley distance of the structural surface.

In Equation (24), the shear properties are mainly determined by the density factor, f_d, and the critical friction angle, φ_c, where φ_c determines the failure stress of loose sand and the residual stress of dense sand. Therefore, the friction parameter, λ_R, is introduced into the expression related to φ_c in order to consider the effect of structural surface roughness. And the expressions of c_1 and c_2 are modified to c_{1R} and c_{2R} as follows:

$$c_{1R} = \sqrt{\frac{3}{8}} \times \frac{3 - \sin(\lambda_R \varphi_c)}{\sin(\lambda_R \varphi_c)} \tag{29}$$

$$c_{2R} = \frac{3}{8} \times \frac{3 + \sin(\lambda_R \varphi_c)}{\sin(\lambda_R \varphi_c)} \tag{30}$$

the strength index, a_1, is modified to a_{1R},

$$a_{1R} = c_{1R} + c_{2R}\|\hat{\sigma}^*\|(1 + \cos\theta)^{-1} \tag{31}$$

Note that a_{1R} is the strength index considering only the effect of structural surface roughness. It is further rewritten to a_{1iR} by substituting Equation (31) into Equation (27), which considers both the initial anisotropy and the effect of structural surface roughness,

$$a_{1iR} = \chi^m [c_{1R} + c_{2R}\|\hat{\sigma}^*\|(1 + \cos\theta)]^{-1} \tag{32}$$

The density factor, f_d, describes the effect of the mean stress and the number of pores on shear properties and determines the response of the nonlinear part of the hypoplasticity model. The modeling experiments of Uesugi [38] showed that the test results of loose sand on rougher structural surfaces are similar with that of dense sand on smoother surfaces. Therefore, the expression of f_d is rewritten to f_{dR} by taking into account the roughness effect:

$$f_{dR} = \left(\frac{e - e_d}{e_c - e_d}\right)^{\alpha \lambda_R^2} \tag{33}$$

Correspondingly, the parameter h_i is rewritten to h_{iR}:

$$h_{iR} = \frac{1}{c_{1R}^2} + \frac{1}{3} - \left(\frac{e_{i0} - e_{d0}}{e_{c0} - e_{d0}}\right)^{\alpha \lambda_R^2} \frac{1}{c_{1R}\sqrt{3}} \tag{34}$$

and the stiffness factor, f_s, is rewritten to f_{sR}:

$$f_{sR} = \frac{h_s}{n h_{iR}} \left(\frac{e_i}{e}\right)^{\beta} \frac{1 + e_i}{e_i} \left(\frac{3p_s}{h_s}\right)^{1-n} \tag{35}$$

By substituting Equations (32)–(35) into Equation (24), the hypoplasticity constitutive model of soil–structure contact surface considering structural surface roughness and initial anisotropy can be obtained as expressed in Equation (36):

$$\begin{bmatrix} \dot{\sigma}_n \\ \dot{\sigma}_p \\ \dot{\tau}_x \\ \dot{\tau}_z \end{bmatrix} = f_{sR} \left[a_{1iR}{}^2 \begin{bmatrix} \dot{\varepsilon}_n \\ 0 \\ \frac{\dot{\gamma}_x}{2} \\ \frac{\dot{\gamma}_z}{2} \end{bmatrix} + \frac{\sigma_n \dot{\varepsilon}_n + \frac{3}{2}\tau_x \dot{\gamma}_x + \frac{1}{2}\tau_z \dot{\gamma}_z}{(\sigma_n + 2\sigma_p)^2} \begin{bmatrix} \sigma_n \\ \sigma_p \\ \tau_x \\ \tau_z \end{bmatrix} + f_{dR} a_{1iR} \frac{\sqrt{\dot{\varepsilon}_n{}^2 + \frac{1}{2}\dot{\gamma}_x{}^2 + \frac{1}{2}\dot{\gamma}_z{}^2}}{\sigma_n + 2\sigma_p} \begin{bmatrix} \frac{2\sigma_n}{\sigma_n + 2\sigma_p} - \frac{1}{3} \\ \frac{2\sigma_p}{\sigma_n + 2\sigma_p} - \frac{1}{3} \\ \frac{2\tau_x}{\sigma_n + 2\sigma_p} \\ \frac{2\tau_z}{\sigma_n + 2\sigma_p} \end{bmatrix} \right] \tag{36}$$

2.5. Model Parameters Determination

The contact surface model presented in this study encompasses 8 fundamental parameters (α, β, e_{i0}, e_{c0}, e_{d0}, h_s, n, and φ_c), 4 initial anisotropy parameters (b_1, b_2, b_3, and m), and 1 roughness parameter (R). The methods for determining these parameters are detailed as follows.

1. Critical friction Angle

Extensive testing has demonstrated that the method of estimating the critical friction angle using the angle of repose is relatively precise, with accuracy sufficient for the majority of sandy soils. Table 2 [39] presents the angle of repose and critical friction angle values obtained from shear tests on various types of sand. The data reveal a minimal discrepancy between these two parameters, indicating that it is entirely viable to approximate the critical friction angle using the angle of repose. However, it is important to note that for particles with a size d $\leq$ 0.1 mm, the error in predicting the critical friction angle using the angle of repose becomes significant, necessitating the use of tests to accurately determine the critical friction angle.

Table 2. Angle of repose and critical friction angle from shear tests.

Materials	Angle of Repose/°	Critical Friction Angle/°	Test
Hochstetten gravel	35.7	36.5	Direct shear
Hochstetten sand	34.0	34.0	Consolidation drained
Hostun sand	32.0	32.0	Consolidation drained
Karlsruhe sand	29.1	30.0	Consolidation drained
Lausitz sand	33.0	32.2	Consolidation drained
Toyoura sand	30.4	30.9	Consolidation undrained
Zbraslav sand	30.8	29.7	Direct shear

2. Particle hardness (h_s) and index n

The values of h_s and n are ascertained through an optimization method that employs local search to identify the best solution, thereby circumventing the issue of excessive error that may stem from the selection of disparate stress state points based on test outcomes.

3. Upper void ratio, e_{i0}, at zero pressure

During the experimental procedures, it is nearly impossible to replicate the upper void ratio; hence, dry sand is commonly poured into a standard mold. During the pouring process, a critical value is derived which exceeds the theoretical maximum void ratio, $e_{\max}$, and it is widely accepted that $e_{i0} \approx 1.2e_{\max}$.

4. Lower bound void ratio, e_{d0}, at zero pressure

According to the compression characteristics of sand, the void ratio, e_d, diminishes as stress increases. Typically, e_d is ascertained from the $e_d - C_u$ relationship curve, which is suggested by Youd [40]. The $e_d - C_u$ relationship is conducted under conditions where the pressure is $p_s = 55$ kPa, $K_0 = 0.40$, and the vertical stress is maintained at 96 kPa. Once the value of e_d is determined, the value of e_{d0} can be deduced by applying the formula that is retroactively derived from Equation (25):

$$e_{d0} = e_d \exp\left[\left(\frac{\sigma_n + 2\sigma_p}{h_s}\right)^n\right] \tag{37}$$

As a general guideline, it is often more favorable to set $e_{d0} \approx e_{\min}$.

5. Critical void ratio, e_{c0}, at zero pressure

Similar to the upper void ratio, the critical void ratio, e_c, also diminishes with increasing stress. The value can likewise be extracted from the expression given in Equation (25):

$$e_{c0} = e_c \exp\left[\left(\frac{\sigma_n + 2\sigma_p}{h_s}\right)^n\right] \tag{38}$$

The results of undrained triaxial test on Sacramento River sand and Monterey sand are summarized, and $e_{c0} \approx e_{\max}$ is generally used.

6. Index α

Parametric equations for two scalars can be obtained from Equation (24):

$$\dot{\sigma}_n = f_s \frac{(\sigma_n + 2\sigma_p)^2}{\sigma_n^2 + 2\sigma_p^2} \left[\dot{\varepsilon}_n + a^2 \frac{\sigma_n \dot{\varepsilon}_n}{(\sigma_n + 2\sigma_p)^2} \sigma_n + f_d \frac{a}{3} \frac{5\sigma_n - 2\sigma_p}{\sigma_n + 2\sigma_p} \dot{\varepsilon}_n \right] \tag{39}$$

$$\dot{\sigma}_p = f_s \frac{(\sigma_n + 2\sigma_p)^2}{\sigma_n^2 + 2\sigma_p^2} \left[a^2 \frac{\sigma_n \dot{\varepsilon}_n}{(\sigma_n + 2\sigma_p)^2} \sigma_p + f_d \frac{a}{3} \frac{4\sigma_p - \sigma_n}{\sigma_n + 2\sigma_p} \dot{\varepsilon}_n \right] \tag{40}$$

When the soil reaches the peak stress state, $\dot{\sigma}_n = 0$, so α can be derived from Equation (23) as follows:

$$\alpha = \frac{\ln\left[6 \frac{(2+K_p)^2 + a^2 K_p (K_p - 1 - \tan v_p)}{a(2+K_p)(5K_p - 2)\sqrt{4 + 2(1 + \tan v_p)^2}} \right]}{\ln\left(\frac{e - e_d}{e_c - e_d} \right)} \tag{41}$$

where K_p is the peak stress ratio:

$$K_p = \frac{\sigma_n}{\sigma_p} = \frac{1 + \sin \varphi_p}{1 - \sin \varphi_p} \tag{42}$$

$$\tan v_p = -\frac{\dot{\varepsilon}_v}{\dot{\varepsilon}_n} \tag{43}$$

$$\sin \varphi_p = \left(\frac{\sigma_n - \sigma_p}{\sigma_n + \sigma_p} \right)_p \tag{44}$$

Since the model uses a reduced component of the stress tensor, $\dot{\varepsilon}_{22} = \dot{\varepsilon}_{33} = 0$, and $\dot{\varepsilon}_v = \dot{\varepsilon}_n$, where $\dot{\varepsilon}_v$ is the volume strain rate, and $\dot{\varepsilon}_n$ is axial strain rate. Hence, $\tan v_p = -1$. The natural logarithm function in the denominator of Equation (41) can be regarded as the relative void ratio related to pressure and can be denoted as r_e; r_e is a function of the void ratio and average stress.

Integrate $\dot{\sigma}_n = 0$ into Equation (40) at the peak stress condition, and the peak dilatancy angle can be obtained:

$$\tan v_p = 2 \frac{K_p - 4 + 5AK_p^2 - 2AK_p}{(5K_p - 2)(1 + 2A)} - 1 \tag{45}$$

$$A = \frac{a^2}{(2 + K_p)^2} \left[1 - \frac{K_p (4 - K_p)}{5K_p - 2} \right] \tag{46}$$

In summary, when the critical friction angle, peak friction angle, and relative void ratio are known, the index α can be calculated according to the above formula.

7. Index β

The index β only plays a role when e is significantly smaller than e_i (dense sand), and the beta has little effect on the model response in the rest of the cases. A large number of test results show that the value of β is roughly 1, so for natural soil, the general method is to assume that $\beta = 1$.

8. Anisotropic parameters: b_1, b_2, b_3, and m

The parameters b_1, b_2, and b_3 can be determined through direct shear tests conducted on the contact surface under normal stress conditions. By performing direct shear tests on the contact surface, we can obtain the shear stress–shear displacement relationship curves at various initial deposition angles. This process then enables the determination of peak shear stress under these different conditions. The peak shear strength coefficient's relationship with the initial deposition angle can be charted by normalizing the peak stress values obtained under isotropic conditions. This process yields a regular variation relationship, as expressed in Equation (26), where b_1, b_2, and b_3 serve as fitting parameters. The exponent m in this context is a dimensionless constant.

9. Roughness, R

For the regular structural planes, R represents the peak-to-valley distance (or depth) of the structural plane, as depicted in Figure 4. In the case of irregular structural planes, an arithmetic mean value derived from multiple measurements can be employed for R.

2.6. Contact Surface Thickness

Within the constitutive model of the contact surface outlined in this document, it is essential to specify the shear zone thickness to determine the relationship between strain and displacement. In this context, we utilize the relationship formula suggested by Gutjahr [23] and Herle and Arnold [24]. The strain at the contact surface is computed on the basis of the predefined shear zone thickness:

$$\tan \gamma_i = \frac{u_i}{d_s} \tag{47}$$

where u_i is displacement at the contact surface, and d_s is the assumed contact surface thickness. The geometric relationship is shown in Figure 5. The value of contact surface thickness depends on factors such as density, average particle size, and contact surface roughness. According to the research of Gutjahr [23], it can be approximated as $5 \sim 10d_{50}$.

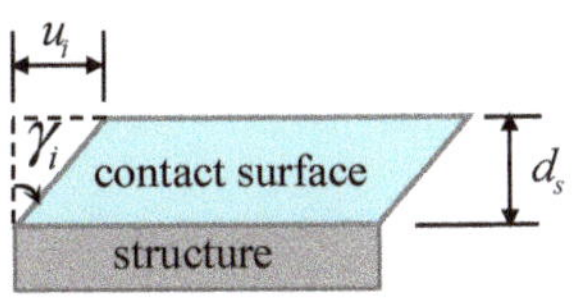

Figure 5. Schematic diagram of strain–displacement relationship at the contact surface.

3. Model Validation

In this section, we utilize the hypoplastic constitutive model for contact surfaces, as it accounts for roughness and initial anisotropy, as detailed in Equation (36). The simulation outcomes are juxtaposed with a range of experimental results from the literature concerning soil–structure contact surfaces to ascertain the model's effectiveness. To align the model's stress–strain relationship curves with the experimental data, these curves are converted into relative displacements of the contact surface, employing an assumed contact surface thickness, as outlined in Equation (48). In the ensuing figures, the data points symbolize the experimental findings, while the continuous lines depict the model's predicted outcomes.

3.1. Experiment on Contact Surfaces of Toyoura Sand and Low Carbon Steel Plate

We evaluated the effectiveness of the hypoplastic model in capturing the stress–deformation behavior of soil–structure contact surfaces with different roughness levels. The model was validated against direct shear test results of Toyoura sand interacting with structural surfaces of varying roughness, as reported by Uesugi et al. [38], under constant normal stress conditions. The material of the structural surface used in the experiments is low-carbon steel, subjected to a normal stress of 78 kPa. The initial void ratio of the sand is 0.68, and the maximum roughness of the structural surfaces is defined as 46 μm, 20.5 μm, and 2.4 μm, corresponding to friction parameters of 0.905, 0.66, and 0.21, respectively. Detailed model parameters are presented in Table 3.

Table 3. Model parameters for the experiment of Toyoura sand—low-carbon steel contact surface.

Parameters	$\varphi_c/^\circ$	h_s/MPa	n	e_{d0}	e_{c0}	e_{i0}	α	β	λ_R
Values	30	2600	0.27	0.61	0.98	1.10	0.25	1.0	0.21/0.66/0.905

The experimental data and model prediction results for the contact surfaces between Toyoura sand and low-carbon steel plates of varying roughness are shown in Figure 6. The findings reveal a robust correlation between the model predictions and the experimental data, both in numerical values and trends, across all three roughness conditions. Moreover, the model adeptly captures the softening phenomenon observed under high roughness conditions, thereby validating its capacity to accurately depict the true shear behavior of the sand–structure contact surface across varying levels of roughness.

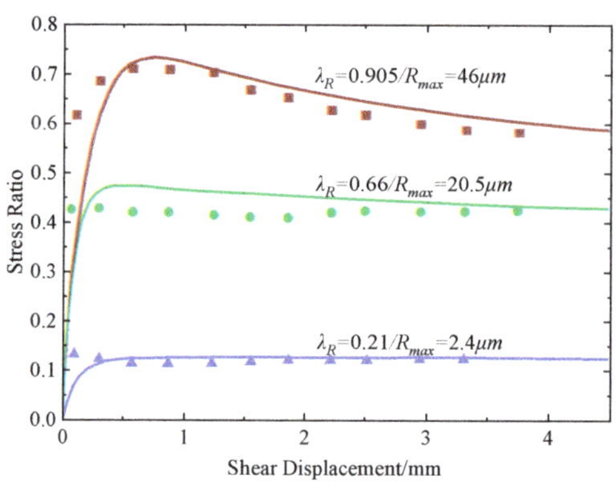

Figure 6. Comparison of experimental results with simulation results for the contact surface between Toyoura sand and low-carbon steel plates.

3.2. Experiment on Contact Surfaces of Changping Composite Gravel/Uniform Gravel and Artificial Steel Plate

To evaluate the effectiveness of the hypoplastic model in capturing the deformation behavior of sand–structure contact surfaces under varying normal stresses and roughness, monotonic shear tests were performed. These tests involved uniform gravel and composite gravel interacting with artificially roughened steel plates, as detailed by Zhang [41]. The experiments were conducted under controlled ideal contact surface boundary conditions, with normal stresses set at 200 kPa and 700 kPa. The composite gravel had a mean particle size of 10.0 mm, and the structural surfaces were artificially roughened steel plates with roughness values of 1 mm and 10 mm. The loading was displacement-controlled at a rate of 1 mm/min. Model parameter values are specified in Table 4, with an initial void ratio of the soil set at 0.75.

Table 4. Model parameters experiment of Changping composite gravel–artificial rough steel plate contact surface.

Parameters	$\varphi_c/°$	h_s/MPa	n	e_{d0}	e_{c0}	e_{i0}	α	β	R
Composite gravel	42	3800	0.2	0.61	0.9	1.08	0.2	1.5	1 mm/10 mm
Uniform gravel	40	3200	0.2	0.61	0.9	1.08	0.2	1.0	1 mm/10 mm

Figures 7 and 8 show the shear stress–shear displacement relationship curves for the two roughness conditions of the artificial steel plates under different normal stresses. The curves predicted by the contact surface hypoplastic model closely match the experimental data points, both in terms of numerical values and overall trends. This alignment suggests that the model is adept at forecasting the mechanical behavior of contact surfaces under a range of normal stress conditions and roughness levels. Additionally, the figures demonstrate that there is no significant softening observed in the shear stress and shear displacement at the contact surface, a phenomenon that the model's predictions accurately capture.

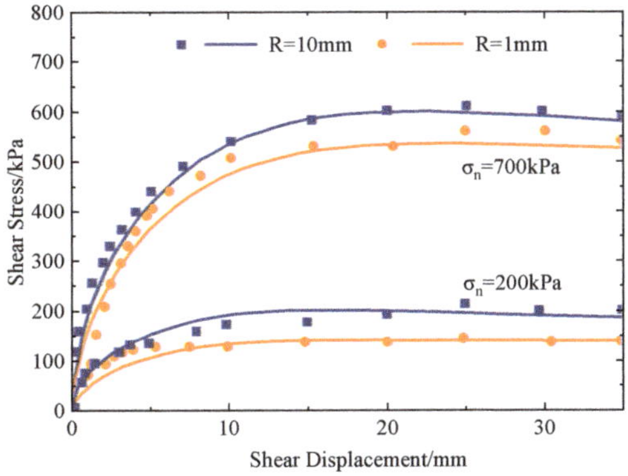

Figure 7. Experimental results and simulation results for Changping composite gravel–artificial rough steel plate contact surface under different roughness conditions.

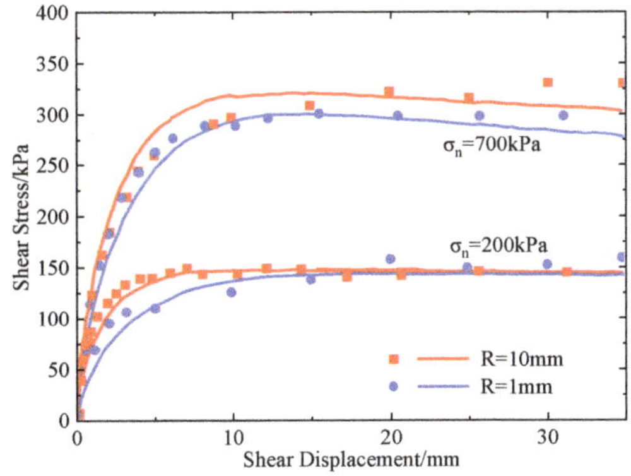

Figure 8. Experimental results and simulation results for Changping uniform gravel–artificial rough steel plate contact surface under different roughness conditions.

3.3. Experiment on Contact Surfaces of Uniform Density Sand and Concrete

This section evaluates the effectiveness of hypoplastic model in capturing the deformation behavior of sand–structure contact surfaces under a staged stress path. The experiments referenced, conducted by Gomez [42], were designed to emulate the real-world behavior of a ship's hull, with a particular emphasis on the interaction between uniform density sand and concrete surfaces. In the experiments, the normal stress was incrementally ramped up from 102 kPa to 274 kPa under meticulously controlled displacement shear conditions, beginning with an initial void ratio of 0.68. The friction parameter was set to 0.85 based on the existing literature, with the specific model parameters outlined in Table 5.

Table 5. Model parameters for uniform density sand–concrete contact surface experiment.

Parameters	$\varphi_c/°$	h_s/MPa	n	e_{d0}	e_{c0}	e_{i0}	α	β	λ_R
Values	30	2600	0.27	0.61	0.98	1.10	0.25	1.0	0.21/0.66/0.905

Figure 9 juxtaposes the experimental outcomes for uniform density sand interfacing with concrete under a progressive stress path against the predicted model curves. The transition point in Figure 9 denotes the juncture at which the normal stress undergoes a change. Initially, a normal stress of 102 kPa was imposed, and upon the shear displacement reaching 0.25 mm, this stress was escalated to 274 kPa. Notably, prior to the transition point, the predicted shear stress–displacement curve closely aligns with the empirical data. Post-transition, although the model's prediction is marginally lower than the experimental results in the lead-up to peak stress, there is a congruence in both the peak stress values and the timing of their manifestation. This concordance suggests that the model adeptly captures the shear behavior of soil–structure interfaces under a staged stress regimen.

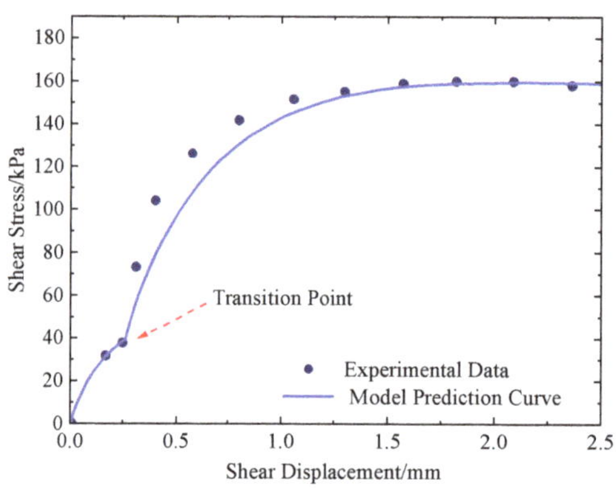

Figure 9. Experimental results and simulation results for uniform density sand–concrete contact surface under staged stress path.

It should be noted that, although the predicted results are in good agreement with the experimental results, minor discrepancies are still observed, which can be attributed to several factors, including inherent model simplifications, assumptions made during model transformation, experimental uncertainties, and differences in boundary and loading conditions. However, the results are within the acceptable error margin for this type of study and that these minor discrepancies do not undermine the validity of the overall findings.

4. Finite Element Implementation of the Interface Constitutive Model

This section presents the finite element implementation of the interface constitutive model in ABAQUS. For the conventional interface modeling, the primary state variables include normal stress, σ_n; normal strain, ε_p; shear stress components (τ_x and τ_y); and shear strain components ($\dot{\gamma}_x$, $\dot{\gamma}_y$). The proposed interface constitutive model introduces an additional state variable, the interface normal stress, σ_p, which must be considered together with void ratio, e, as an auxiliary state variable, The algorithm flowchart is shown in Figure 10.

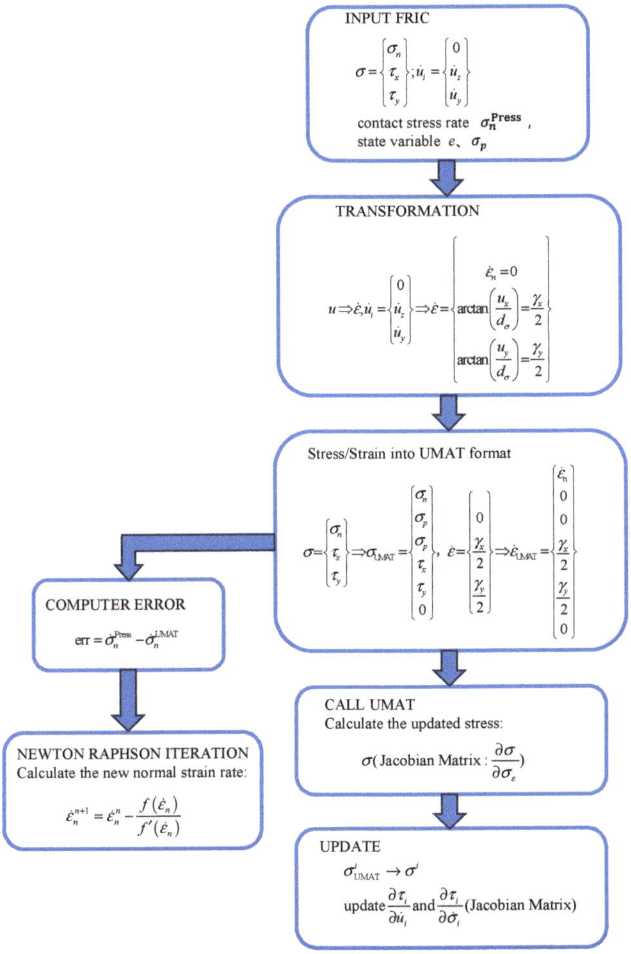

Figure 10. Basic algorithm flow of the finite element method for the hypoplasticity model of contact surfaces.

The implementation procedure follows these sequential steps:

Step 1: At each time step, utilize the user-defined FRIC subroutine within the ABAQUS interface to input model parameters and to retrieve response and displacement data.
Step 2: Utilize the apparent shear modulus to convert shear displacement, u, to shear strain, ε.
Step 3: Convert the inputs obtained from the FRIC subroutine into a format that is compatible with the UMAT subroutine framework.
Step 4: Apply the UMAT subroutine to compute updated stress components.
Step 5: Employ the Newton–Raphson iteration method to determine the normal strain rate, $\dot{\varepsilon}_n$, using the normal contact stress rate, $\dot{\sigma}_n^{Press}$. After the UMAT subroutine execution, compute the difference in normal stress ratios.

Step 6: If the error exceeds the tolerance threshold, implement the Newton–Raphson method to recalculate the stress ratio and return to the UMAT subroutine for further iterations, continuing this process until the error is reduced to an acceptable range.
Step 7: When the error meets the tolerance criteria, update the interface pressure, σ_{UMAT}, and return to the FRIC subroutine to compute the corresponding stress components, σ_{FRIC}, and initiate the next calculation cycle.

As Equation (47) mentions, the apparent shear modulus is employed to correlate the relative displacement and shear strain transformation. Under shear conditions, the shear strain is given by the following:

$$\tan \dot{\gamma}_i = \frac{u_i}{d_v^s} \tag{48}$$

where d_v^s represents the apparent shear thickness.

4.1. Numerical Model

Figure 11 illustrates the geometric configuration and boundary conditions of the direct shear test model. The upper soil specimen utilizes sand from the river sand of the Karnaphuli River tunnel project in Chittagong, Bangladesh, with dimensions of 25 × 25 × 5 cm. The structural component consists of C30 concrete with dimensions of 35 × 25 × 5 cm, modeled as a linear elastic material with a Young's modulus of 1×10^6 kPa and a Poisson's ratio of 0.25. Both the soil and structural elements are discretized using eight-node linear elements (C3D8). The soil–structure interface is modeled using a surface-to-surface contact, with the master surface being the concrete surface and the slave surface being the soil surface. The interface behavior is described using both the Mohr–Coulomb model and the proposed constitutive model for comparative analysis. The model parameters and boundary conditions are detailed in Tables 6 and 7, respectively.

Note that, in Table 6, the critical friction angle φ_c was estimated from the angle of repose of the sandy soils in the site; the particle hardness h_s and index n are obtained from the material compression curve; the value of e_{d0} can be computed by Equation (49):

$$e_{d0} = e_d \exp\left[\left(\frac{\sigma_n + 2\sigma_p}{h_s}\right)^n\right] \tag{49}$$

where the value of e_d can be obtained by the relationship of $e_d - C_u$, according to the experimental results in Youd [41].

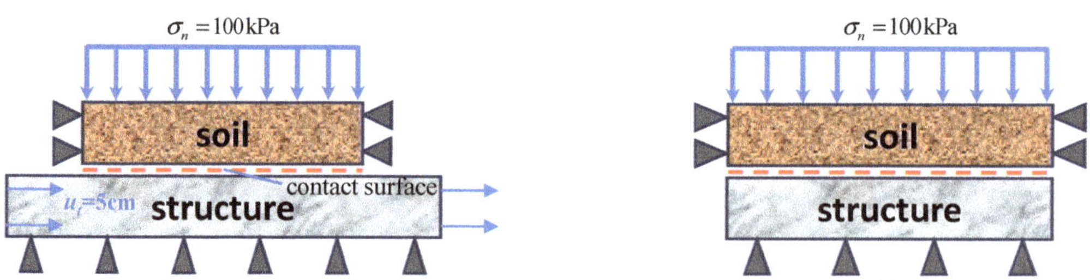

Figure 11. Schematic diagram of the geometric structure and boundary conditions of the finite element model for direct shear test: (**a**) front view and (**b**) side view.

Table 6. Parameter values for the hypoplasticity model of contact surfaces.

Initial Parameters	$\varphi_c/^\circ$	h_s/MPa	n	e_{d0}	e_{c0}	e_{i0}	α	β
	31	1000	0.29	0.61	0.96	1.09	0.13	2.0
Anisotropic Parameters	b_1	b_2	b_3	m	Roughness Parameters			R
	1.15	−0.28	42^2	2				0.3–1.5

Table 7. Finite element direct shear test simulation scheme.

Number	Normal Stress	Roughness	Angle of Repose	Load Amplitude/cm	Shear Path
1	100	0.3	20	5	Monotonic linear shear
2	100	0.5	20	5	Monotonic linear shear
3	100	0.8	20	5	Monotonic linear shear
4	100	1.0	20	5	Monotonic linear shear
5	100	1.5	20	5	Monotonic linear shear
6	100	0.8	0	5	Monotonic linear shear
7	100	0.8	30	5	Monotonic linear shear
8	100	0.8	90	5	Monotonic linear shear

The model boundary conditions are depicted in Figure 11, with the corresponding finite element mesh shown in Figure 12. As illustrated in Figure 11, the model configurations include the following boundary conditions: during the shear, the horizontal displacement of the soil is fixed, and the vertical displacement at the bottom of the structure is also constrained. In the front view (in Figure 11a), the bottom of the structure is assigned a free boundary in the horizontal direction. During the calculation, a continuous horizontal displacement is applied to the structure to perform a soil–structure interface shear test. In contrast, in the side view (in Figure 11b), the horizontal displacement of the structure is fixed. These configurations ensure that the test accurately simulates the interaction between the soil and structure under shear loading conditions. The testing procedure follows these steps:

(1) Application of gravity loading;
(2) Application of normal stress, specified as 100 kPa;
(3) Progressive application of shear displacement up to a maximum of 5 cm.

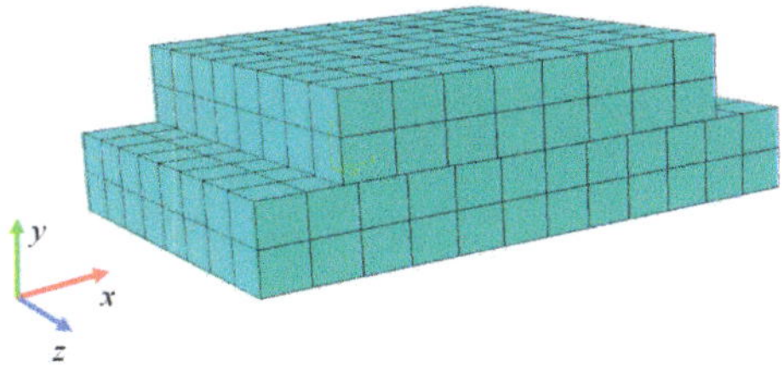

Figure 12. Model grid schematic diagram.

This study conducts a comprehensive analysis of the response characteristics under varying normal stresses, shearing velocities, and material properties of both the soil mass and the interface, utilizing direct shear testing.

4.2. Results Analysis

Based on the numerical model shown in Figure 12, direct shear tests were conducted under different roughness and initial deposition angle conditions to verify the correctness of the FRIC subroutine and the rationality and superiority of the proposed hypoplasticity model for contact surfaces. The test results are presented in Figures 13–15.

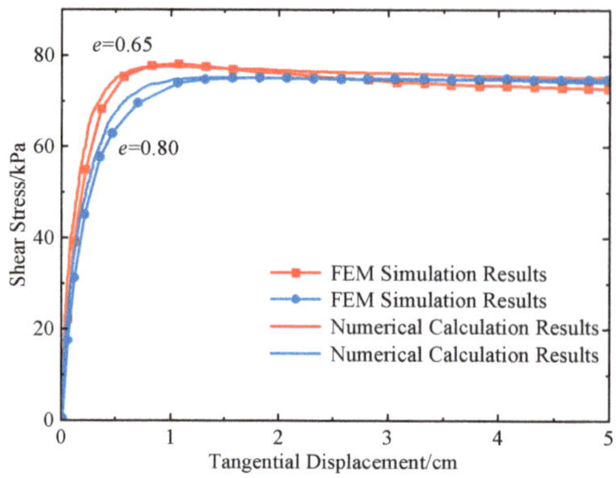

Figure 13. Comparison between the numerical results of the hypoplastic model for contact surfaces and the theoretical solution.

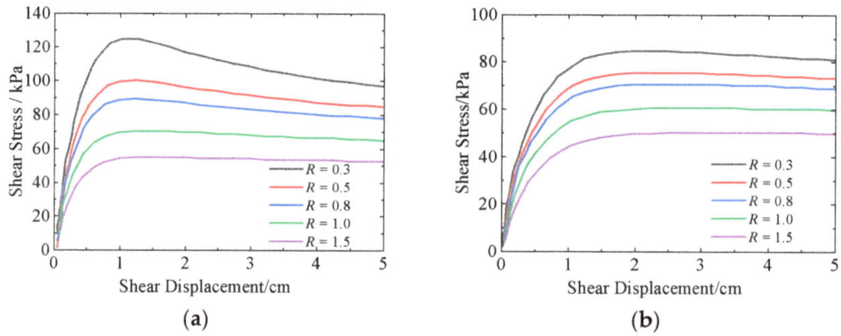

Figure 14. Shear stress–displacement relationship curves for contact surfaces under different interface roughness conditions: (**a**) e = 0.65 and (**b**) e = 0.95.

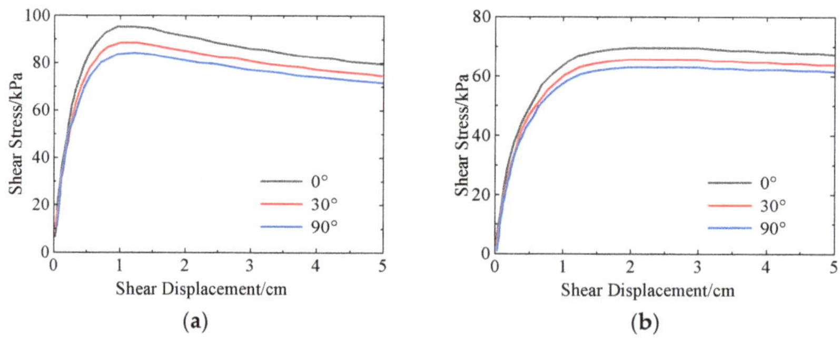

Figure 15. Shear stress–displacement relationship curves for contact surfaces under different initial deposition angle conditions: (**a**) e = 0.65 and (**b**) e = 0.90.

Figure 13 presents a comparison of the numerical calculation results from the hypoplastic model for contact surfaces with the theoretical analytical results under various initial void ratio conditions. It is evident that the stress levels derived from numerical calculations are marginally lower than those predicted by the theoretical solution, with the overall discrepancy being quite minor. Notably, at higher void ratios, the theoretical and numerical results are almost indistinguishable, making the error effectively negligible. This close alignment between the two sets of results confirms the accuracy of the FRIC subroutine.

Figure 14 shows the shear stress–displacement relationship curves for dense sand (e = 0.65) and loose sand (e = 0.90) under different interface roughness conditions.

It is noticeable that, with an increase in the roughness, R, of the structural surface, the peak shear strengths for both loose and dense sands correspondingly increase. Moreover, the softening behavior in dense sand becomes more significant as roughness increases; when the structural surface is relatively smooth, dense sand exhibits minimal signs of softening. In contrast, the shear stress–strain relationship curve for loose sand remains largely unaffected by variations in roughness.

Figure 15 shows the shear stress–displacement relationship curves for the aforementioned dense sand and loose sand under different initial deposition angle conditions. Notably, when the initial deposition angle of the sand approximately 30°, the peak shear stress at the contact surface is minimized. Additionally, the deposition angle impacts only the peak stress and residual strength, leaving the shape and trend of the curves unaltered. The finding is consistent with the observations from Figure 3, which also demonstrate the influence of the deposition angle on the shear behavior of the contact surface.

In addition, Figure 16 compares the shear stress–displacement curves obtained from the contact surface model proposed in this study and from the Mohr–Coulomb contact surface model commonly used in engineering. It can be observed that the maximum shear stress is limited when using the Mohr–Coulomb model; following that, the yield stress limit is achieved, and the shear stress keeps constant. In this sense, the predicted results by Mohr–Coulomb model do not accurately describe the experimental results, even though the Mohr–Coulomb model can simply reflect the tangential friction behavior at the contact surface. Meanwhile, the hypoplasticity contact surface model proposed in this study can more accurately capture the nonlinear behavior of the contact surface, suggesting that the proposed model can better predict the mechanical behaviors of the contact surfaces with large deformation, sliding, and cohesion.

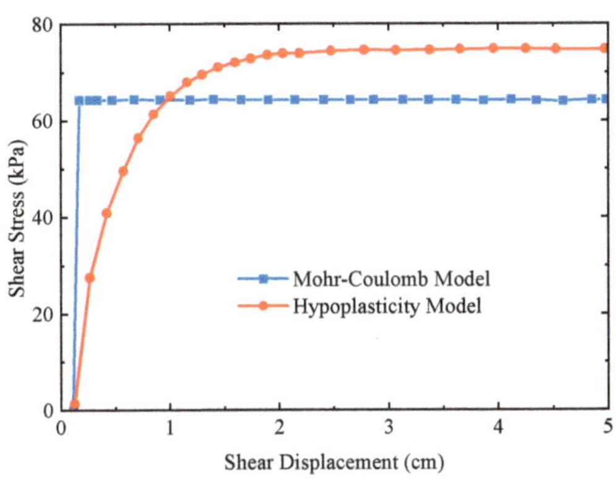

Figure 16. Comparison of shear stress–displacement relations between the Mohr–Coulomb model and the hypoplasticity model.

In summary, the analysis indicates that the direct shear test finite element model using the hypoplastic model can effectively capture the influence of different roughness and the initial deposition angle of sand on the mechanical properties of the contact surface. This validates the rationale of the model proposed in this paper, which considers initial anisotropy and roughness. The proposed model is expected to have a broad applicability in various engineering projects, including tunnel engineering, foundation design, and slope stability analysis. We are currently exploring the application of the constitutive model proposed in this paper to practical engineering projects, including the prediction of frictional resistance between tunnel segments and slurry during the VSM (Vertical Shaft Sinking Machine) construction method.

It is worth mentioning that two assumptions of the whole methodology are considered in this study to establish the hypoplasticity contact surface model: (1) the shear action and control mechanism at the contact surface is consistent with that of soil, and the contact surface can be regarded as a tiny continuous soil unit with the same deformation pattern as that of soil; and (2) an isotropic stress state is assumed at the contact surface (i.e., $\sigma_{22} = \sigma_{33} \neq \sigma_{11}$), which develops independently of the normal stress during shear. In this sense, the model proposed in this study has some limitations for predicting stress-induced anisotropic and cohesive soil–structure contact surfaces, which should be focused on in future work.

5. Conclusions

Drawing from the foundational equations of the Gudehus–Bauer eight-parameter hypoplasticity model, this paper examines the influences of initial anisotropy and surface roughness on the mechanical behavior of sandy soils and develops a hypoplasticity constitutive model for the sandy soil–structure contact surface that incorporates these factors. The model is subsequently validated against results from various contact surface experiments. The principal findings are summarized as follows:

(1) A new hypoplasticity constitutive model for sandy soil–structure contact surfaces is proposed by incorporating the effects of initial anisotropy and surface roughness into the Gudehus–Bauer eight-parameter hypoplasticity model. This model consists of 13 fundamental parameters, which can be determined based on the particle size characteristics of granular materials.

(2) The proposed model is validated by comparing its predictions with experimental results from various contact surface conditions, including Toyoura sand–steel plate, Changping composite gravel/uniform gravel–steel plate, and dense sand–concrete contact surfaces under different roughness conditions. The predicted results align well with experimental data, confirming that the model accurately captures the mechanical behavior of contact surfaces under varying roughness conditions.

(3) A FRIC-UMAT subroutine was developed to implement the hypoplasticity model in finite element simulations. The modified subroutine was applied to typical direct shear tests on contact surfaces, and the simulation results were compared with experimental data. Additionally, the performance of the proposed model was compared with that of the Mohr–Coulomb contact surface model under varying initial deposition angles and surface roughness conditions. The results demonstrate the accuracy of the FRIC subroutine and validate the rationality and superiority of the proposed model.

As far as the isotropic condition is considered, more influencing factors, such as stress-induced anisotropy and temperature, should be introduced into the proposed hypoplasticity contact surface model for the practical engineering. This will enhance the reliability of the model. At the same time, the hypoplasticity model of cohesive soil will be further adapted to the cohesive soil–structure contact surface. This innovation work will expand the applicability of the hypoplasticity model, enabling it to be used for both cohesive and non-cohesive soil contact surfaces. The development of this hypoplasticity contact surface model offers potential advantages for practical engineering applications, particularly in analyzing large deformations of soil foundations, such as the floating or settling of pile foundations.

Author Contributions: Conceptualization, Z.C.; data curation, J.Y. and J.Z.; investigation, J.Y. and J.C.; methodology, Z.C.; software, Z.C. and Y.Z.; supervision, P.Y.; validation, Z.C.; visualization, J.Z.; writing—original draft, J.Y., J.C. and Z.C.; writing—review and editing, J.Z., Y.Z. and P.Y. All authors have read and agreed to the published version of the manuscript.

Funding: This research received no external funding.

Institutional Review Board Statement: Not applicable.

Informed Consent Statement: Not applicable.

Data Availability Statement: Details of the analytical model presented in this study are available upon request from the corresponding author. The data are not publicly available due to privacy.

Conflicts of Interest: The authors declare no conflicts of interest.

References

1. Clough, G.W.; Duncan, J.M. Finite element analyses of retaining wall behavior. *J. Soil Mech. Found. Div.* **1971**, *97*, 1657–1673. [CrossRef]
2. Lashkari, A. Prediction of the shaft resistance of nondisplacement piles in sand. *Int. J. Numer. Anal. Methods Geomech.* **2013**, *37*, 904–931. [CrossRef]
3. Long, S.; Xu, S.; Zhang, Y.; Li, B.; Sun, L.; Wang, Y.; Wang, J. Method of soil-elastoplastic dem parameter calibration based on recurrent neural network. *Powder Technol.* **2023**, *416*, 118222. [CrossRef]
4. Mortara, G.; Ferrara, D.; Fotia, G. Simple model for the cyclic behavior of smooth sand-steel interfaces. *J. Geotech. Geoenviron. Eng.* **2010**, *136*, 1004–1009. [CrossRef]
5. Stutz, H.H. Hypoplastic Models for Soil-Structure Interfaces—Modelling and Implementation. Ph.D. Thesis, Kiel University, Kiel, Germany, 2016.
6. ur Rehman, Z.; Zhang, G. Three-dimensional elasto-plastic damage model for gravelly soil-structure interface considering the shear coupling effect. *Comput. Geotech.* **2021**, *129*, 103868. [CrossRef]
7. Zhou, Z.; Li, Z.; Gao, C.; Zhang, D.; Wang, M.; Wei, C.; Bai, S. Peridynamic micro-elastoplastic constitutive model and its application in the failure analysis of rock masses. *Comput. Geotech.* **2021**, *132*, 104037. [CrossRef]
8. Kolymbas, D. A rate-dependent constitutive equation for soils. *Mech. Res. Commun.* **1977**, *4*, 367–372. [CrossRef]
9. Dafalias, Y.F. Bounding surface plasticity. I: Mathematical foundation and hypoplasticity. *J. Eng. Mech.* **1986**, *112*, 966–987. [CrossRef]
10. Kolymbas, D. An outline of hypoplasticity. *Arch. Appl. Mech.* **1991**, *61*, 143–151. [CrossRef]
11. Wu, W.; Bauer, E. A simple hypoplastic constitutive model for sand. *Int. J. Numer. Anal. Methods Geomech.* **1994**, *18*, 833–862. [CrossRef]
12. Gudehus, G. A comprehensive constitutive equation for granular materials. *Soils Found.* **1996**, *36*, 1–12. [CrossRef]
13. Bauer, E. Calibration of a comprehensive hypoplastic model for granular materials. *Soils Found.* **1996**, *36*, 13–26. [CrossRef]
14. Von Wolffersdorff, P.A. A hypoplastic relation for granular materials with a predefined limit state surface. *Mech. Cohesive-Frict. Mater* **1996**, *1*, 251–271. [CrossRef]
15. Wu, W.; Lin, J.; Wang, X. A basic hypoplastic constitutive model for sand. *Acta Geotech.* **2017**, *12*, 1373–1382. [CrossRef]
16. Qian, H.Y.; Wu, W.; Du, X.L.; Xu, C.S. A Hypoplastic Constitutive Model for Granular Materials with Particle Breakage. *Int. J. Geomech.* **2023**, *23*, 04023065. [CrossRef]
17. Qian, H.Y.; Wu, W.; Xu, C.S.; Liao, D.; Du, X.L. An extended hypoplastic constitutive model considering particle breakage for granular material. *Comput. Geotech.* **2023**, *159*, 105503. [CrossRef]
18. Subair, A.; Aljorany, A. Experimental and Numerical Analysis of Axially Loaded Slender Piles Implementing a Hypoplastic Model. In Proceedings of the IOP Conference Series: Materials Science and Engineering, Kerbala, Iraq, 5–6 December 2020; p. 012069.
19. Tang, Y.L.; Wang, S.; Zhang, C.S. A gradation-dependent hypoplastic model for crushable sands. *Acta Geotech.* **2024**, *19*, 6937–6953. [CrossRef]
20. Tian, H.; Zeng, K.; Zhang, H.; Lu, S. Research on Extended Hypoplastic Model and Its Verification for Deposits Soil. In Proceedings of the IOP Conference Series: Earth and Environmental Science, Beijing, China, 23–26 October 2020; p. 062034.
21. Wang, S.; Fang, H.-J.; Kang, X.; Li, D.-Q.; Wu, W. Simhypo-sand: A simple hypoplastic model for granular materials and SPH implementation. *Acta Geotech.* **2024**, *19*, 4533–4555. [CrossRef]
22. Herle, I.; Nübel, K. Hypoplastic description of the interface behaviour. In *Numerical Models in Geomechanics*; Pande, G.N., Pietruszczak, S., Schweiger, H.F., Eds.; CRC Press: London, UK, 2020; pp. 53–58. [CrossRef]

23. Gutjahr, S. Optimierte Berechnung von Nicht Gestützten Baugrubenwänden in Sand. Doctor's Thesis, Universität Dortmund, Dortmund, Germany, 2003. [CrossRef]
24. Herle, I.; Arnold, M. Hypoplastic description of the frictional behaviour of contacts. In *Numerical Methods in Geotechnical Engineering*; Schweiger, H.F., Ed.; Taylor & Francis: Graz, Austria, 2006; pp. 101–106. [CrossRef]
25. Stutz, H.; Mašín, D.; Wuttke, F. Enhancement of a hypoplastic model for granular soil–structure interface behaviour. *Acta Geotech.* **2016**, *11*, 1249–1261. [CrossRef]
26. Stutz, H.; Mašín, D. Hypoplastic interface models for fine—Grained soils. *Int. J. Numer. Anal. Methods Geomech.* **2017**, *41*, 284–303. [CrossRef]
27. Jong, Y.; Ma, Y.; Liu, Y. Granular Hypoplastic Interface Model Considering Surface Roughness. In Proceedings of the Journal of Physics: Conference Series, Busan, Republic of Korea, 14–15 November 2020; p. 042004.
28. Jong, Y.-G.; Liu, Y.; Chen, Z.; Mabe Fogang, P. Hypoplastic interface model considering plane strain condition and surface roughness. *Adv. Civ. Eng.* **2021**, *2021*, 1473181. [CrossRef]
29. Staubach, P.; Machaček, J.; Wichtmann, T. Novel approach to apply existing constitutive soil models to the modelling of interfaces. *Int. J. Numer. Anal. Methods Geomech.* **2022**, *46*, 1241–1271. [CrossRef]
30. Staubach, P.; Machaček, J.; Wichtmann, T. Mortar contact discretisation methods incorporating interface models based on Hypoplasticity and Sanisand: Application to vibratory pile driving. *Comput. Geotech.* **2022**, *146*, 104677. [CrossRef]
31. Wu, W. Rational approach to anisotropy of sand. *Int. J. Numer. Anal. Methods Geomech.* **1998**, *22*, 921–940. [CrossRef]
32. Yang, Z.; Liao, D.; Xu, T. A hypoplastic model for granular soils incorporating anisotropic critical state theory. *Int. J. Numer. Anal. Methods Geomech.* **2020**, *44*, 723–748. [CrossRef]
33. Bauer, E.; Huang, W.; Wu, W. Investigations of shear banding in an anisotropic hypoplastic material. *Int. J. Solids Struct.* **2004**, *41*, 5903–5919. [CrossRef]
34. Saberi, M.; Annan, C.-D.; Konrad, J.-M. On the mechanics and modeling of interfaces between granular soils and structural materials. *Arch. Civ. Mech. Eng.* **2018**, *18*, 1562–1579. [CrossRef]
35. Stutz, H.H.; Martinez, A. Hypoplastic simulation of axisymmetric interface shear tests in granular media. In Proceedings of the China-Europe Conference on Geotechnical Engineering: Volume 1, Vienna, Austria, 3 August 2018; pp. 69–73.
36. Wu, W.; Kolymbas, D. Numerical testing of the stability criterion for hypoplastic constitutive equations. *Mech. Mater.* **1990**, *9*, 245–253. [CrossRef]
37. Li, S. The Macro-Micromechanical Analysis and Numerical Simulation of Shear Anisotropy in the Sand-Structure Contact Zone. Master's Thesis, University of Science and Technology Beijing, Beijing, China, 2016.
38. Uesugi, M.; Kishida, H.; Tsubakihara, Y. Behavior of sand particles in sand-steel friction. *Soils Found.* **1988**, *28*, 107–118. [CrossRef]
39. Herle, I.; Gudehus, G. Determination of Parameters of a Hypoplastic Constitutive Model from Properties of Grain Assemblies. *Mech. Cohesive-Frict. Mater* **1999**, *4*, 461–486. [CrossRef]
40. Youd, T. Factors Controlling Maximum and Minimum Densities of Sands. *ASTM Spec. Tech. Publ.* **1972**, *523*, 98–112. [CrossRef]
41. Zhang, G.; Zhang, J.M. Unified modeling of soil-structure interface and its test confirmation. *Chin. J. Geotech. Eng.* **2005**, *27*, 1175–1179. [CrossRef]
42. Gómez, J.E.; Filz, G.M.; Ebeling, R.M. Extended hyperbolic model for sand-to-concrete interfaces. *J. Geotech. Geoenviron. Eng.* **2003**, *129*, 993–1000. [CrossRef]

applied sciences

MDPI

Article

Effect of Design Parameters of Supporting Structure on the Energy Evolution Characteristic of Surrounding Rock

Ying Chen, Qi Da, Lei Zhang *, Danli Li and Bing Dai

School of Resource Environment and Safety Engineering, University of South China, Hengyang 421001, China; usc_chen@usc.edu.cn (Y.C.); 20222002110041@stu.usc.edu.cn (Q.D.); 20212002110041@stu.usc.edu.cn (D.L.); daibingusc@usc.edu.cn (B.D.)
* Correspondence: zhagleicpu1@usc.edu.cn

Abstract: A reasonable support parameter design scheme is the key to ensuring the stability of the roadway. This study established 217 models using FLAC3D to analyze the evolution characteristics of elastic strain energy and plastic dissipation energy of surrounding rock under different shotcrete and rockbolt support structures. Additionally, five single models (BP, DT, ELM, RF, SVM) were introduced to explore the application of machine learning in predicting the stability of the roadway. The study found that in the parameters of the shotcrete layer support structure, the energy evolution of the surrounding rock is more sensitive to isotropic and thickness; in the parameters of the anchor rod support structure, the energy evolution of the surrounding rock is more sensitive to Young's modulus, cross-sectional area young, and grout stiffness. Additionally, the parameters of the shotcrete layer support structure are not necessarily the larger the better. When isotropic is 100 GPa, both the dissipated energy and the elastic strain energy are higher than that at 25 GPa. The results of the single model test indicate that machine learning is relatively accurate in predicting different shotcrete and anchor support structures. The runtime difference between traditional methods and machine learning models highlights the potential advantages of machine learning.

Keywords: surrounding rock; design parameters of supporting structure; FLAC3D; energy evolution; machine learning

1. Introduction

In recent years, as mining activities proceed, shallow and easily accessible resources have gradually become depleted. Consequently, exploring resources in the depths of the earth has emerged as a major mainstream trend in the field of mineral resources [1–4]. In the deep mining area, the safety of the roadway will be exposed to more risks, such as roof falling [5,6], wall caving [7,8], water gushing [9,10], and particularly the rockburst disaster [11–13], which is challenging to predict and prevent. When all the elastic strain energy (ESE) stored in the rock mass is suddenly and violently released under certain circumstances, leading to brittle fracture of the rock mass and causing it to explode like a detonation, a phenomenon known as rockburst occurs [14,15]. The entire process is accompanied by the accumulation, dissipation, and release of energy. To protect the stability of the roadway, it is crucial to understand the energy evolution characteristics (EEC). Results from references [16,17] indicate that supporting the tunnel can effectively enhance its safety, thereby reducing the likelihood of engineering accidents. Therefore, conducting research on optimizing support parameters holds significant importance. Figure 1 shows the keyword contribution map. The keywords used to generate the word cloud were extracted from the Web of Science database, focusing on articles related to the topics "Numerical Simulation," "Energy Evolution Characteristics", and "Surrounding Rock". This approach ensures comprehensive coverage of relevant terms and concepts within the field. The word cloud reflects the frequency and significance of these terms, highlighting the core research focus areas.

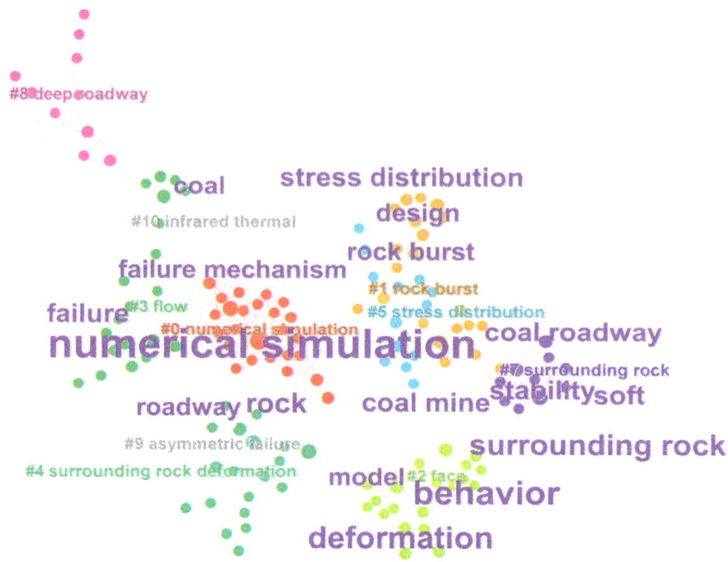

Figure 1. Keyword contribution map.

Scholars have carried out a series of studies on the EEC of the surrounding rock. Zhu et al. [18] calculated the stability of the elastic-plastic and rheological state of the rock mass and analyzed the energy dissipation (ED) characteristics during the excavation process based on an example from the Three Gorges Project in China. Xie et al. [19] used the damage evolution equation to describe the variation law of the damage variable and its associated generalized thermodynamic force-damage energy release rate from a macroscopic perspective, and further revealed the internal mechanism of ED during rock deformation and failure through the study of mesoscopic damage mechanics. They then established a macro-meso-micro scale coupling rock mechanics system based on damage evolution and ED. Liang et al. [20] carried out an acoustic emission (AE) monitoring experiment of roadway rockburst under biaxial loading, analyzed the macro failure characteristics of roadway rockburst process, and conducted a comparative study with the on-site rockburst occurrence process, and finally obtained the accumulation and release laws of energy and time effect laws of rockburst in the macro-damage process. Using Particle Flow Code (PFC), Yu et al. [21] obtained the macroscopic stress field and full-section deformation dynamic evolution curves of the surrounding rock in a semicircular arch roadway (mudstone), revealed the AE characteristics and EEC of the roadway excavation, and explored the influence of vertical stress and lateral pressure coefficients on EEC. Dai et al. [22,23] conducted tests on rock specimens with different crack distribution angles using a Split Hopkinson Pressure Bar (SHPB); they found that the ED was positively correlated with the dynamic strength, and established a damage variable model for describing the damage evolution of granite under dynamic and static loading. Zhang et al. [24], through PFC2D simulation tests, found that when the coal is destroyed, the hard top plate will produce a rebound "energy supply" phenomenon. They then analyzed the effect of rock strength on the EEC of coal-rock combination, studied the EEC of the roadway perimeter rock under different roof strength, and preliminarily established the roadway perimeter rock zoned energy storage model. Yi et al. [25] analyzed the transfer and dissipation of ESE in the surrounding rock of a deep roadway by considering the strain-softening and expand properties of the rock mass. He et al. [26,27] studied the influence of in situ stress value, joint angle, and joint spacing on tunnel rockburst under deep buried strong dynamic disturbance, and discussed the failure zone, deformation process, and stress change of surrounding rock, the velocity of the damaged rock mass, and kinetic energy of failure under different schemes. Table 1 provides an incomplete summary of recent studies on the energy evolution characteristics (EEC) of surrounding rock.

Table 1. Summary of Research on Energy Evolution Characteristics (EEC) of Surrounding Rock.

Research Direction	Authors and References	Key Findings
Energy Dissipation Characteristics	Zhu et al. [18]	Investigated the stability of elastic-plastic and rheological states of rock masses.
Macro-Micro Coupled Models	Xie et al. [19]	Developed a macro-meso-micro coupled model for damage evolution and energy release.
Acoustic Emission Characteristics	Liang et al. [20]	Analyzed failure characteristics of roadway rockburst through acoustic emission.
Numerical Simulation and Energy Evolution Laws	Yu et al. [21]	Used particle flow methods to reveal stress field evolution in surrounding rock.
Damage Models under Dynamic and Static Loading	Dai et al. [22,23]	Established a damage variable model through dynamic loading experiments.
Energy Transfer and Zonal Energy Models	Zhang et al. [24]	Proposed zonal energy storage models for surrounding rock.
Elastic Energy Transfer and Dissipation	Yi et al. [25]	Proposed an energy dissipation zoning pattern considering strain-softening and expansion properties.
Dynamic Disturbances in Rockbursts	He et al. [26,27]	Explored effects of in situ stress, joint angles, and spacing on deep roadway rockbursts.

In recent years, the field of geotechnical engineering has witnessed the rapid integration of artificial intelligence (AI) and machine learning (ML), especially in the field of data mining and predictive modeling. Wang et al. [28] combined finite element simulation with machine learning, obtained meso mechanical properties through simulation, and put them into machine learning algorithm for training. Taking uniaxial elastic modulus, Poisson's ratio, tensile strength, compressive strength, and confining pressure of rock as characteristic variables, they carried out multi-objective machine learning on triaxial elastic modulus, Poisson's ratio, and compressive strength of rock. Li et al. [29] simulated the slope through FLAC to obtain the parameters, and used machine learning to develop the agent model. Table 2 shows the application of machine learning in the field of geotechnical engineering.

Table 2. Applications of machine learning.

Reference	Time	Model	Data Number	Research Object
Wang et al. [28]	2023	ANN, SVR, RF	384	Mechanical property
Li et al. [29]	2023	DNN	880	Slope stability
Lu et al. [30]	2024	ANN	30	Shear strength

In summary, although scholars have extensively studied the energy evolution characteristics (EEC) of surrounding rock and explored the effects of support systems on roadway stability, the specific mechanisms by which different support parameters influence the EEC remain inadequately investigated. Furthermore, while artificial intelligence (AI) and machine learning (ML) techniques have been applied in geotechnical engineering, there has been no systematic research utilizing machine learning methods to predict the impact of various support parameters on the EEC. To effectively address these challenges, this study takes the deep mining area of the Sanshandao Gold Mine as its research context. The primary objectives are to analyze the mechanisms through which different support parameters affect the EEC of the surrounding rock, optimize roadway support design from an energy perspective, and provide theoretical guidance. Additionally, five commonly used regression models (Back propagation (BP) [30], Decision Tree (DT) [31], Extreme Learning Machine (ELM) [32], Random Forests (RF) [33], Support Vector Machines (SVM) [34]) are introduced to investigate the application of machine learning in predicting roadway stability and evaluate the predictive performance of different models for the EEC influenced by various support parameters.

2. Materials and Methods

Fast Lagrangian Analysis of Continua in three dimensions (FLAC3D) can simulate plastic failure and flow of materials very accurately. It features fast computation speed, high accuracy, and a wide range of applicability, and has been widely used in geotechnical engineering analysis [35,36]. Our work, presented in this article, utilizes this software to conduct subsequent research.

Figure 2 illustrates the composition of the "cable" and "shell" structural elements in FLAC3D. These two structural elements are widely used in engineering simulations, representing rock bolts and shotcrete structures, respectively. Figure 3 presents the support system, stress distribution, and energy evolution characteristics (EEC) during roadway excavation. As the excavation progresses, stress in the surrounding rock undergoes redistribution. With increasing distance from the excavation face, the stress field of the surrounding rock can be classified into three distinct zones: the stress relaxation zone (SRZ), the stress concentration zone (SCZ), and the in situ stress zone (ISZ). Simultaneously, stress redistribution induces strain variations, accompanied by the accumulation and dissipation of energy. Excessive energy concentration, characterized by elastic strain energy (ESE), can result in frequent geological hazards such as rockbursts. Conversely, excessive energy dissipation, defined as plastic dissipation energy (PDE), deteriorates the physical and mechanical properties of the rock mass, compromising roadway stability. Therefore, determining the effects of different support parameters on the EEC of the surrounding rock holds significant practical importance for ensuring roadway safety.

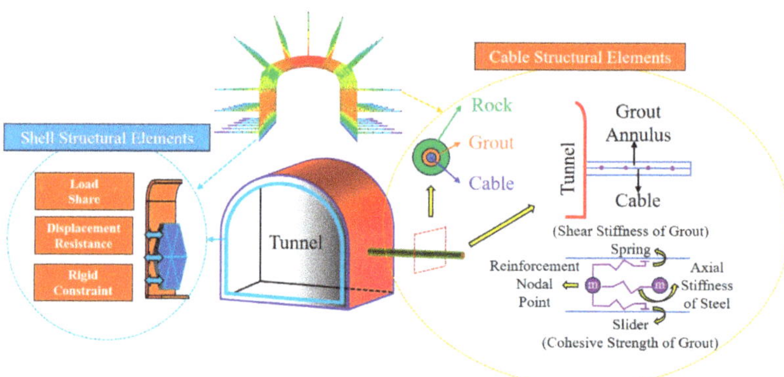

Figure 2. Structural elements action principal schematic.

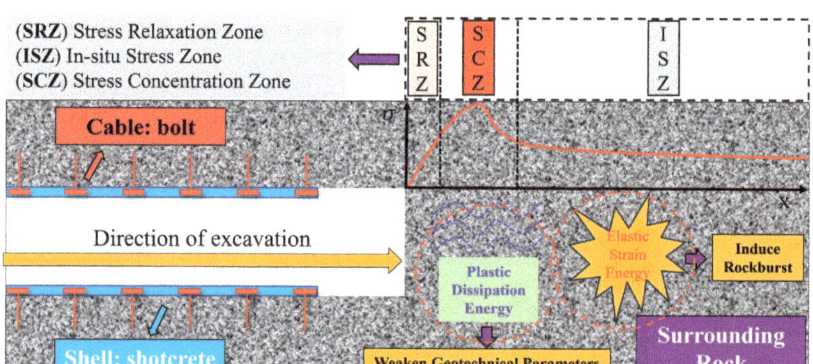

Figure 3. Roadway excavation, support, stress distribution, and EEC.

2.1. Model Building

In this study, the roadway in the 945 m deep buried area of Sanshandao gold mine is taken as the background [37]. The cross-section shape of this roadway is a straight wall arch with a span width of 4.4 m, wall height of 2.75 m, and arch height of 1.15 m. The roadway support scheme is constructed from shotcrete combined with rock bolt support structure. The rock bolts are resin bolts, 2.2 m in length, and spaced 1 m apart in rows. According to the results of in situ measurements, the in situ stress at the roadway depth of 945 m is 31.0 MPa~44.1 MPa, and in this study, the vertical stress of the model is 40 MPa, and the lateral stress coefficient is taken to be 1.0. The model has a specification of 50 m × 50 m × 6 m (W × H × L), and there are a total of 34,884 zones and 26,793 grid-points in the model. Additionally, the mesh elements of the roadway attachments are encrypted. After the model is established, displacement and velocity constraints are applied to the front, back, left, and right surfaces and lower surfaces of the model, and the top of the model is a free surface. In the model used in this study, the value of gravitational acceleration has a minor impact on the results. The primary focus of the simulation is to investigate the influence of anchor rod parameters and shotcrete layer parameters on the energy evolution characteristics of the surrounding rock. Therefore, in this study, numerical simulations were conducted based on the deep-buried area of the Sanshandao gold mine at a depth of 945 m, utilizing in situ measured data to construct a mechanical model of the surrounding rock. The selected geotechnical parameters include a bulk modulus of 21.08 GPa, a shear modulus of 9.73 GPa, cohesion of 4 MPa, a Poisson's ratio of 0.3, and an internal friction angle of 30°. These parameters, derived from field measurements or relevant literature, are representative and reliable. The model dimensions are 50 m × 50 m × 6 m, and boundary conditions were applied with displacement and velocity constraints on the front, back, left, right, and bottom surfaces, while the top surface was set as a free boundary. This configuration ensures the completeness of stress distribution and energy evolution characteristics within the study area. Furthermore, to enhance computational accuracy, the mesh around the support structures was refined to more precisely capture the energy variation in stress concentration zones. The gravitational acceleration was set to 10 m/s^2, considering its minor impact in deep rock environments, allowing the study to focus on the influence of support parameters on the energy evolution of surrounding rock. A vertical stress of 40 MPa was selected, reflecting a representative value within the in situ measured range of 31.0 MPa to 44.1 MPa at a depth of 945 m. A lateral pressure coefficient of 1.0 was used to simplify the complexity of horizontal stress distribution for analytical purposes. The Mohr–Coulomb model [38], widely utilized in geomechanical analysis, was adopted to describe the elastic-plastic behavior and failure characteristics of the rock mass. The assumptions of the Mohr–Coulomb model make it particularly relevant for this study. While the model simplifies the complex behavior of real rock masses by assuming homogeneity and isotropy, these assumptions are appropriate for analyzing the large-scale energy evolution characteristics of surrounding rock in deep mining conditions. The model effectively captures the transition from elastic energy accumulation to plastic energy dissipation, which aligns with the primary focus of this research. The simulation process and specific design scheme are shown in Figures 4 and 5.

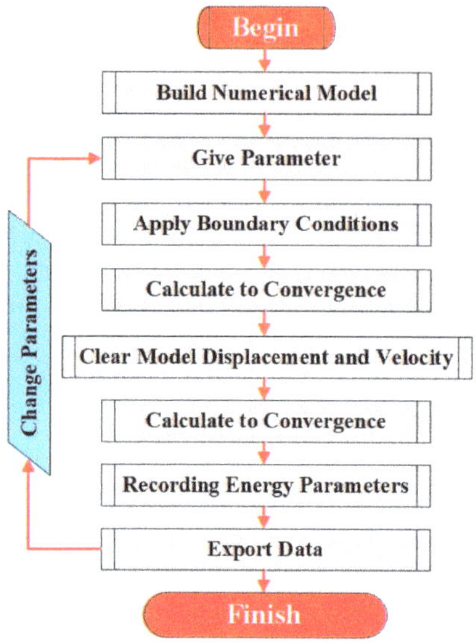

Figure 4. Simulation process.

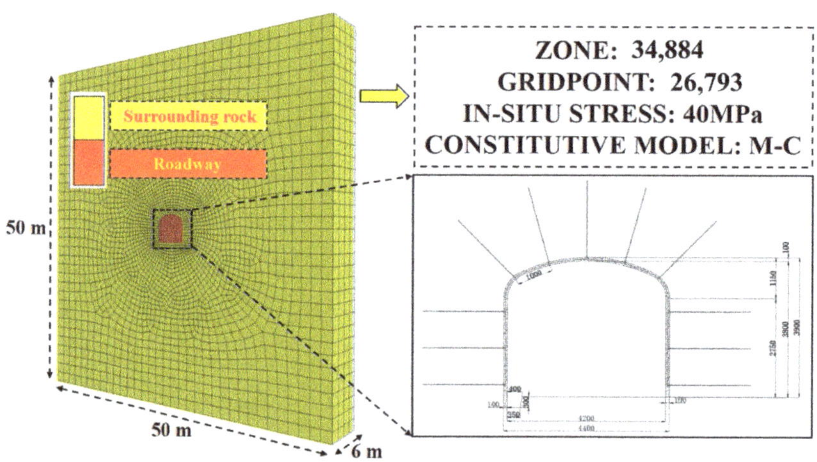

Figure 5. Schematic diagram of the model and supporting structure.

2.2. Scheme Design

Sanshandao Gold Mine adopts shotcrete combined with rock bolt support structures as the perimeter rock stability control program, which significantly reduces the probability of engineering disasters. To study the EEC of the surrounding rock, the support parameter scheme shown in Appendix A is constructed, where A1: Isotropic elastic modulus, A2: Poisson's Ratio, A3: Thickness (m), B1: Young's modulus (GPa), B2: Yield-tension (MPa), B3: Grout-cohesion (MPa), B4: Grout-stiffness (GPa), B5: Grout-friction (°), B6: Grout-perimeter (m), B7: Cross-sectional-area (m^2), and B8: Pre-tensioning Force (kN). Isotropic elastic modulus is abbreviated as Isotropic and Young's modulus is abbreviated as Young. Additionally, specific parameters for rock bolts and the shotcrete layer are listed in Tables 3 and 4, respectively.

Table 3. Parameter of the shotcrete layer.

Scheme	Isotropic (GPa)	Poisson's Ratio	Thickness (m)
A-number	35	0.25	0.1

Table 4. Parameter of the rock bolt.

Scheme	Young (GPa)	Yield-Tension (MPa)	Grout-Cohesion (MPa)	Grout-Stiffness (GPa)	Grout-Friction (°)	Grout-Perimeter (m)	Cross-Sectional-Area (m^2)	Pre-Tensioning Force (kN)
B-number	200	50	6	3	40	0.1884	0.002826	150

Numerical simulation can effectively track elastic stored energy (ESE) and plastic dissipated energy (PDE). FLAC 3D employs the incremental method for continuous iterative solution, where the motion equation is solved at grid nodes, while the stress-strain equation is applied to each time step in the model region. During the calculation of the stress-strain equation, as the system approaches an equilibrium state, the incremental changes in energy components are determined and accumulated. Additionally, FLAC3D provides a variety of plastic models to characterize the deformation capacity of a region. When a region undergoes irreversible deformation, energy is dissipated in the form of plastic work. The strain within the region can typically be decomposed into two parts: elastic strain and plastic strain [39]. In this study, the energy calculation module of FLAC3D was employed, and the equivalence between elastic and plastic energy was explored through the utilization of relevant command workflows.

2.3. Correlation

Studying parameter correlation is crucial when examining how the design parameters of a supporting structure affect the energy evolution of the surrounding rock. Each index is analyzed using Kendall correlation to take the correlation between characteristic parameters into account. The correlation coefficient is a real number between (−1, +1). When the correlation coefficient is between −1 and 0, it indicates that there is a negative correlation between the variables [40]. When the correlation coefficient is between 0 and 1, it indicates that there is a positive correlation between the variables. When the correlation coefficient is 0, there is no correlation between the two variables. Talking about the correlation coefficient alone without mentioning the *p*-value is pointless when attempting to determine whether the two variables are correlated. Instead, we need to talk about the significance level. Determining the correlation's significance level is important because the correlation between the two variables could only be the result of coincidental factors.

3. Results and Analysis

3.1. Correlation Analysis

In conjunction with Figure 6 and Table 5, the phase relationship table reveals that the only variable that exhibits a weak negative correlation with isotropy is the thickness of the shotcrete layer. However, the *p* value of both variables is 0.957 in the *p*-value table, exceeding the predetermined significance level (0.05), indicating that there is no correlation between the two variables.

Table 5. Kendall correlation *p*-value table (A).

Variable	Isotropic	Poisson's Ratio	Thickness
Isotropic	0	1	0.957
Poisson's Ratio	1	0	1
Thickness	0.957	1	0

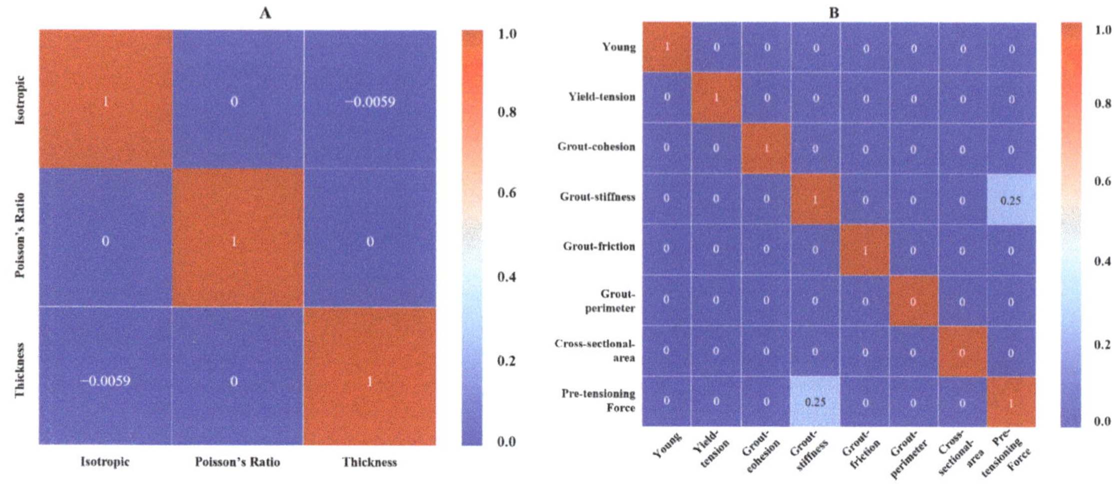

(**a**) Parameters of the jet layer (**b**) Parameters of bolt

Figure 6. Kendall correlation coefficient.

As shown in Figure 6 and Table 6, the correlation table indicates that only Grout-stiffness and Pre-tensioning Force exhibit a positive correlation. Moreover, the *p*-value for both variables is 0.001, which is well below the predefined significance threshold of 0.05. This result suggests a statistically significant correlation between the two variables.

Table 6. Kendall correlation *p*-value table (B).

Variable	Young's	Yield-Tension	Grout-Cohesion	Grout-Stiffness	Grout-Friction	Grout-Perimeter	Cross-Sectional-Area	Pre-Tensioning Force
Young	0	1	1	1	1	1	1	1
Yield-tension	1	0	1	1	1	1	1	1
Grout-cohesion	1	1	0	1	1	1	1	1
Grout-stiffness	1	1	1	0	1	1	1	0.001
Grout-friction	1	1	1	1	0	1	1	1
Grout-perimeter	1	1	1	1	1	0	1	1
Cross-sectional-area	1	1	1	1	1	1	0	1
Pre-tensioning Force	1	1	1	0.001	1	1	1	0

3.2. Model Analysis

Figure 7 illustrates the energy analysis results under different parameter design schemes for the shotcrete layer. As shown in Figure 7a, with the exception of an isotropic modulus of 1 GPa, the maximum plastic dissipated energy (PDE) of the surrounding rock initially decreases and then increases as the isotropic modulus of the shotcrete layer increases, while the elastic strain energy (ESE) exhibits a gradual increase. When the isotropic modulus is 1 GPa, the structural stiffness of the shotcrete layer is insufficient to provide reliable support for the roadway. This inadequacy leads to damage and energy dissipation in the surrounding rock, resulting in increased stress concentration and significant energy accumulation. At an isotropic modulus of 100 GPa, the shotcrete layer exhibits high structural stiffness, which limits the deformation of the surrounding rock. As a result, elastic strain energy accumulates within the surrounding rock due to the inability to dissipate energy through displacement, reaching approximately 185 kJ. This accumulation, often accompanied by stress concentration, may lead to damage in the rock mass, with PDE values exceeding 22 kJ. In contrast, a moderate isotropic modulus (e.g., 25 GPa) effectively facilitates the release of accumulated ESE in the surrounding rock, thereby reducing the extent of damage and achieving significant displacement control. As depicted in Figure 7b, Poisson's ratio has a comparatively weaker influence on the energy evolution characteristics (EEC) of the surrounding rock than isotropic modulus and thickness. The PDE increases with higher Poisson's ratio, whereas ESE decreases correspondingly. Figure 7c illustrates that the thickness of the shotcrete layer also affects the energy dynamics. Specifically, as thickness increases, PDE initially decreases, then increases, while ESE follows a trend of increasing, subsequently decreasing, and eventually stabilizing. At a thickness of 0.01 m, the shotcrete layer's structural stiffness is notably insufficient, failing to provide stable roadway support. Under such conditions, the surrounding rock is prone to destructive deformation and energy dissipation, further exacerbating stress concentration and significantly increasing the accumulated energy.

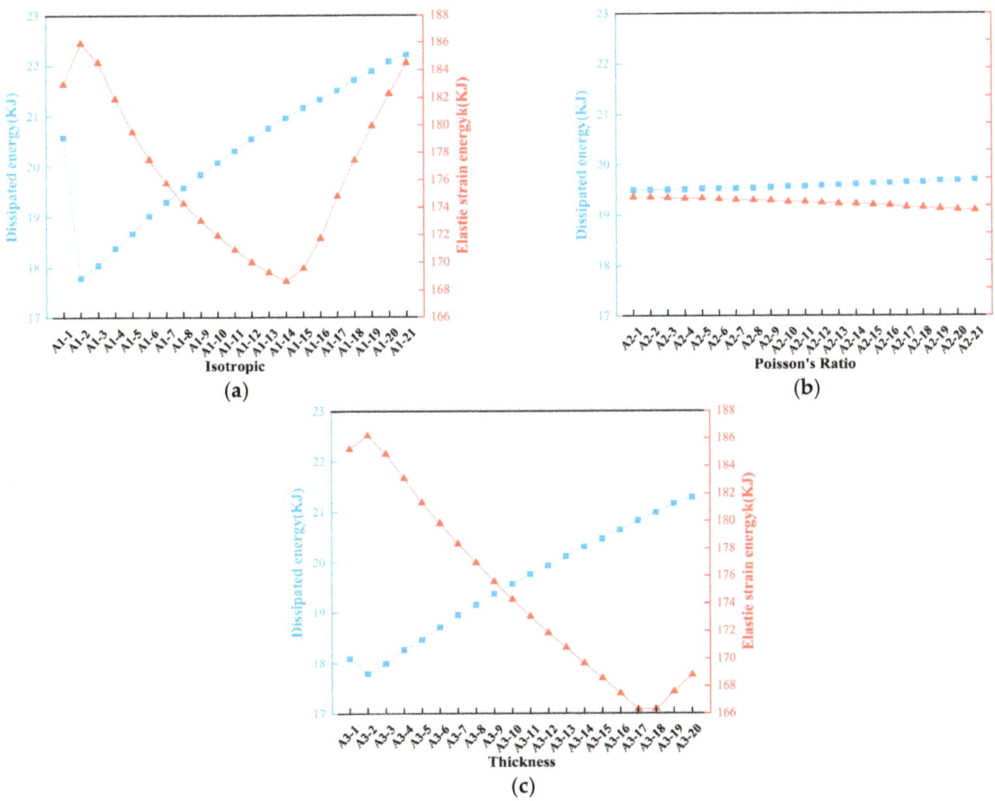

Figure 7. Energy analysis of parameters design schemes of shotcrete layer. (**a**) Isotropic; (**b**) Poisson' ratio; (**c**) Thickness.

The energy analysis results under different bolt parameter design schemes are presented in Figure 8. Compared to the shotcrete layer, the influence of bolt parameters on the plastic dissipated energy (PDE) of the surrounding rock is relatively minor. Most bolt parameters exhibit a slight increasing trend in PDE as their values increase, but the overall variation remains insignificant. For elastic strain energy (ESE), the Yield-tension and Pre-tension force parameters of the bolt have minimal impact. However, as the Grout-cohesion, Grout-friction, and Grout-perimeter parameters increase, ESE shows a slight decrease before stabilizing. Additionally, it can be observed that continuous increases in Young's modulus and Grout-stiffness of the bolts effectively reduce energy dissipation in the surrounding rock. Similarly, increasing the cross-sectional area of the bolt initially reduces energy dissipation. However, when the cross-sectional area exceeds a threshold value of 0.006079 m^2, the dissipation energy begins to rise instead of further decreasing. This suggests that there is an optimal range for the bolt cross-sectional area in minimizing energy dissipation, beyond which the effect may become counterproductive.

The plastic dissipation energy (PDE) distribution for the shotcrete layer parameter design scheme is illustrated in Figure 9. The diagram indicates that the dissipation energy is primarily concentrated near the sidewalls and floor of the roadway. The depth of energy dissipation at the sidewalls extends to approximately 2 m, while at the floor, the dissipation depth reaches nearly twice this value, reflecting a pronounced disparity in energy distribution between these zones. In cases where the shotcrete layer is characterized by low isotropy and minimal thickness, the dissipation energy tends to propagate towards the roof of the roadway. The maximum dissipation energy is observed at the junction between the sidewall and floor, which emerges as the primary zone of energy concentration. As the isotropy and thickness of the shotcrete layer increase, the distribution range of dissipation energy gradually diminishes, indicating an enhancement in the structural ability to contain energy dissipation. Conversely, Poisson's ratio exhibits minimal influence on the spatial distribution range of dissipation energy.

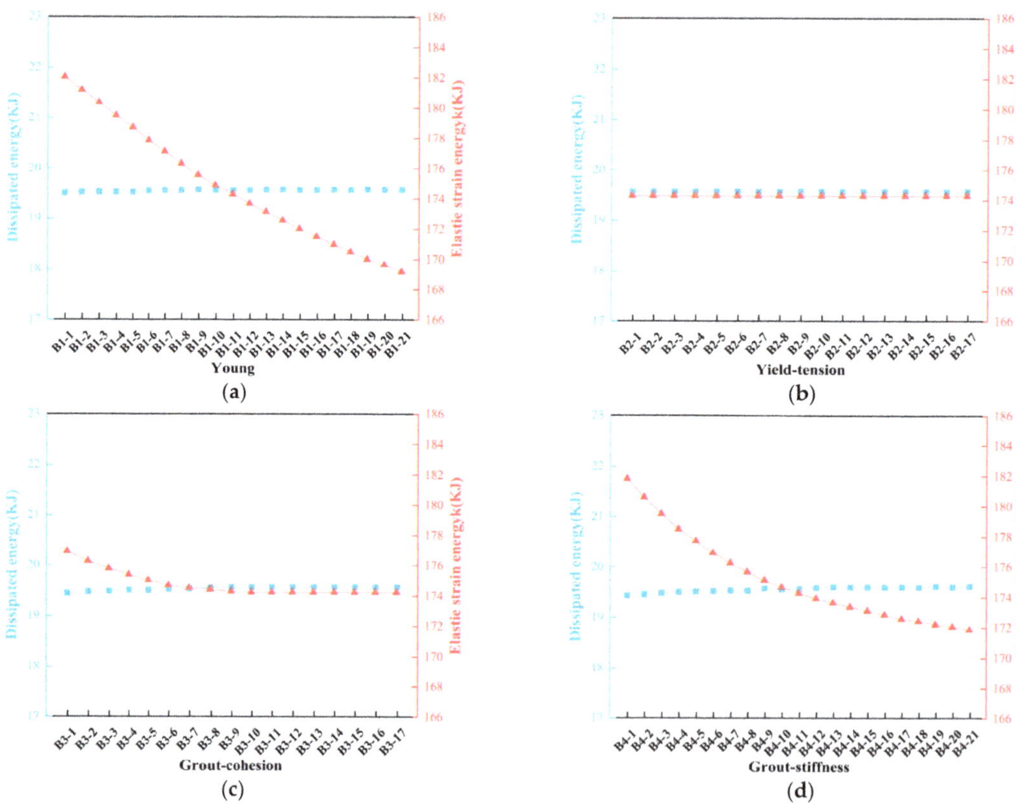

Figure 8. *Cont.*

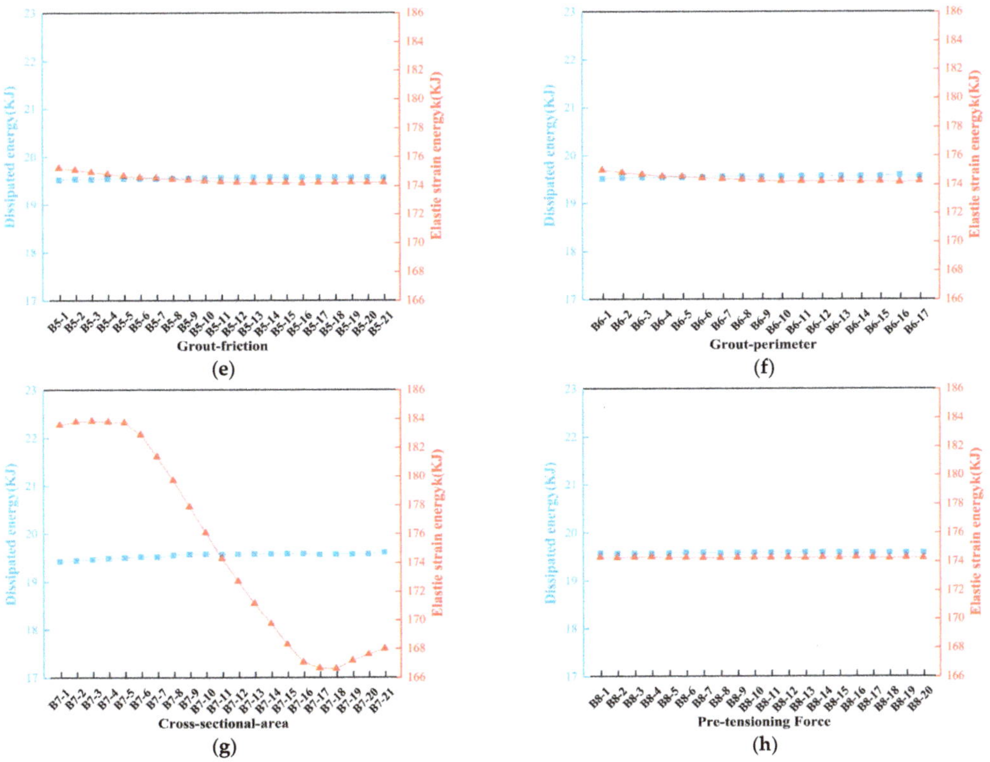

Figure 8. Energy analysis of rock bolt parameter design schemes. (**a**) Young; (**b**) Yield-tension; (**c**) Grout-cohesion; (**d**) Grout-stiffness; (**e**) Grout-friction; (**f**) Grout-perimeter; (**g**) Cross-sectional-area; (**h**) Pre-tension force.

The parameter design scheme for the shotcrete layer and the corresponding elastic strain energy (ESE) distribution cloud diagram are presented in Figure 10. The diagram illustrates that the elastic strain energy is predominantly concentrated within a vertical band extending approximately 1 m from the tunnel section in all directions. As the support parameters increase, the distribution of elastic strain energy exhibits a tendency to shift toward the inner wall of the tunnel. Notably, when the isotropic parameter reaches 100 GPa, a significant change is observed in the distribution pattern of elastic strain energy, suggesting an enhanced capacity of the support structure to constrain energy accumulation. In contrast, variations in Poisson's ratio appear to have a negligible effect on the spatial distribution of elastic strain energy.

Analysis of the results reveals that the isotropic properties and thickness of the concrete layer have a significant influence on both elastic strain energy (ESE) and plastic dissipated energy (PDE), whereas the energy distribution shows minimal sensitivity to variations in Poisson's ratio. In contrast, the distribution of energy is notably affected by parameters such as Young's modulus, cross-sectional area, and grout stiffness of the anchor rod. Furthermore, the distribution patterns observed in the cloud maps appear to be correlated with the maximum energy values. Changes in the maximum energy values are accompanied by corresponding alterations in the energy distribution. Notably, the cloud map distribution patterns under different anchor rod parameter design schemes exhibit a high degree of consistency with those observed for different shotcrete layer parameter design schemes. Therefore, detailed descriptions of these similarities are omitted for brevity.

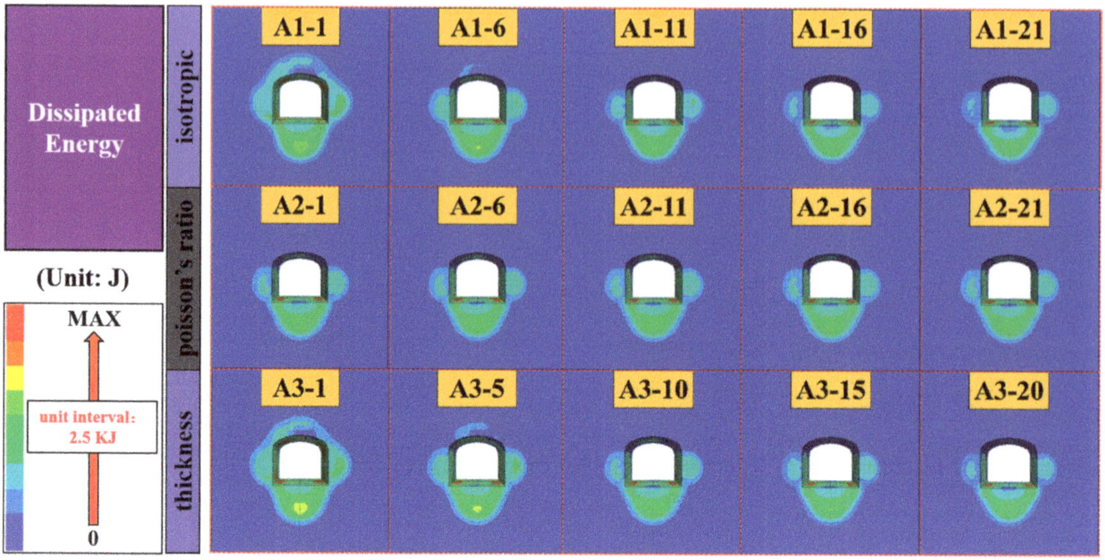

Figure 9. Plastic dissipation energy contours diagram of different shotcrete layer scheme.

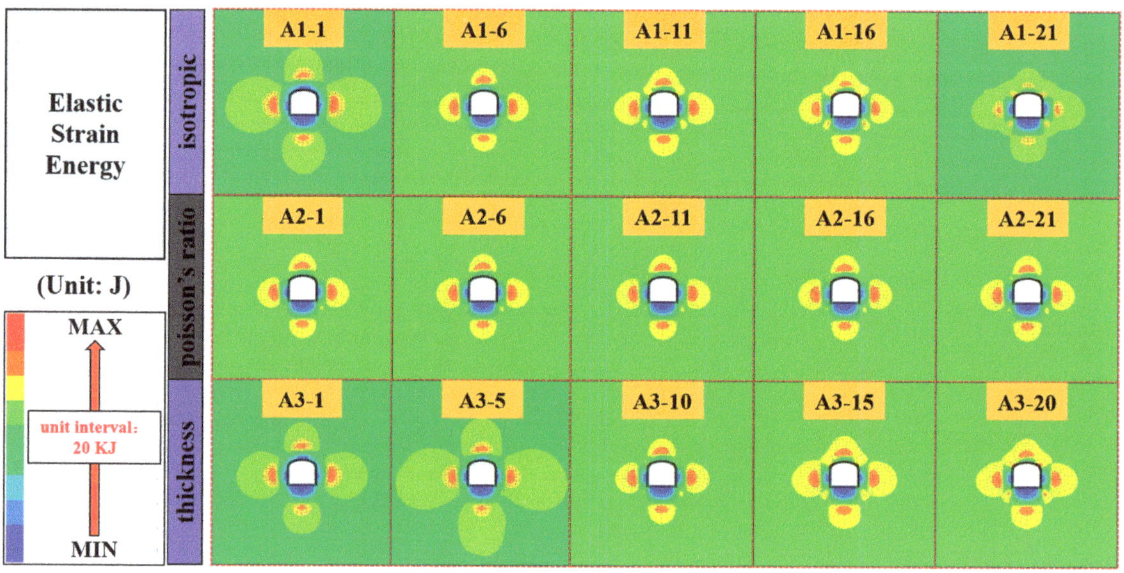

Figure 10. Elastic strain energy contours diagram of different shotcrete layer scheme.

4. Model Development

This study employs multiple numerical simulations using FLAC3D to analyze the evolution characteristics of elastic strain energy (ESE) and plastic dissipated energy (PDE) in surrounding rock under various bolt-shotcrete support configurations. Meanwhile, using ESE and PDE as outputs, a dataset of 217 groups is generated, as shown in Appendix A. To explore the application of machine learning in predicting the stability of tunnels, this dataset is used to train five commonly used machine learning models: BP, DT, ELM, RF, and SVM. It is worth noting that since this article changes a single factor, the data are randomly sorted before model training. The dataset is divided into an 80% training set and a 20% test set, and MAE, MAPE, MSE, RMSE, and R^2 are used to comprehensively evaluate each model. The flow chart is shown in Figure 11. Table 7 shows the detailed parameters of different models.

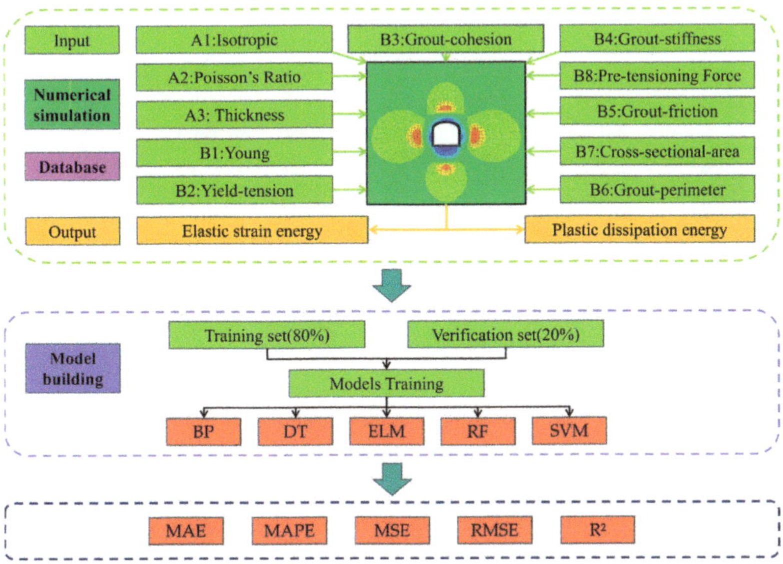

Figure 11. Overall flow chart.

Table 7. Specific structural parameters of different models.

Model	Parameter Name	Default Value	Model	Parameter Name	Default Value
BP	Hidden layer	2	ELM	Number of Hidden Neurons	100
	cell	64		Activation Function	sigmoid
	Output layer	1		Regularization Parameter	0.001
	activation function	Sigmoid		Random Seed	1
	optimizer	SGD		Input Weights Initialization	uniform distribution
	learning rate	0.01	RF	Number of Trees	100
	batch size	16		Max Depth	
	Maximum iterations	1000		Min Samples Split	2
	momentum	0.9		Min Samples Leaf	1
	lambda	0.0001		Max Features	sqrt
	Seed	42		Bootstrap	TRUE
DT	max depth	None	SVM	Kernel Function	Gaussian (RBF)
	min samples split	2		Kernel Scale	1
	min samples leaf	1		Box Constraint	1
	max features	None		Epsilon	0.1
	max leaf nodes	None		Standardize	TRUE
	random state	1		Convergence Criterion	1.00×10^{-3}
	min impurity decrease	0		Max Iterations	1000

Traditional point plots are insufficient for detailed analysis of each model's predictions, making it challenging to distinguish differences in performance. To facilitate a more effective analysis and discussion, this study utilizes thermal maps to visualize the prediction results of each model. As shown in Figures 12 and 13, under the concrete parameter design scheme, the DT model demonstrates the best predictive performance for elastic strain energy, achieving an MAE of 2.14, an MSE of 12.7, and an R^2 value of 0.432. The BP model follows closely with an MAE of 2.38, an MSE of 13.089, and an R^2 value of 0.415. In contrast, the other three models exhibit R^2 values below 0.3, indicating poor data fit, weak explanatory power for variable changes, and limited predictive accuracy. This underperformance can be attributed to the use of default model parameters during the training process. Future studies should aim to enhance model performance by optimizing hyperparameters, integrating models, and employing ensemble learning techniques.

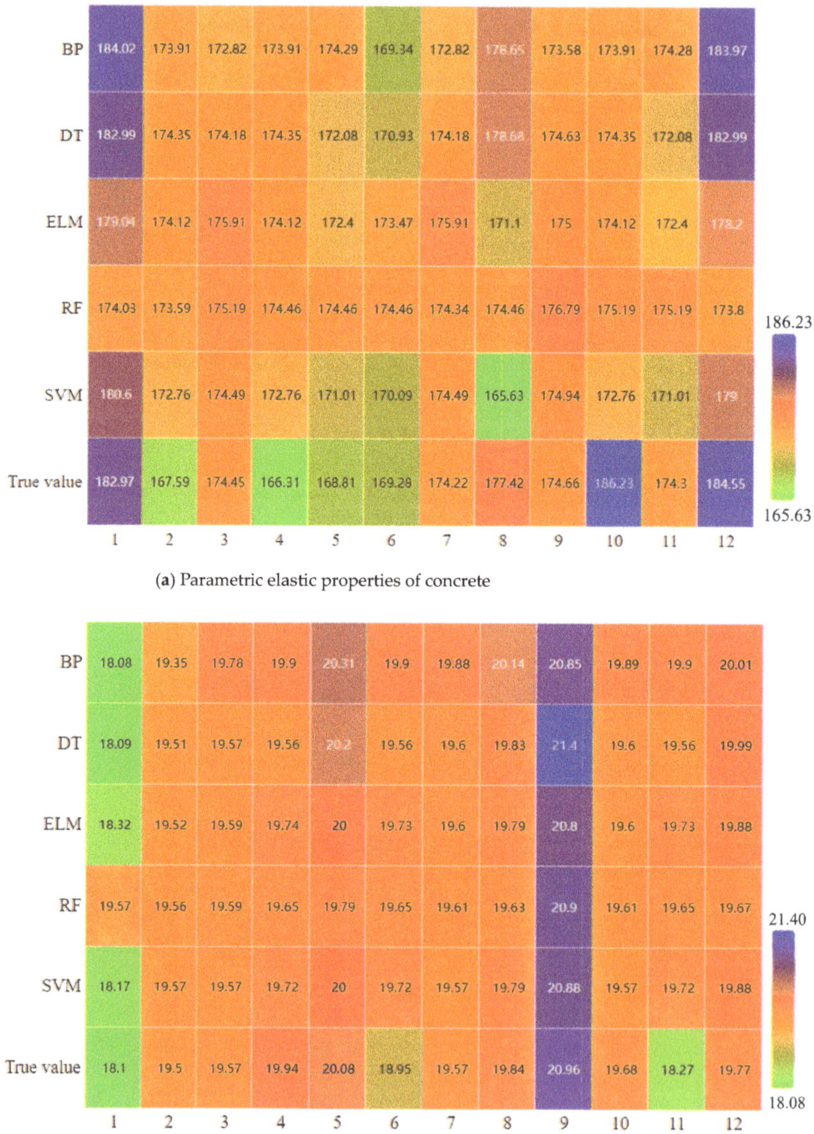

(**a**) Parametric elastic properties of concrete

(**b**) Parametric dissipative energy of concrete

Figure 12. Comparison of concrete prediction results.

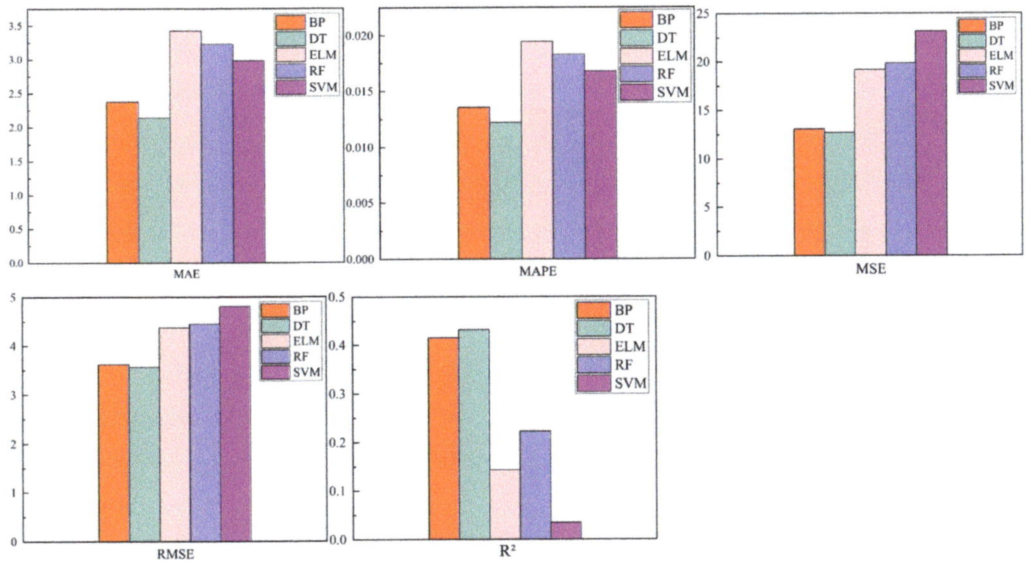

(**a**) Parametric elastic properties of concrete.

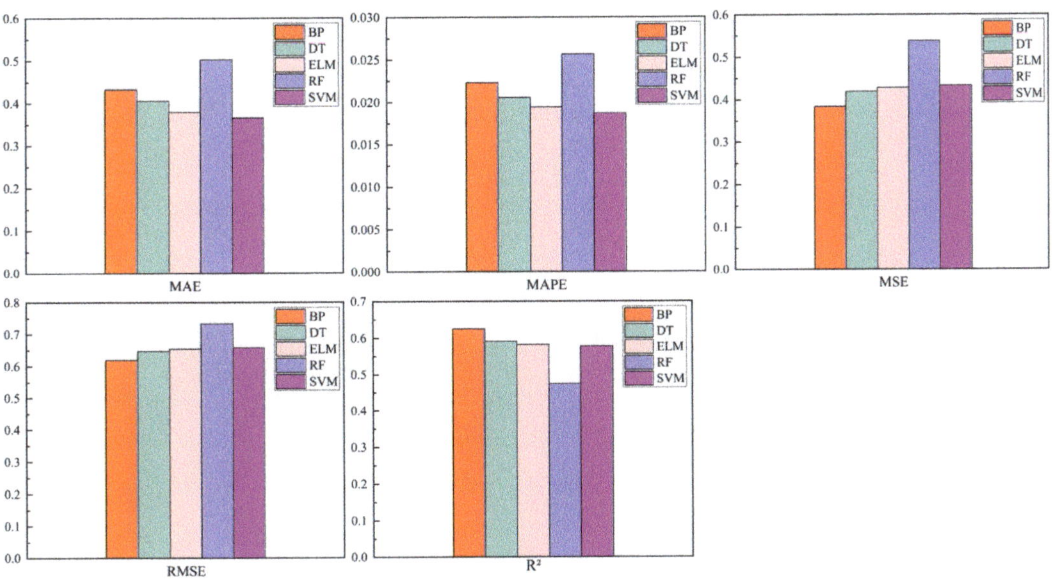

(**b**) Parametric dissipative energy of concrete

Figure 13. Comparison of concrete parameters.

For the dissipative energy prediction model, the SVM model exhibits the best predictive performance, with an MAE of 0.37, an MSE of 0.38, and an R^2 value of 0.58. Analysis of the simulation results reveals that dissipative energy is nearly linearly correlated with variations in the design parameters. Since the SVM model utilizes a linear kernel function by default, it achieves superior performance under these conditions. Additionally, other models also demonstrate satisfactory performance in predicting dissipated energy within the concrete parameter design schemes.

Combined with Figures 14 and 15, the DT model has the best comprehensive performance of all the indices among the prediction models of bolt parametric elastic energy, with an R^2 of 0.77, an MAE of 0.83, and an MSE of 2.32, indicating that the DT model has a small prediction error and a good fitting effect for the bolt parametric elastic energy index. This was followed by the RF model with an MAE of 1.03, an MSE of 3.26, and an R^2 of 0.68. Among them, the BP model shows the worst performance, which may be due to the default model parameters and improper parameter settings, resulting in the model's failure to converge. Among the models of energy dissipated by bolt parameters, the RF model shows better fit and accuracy.

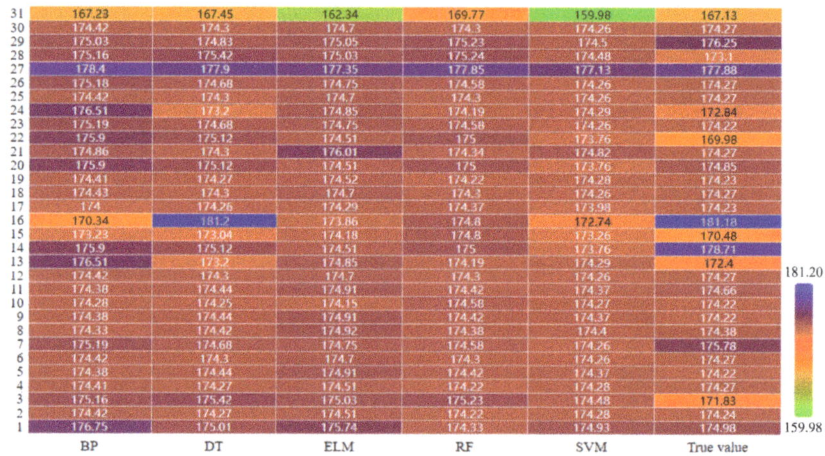

(**a**) Parameter elastic performance of bolt

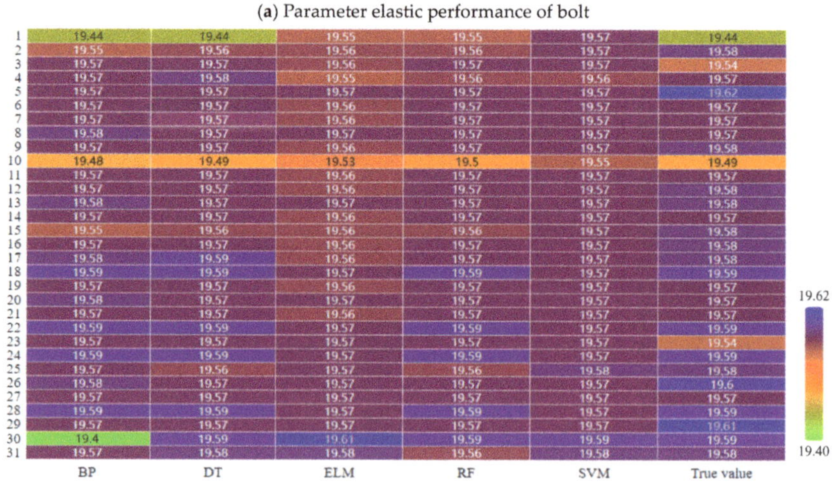

(**b**) Parameter dissipated energy of bolt

Figure 14. Comparison of concrete prediction results.

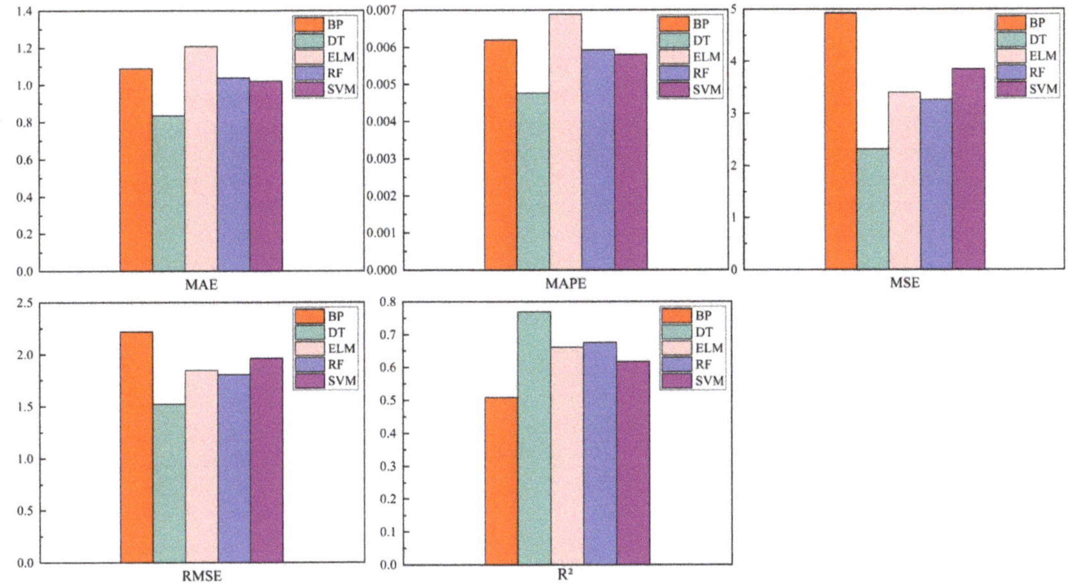

(**a**) Parameter elastic performance of bolt

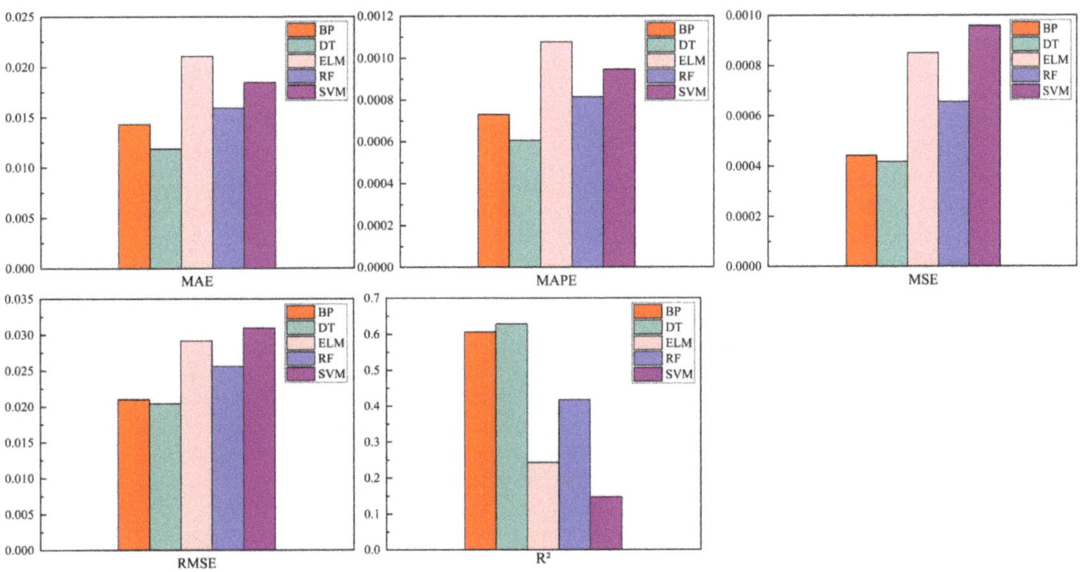

(**b**) Parameter dissipated energy of bolt

Figure 15. Comparison of bolt parameters.

Based on the research above, it is evident that the DT model performs well in predicting elastic energy for the design parameters of the supporting structure, while the SVM model excels in predicting parametric dissipative energy for concrete. The RF model is considered superior to other models in terms of bolt parametric dissipated energy.

Table 8 presents the time required for various calculation methods. Utilizing FLAC3D to compute ESE and PDE under different parameter configurations requires approximately

90 s per simulation. In contrast, employing machine learning models for prediction significantly reduces computation time, with the regression process for a training set and test set comprising 155 datasets requiring only 1.52367 s and 0.00318 s, respectively. These findings underscore the potential advantages of machine learning-based surrogate models in predicting the evolution of ESE and PDE in surrounding rock. The substantial reduction in computational time highlights the feasibility of machine learning as an efficient alternative to traditional numerical methods. By enabling rapid and accurate predictions, machine learning models empower decision-makers to implement timely measures to mitigate geological hazards, thereby reducing the likelihood of accidents and associated losses. Consequently, the integration of machine learning models can complement traditional computational approaches, enhance the efficiency of ESE and PDE predictions, and support the development of more effective risk management strategies.

Table 8. Time consumption of various calculation methods.

Methods Set	Numerical Simulation	Regression	
		Training	Test
Case No.	1	124 (80%)	31 (20%)
Time/s	90	1.52367	0.00318

By combining the results of machine learning and numerical simulation, the efficiency of studying the energy evolution characteristics of surrounding rock with design parameters of supporting structure is effectively improved. Machine learning-based methods can assist researchers in conducting a more comprehensive analysis and gaining a deeper understanding of support selection. This, in turn, can improve engineering practice and decision-making in mining engineering. Additionally, it provides valuable guidance for subsequent research on the influence of support parameters on the energy evolution characteristics of surrounding rock.

5. Discussion

This study systematically investigates the energy evolution characteristics (ESE and PDE) of surrounding rock in deep mining areas by integrating numerical simulation and machine learning techniques. The findings reveal the nonlinear effects of support parameters on the dynamic energy behavior of surrounding rock. The analysis demonstrates significant variations in the influence of support stiffness on elastic strain energy and plastic dissipation energy across different support designs. For instance, a moderate isotropic modulus (e.g., 25 GPa) effectively balances energy accumulation and dissipation, thereby mitigating rock damage and achieving enhanced displacement control. In terms of bolt design parameters, grout-related factors (e.g., cohesion and stiffness) play a pivotal role in minimizing energy dissipation. However, when the bolt cross-sectional area exceeds a certain threshold, energy dissipation increases instead of stabilizing, emphasizing the necessity of balancing stiffness and flexibility in support design.

The findings of this study align with existing research on the optimization of support parameters and their impact on the stability of surrounding rock. For example, Yu et al. [41] demonstrated through numerical simulations and orthogonal experiments that optimized support parameters, such as anchor length and spacing, significantly improved the stability of soft rock roadways. Similarly, Li et al. [42] employed a displacement control-based optimization method and highlighted the effectiveness of adjusting anchor cable length and pre-stress in controlling deformation, which corroborates this study's results showing the influence of support parameters on energy evolution in the surrounding rock. Moreover, Yuan et al. [43] emphasized the critical role of support parameters in controlling the plastic zone distribution in high-stress conditions, which is consistent with the energy evolution trends observed in this study. While previous research has primarily focused on plastic zone control, this study extends the understanding by incorporating the perspective of energy evolution, providing a more comprehensive framework for analyzing the effects of support design. These comparisons highlight the practical implications of optimizing support parameters for enhancing the stability of underground structures. Future work could further validate these findings through field measurements, as demonstrated by Li et al. [42], who combined numerical simulations with site observations to refine their optimization model. This approach would strengthen the applicability of the presented results in real-world engineering scenarios.

Building on these findings, machine learning techniques were applied to predict energy data generated from numerical simulations. The results indicate that the decision tree (DT) model performs optimally in predicting elastic energy for support parameter designs, particularly for bolt configurations, achieving an R^2 value of 0.77. The support vector machine (SVM) model demonstrates superior performance in predicting dissipation energy, especially for shotcrete parameters, owing to the near-linear relationship between dissipation energy and support parameters. Additionally, the study highlights the computational efficiency of machine learning models. While FLAC3D numerical simulations require approximately 90 s to compute energy data for each case, machine learning models achieve comparable predictions with a training time of only 1.5 s and a testing time of 0.01 s. This substantial reduction in computational time enables real-time predictions and rapid decision-making, facilitating timely prevention and mitigation of geological hazards.

Despite the significant theoretical and practical contributions of this study, several limitations must be acknowledged. First, the lack of validation with field data remains a major constraint. The machine learning models developed in this study rely entirely on simulation data generated by FLAC3D, which have not been validated through field monitoring. Future research should prioritize the collection of field data, such as elastic strain energy and dissipation energy, to validate and refine the simulation results, thereby enhancing the practical applicability of the models. Second, the study assumes that the rock mass is homogeneous and isotropic, which may not adequately reflect the complexities of real geological conditions. Incorporating anisotropic and heterogeneous properties into future models could improve the accuracy and reliability of the simulations. Additionally, this study focuses primarily on static loading conditions and does not account for dynamic factors, such as seismic activities, or time-dependent effects, such as creep, which could significantly influence the energy evolution process. Furthermore, the use of default parameters in the machine learning models has resulted in suboptimal performance for certain models, such as the BP model. Future work should focus on hyperparameter tuning, ensemble learning, and model integration to enhance predictive accuracy and robustness. Lastly, the study does not consider dynamic loading conditions, such as seismic events, or time-dependent effects, such as creep. Integrating these dynamic factors into the modeling framework could provide a more comprehensive understanding of energy evolution in complex geological systems.

In conclusion, this study combines numerical simulation and machine learning techniques to establish a novel theoretical framework and practical guidance for optimizing support parameter design and understanding the energy evolution characteristics of surrounding rock. The demonstrated efficiency and reliability of machine learning models in energy prediction highlight their potential to complement traditional computational methods, particularly in improving predictive efficiency and supporting rapid decision-making. To further enhance the robustness of the results, future research should incorporate field data and address the complexities of dynamic geological environments and conditions.

6. Conclusions

This study utilizes FLAC3D to establish 217 models, investigates the influence of anchor rod parameters and shotcrete layer parameters on the energy evolution characteristics of surrounding rock, and fits the results through machine learning models. The main conclusions are as follows:

(1) The energy evolution characteristics of the surrounding rock are more sensitive to the isotropic and thickness parameters of the shotcrete layer support structure, and the

cross-sectional-area, Young's modulus, and grout-stiffness parameters of the rock anchor support structure, respectively.

(2) The parameters of shotcrete layer support structure are not the bigger the better. When isotropic is 100 GPa, the peak values of elastic strain energy and plastic dissipation energy are higher than the value when isotropic is 25 GPa. In addition, for rock anchors, when the value of cross-sectional-area (more than 0.005024 m^2) is too large, the value of plastic dissipative energy in the surrounding rock starts to show an increasing trend.

(3) Appropriate support design parameter schemes can reduce the peak value of energy (elastic strain energy, dissipated energy) and at the same time change the distribution characteristics of energy. Therefore, the design of the support program from the perspective of energy may be able to improve the stability control effect from the root. At the same time, analysis based on the Kendall correlation coefficient found that there was a significant positive correlation between Grout-stiffness and Pre-tensioning Force.

(4) By introducing BP, DT, ELM, RF, and SVM, the prediction models of elastic strain energy and plastic dissipative energy of surrounding rock are constructed. In the concrete parameter design scheme, the decision tree (DT) model performs best in predicting elastic energy, while the support vector machine (SVM) model performs best in predicting dissipative energy. For predicting bolt parameters, the DT model performs well in predicting elastic energy, while the Random Forest (RF) model performs well in predicting dissipative energy. Future research could enhance the model's performance by adjusting the model parameters, optimizing the fusion model, and utilizing ensemble learning to more accurately predict the performance indicators of design parameters of supporting structure.

Meanwhile, it is important to recognize the limitations of this study. First, when establishing the model in this article, the influence of a single variable on the dissipation energy and elastic performance of the surrounding rock is studied by controlling it. In subsequent research, orthogonal experiments should be used to control multiple variables and explore the impact of multivariate changes on the dissipation energy and elastic performance of the surrounding rock. Second, there is a lack of empirical validation for the proxy model, particularly in terms of the prediction accuracy of ESE and PDE. Future research should focus more on introducing conditions that are closer to actual operating conditions. Although the proxy model exhibits good predictive capabilities in simulations, its applicability in practical engineering still requires further verification. To enhance its practicality, research could further consider the anisotropic characteristics under complex geological conditions, the dynamic changes in stress-strain during construction processes, and the nonlinear effects caused by environmental factors and material property fluctuations. Additionally, it is recommended to adopt a more diverse dataset and input parameters to conduct an in-depth analysis of the impact of support structure design on the energy evolution of surrounding rock in order to optimize algorithms and identify key factors affecting slope stability. In summary, although this study provides important insights into slope stability analysis, future research needs to address the aforementioned limitations to further refine the model and enhance its practicality and reliability.

Author Contributions: Conceptualization, Y.C., Q.D., L.Z., D.L. and B.D.; methodology, Y.C., Q.D. and L.Z.; software, Y.C., Q.D. and L.Z.; validation, D.L. and B.D.; investigation, D.L. and B.D.; resources, L.Z., D.L. and B.D.; data curation, Q.D. and L.Z.; writing—original draft preparation, Y.C., Q.D. and L.Z.; writing—review and editing, Y.C., Q.D., L.Z., D.L. and B.D.; visualization, Y.C. and Q.D.; supervision, Y.C. and L.Z.; project administration, Y.C. and L.Z. All authors have read and agreed to the published version of the manuscript.

Funding: This project was sponsored by the National Natural Science Foundation of China (No. 151374244); Key Project of Education Department of Hunan Province (22A0293); National Nature Science Foundation of Hunan Province (2022JJ40373); General Project of Education Department of Hunan Province (22C0235).

Institutional Review Board Statement: Not applicable.

Informed Consent Statement: Not applicable.

Data Availability Statement: The original contributions presented in the study are included in the article, further inquiries can be directed to the corresponding author.

Conflicts of Interest: The authors declare no conflicts of interest.

Appendix A

Option	A1	A2	A3	B1	B2	B3	B4	B5	B6	B7	B8	PDE/KJ	ESE/KJ
1	1	0.25	0.1	200	50	6	3	40	0.1884	0.002826	150	20.580	182.970
2	5	0.25	0.1	200	50	6	3	40	0.1884	0.002826	150	17.785	185.920
3	10	0.25	0.1	200	50	6	3	40	0.1884	0.002826	150	18.038	184.540
4	15	0.25	0.1	200	50	6	3	40	0.1884	0.002826	150	18.375	181.910
5	20	0.25	0.1	200	50	6	3	40	0.1884	0.002826	150	18.670	179.520
6	25	0.25	0.1	200	50	6	3	40	0.1884	0.002826	150	19.021	177.500
7	30	0.25	0.1	200	50	6	3	40	0.1884	0.002826	150	19.290	175.770
8	35	0.25	0.1	200	50	6	3	40	0.1884	0.002826	150	19.570	174.300
9	40	0.25	0.1	200	50	6	3	40	0.1884	0.002826	150	19.835	173.040
10	45	0.25	0.1	200	50	6	3	40	0.1884	0.002826	150	20.083	171.950
11	50	0.25	0.1	200	50	6	3	40	0.1884	0.002826	150	20.310	170.930
12	55	0.25	0.1	200	50	6	3	40	0.1884	0.002826	150	20.545	170.040
13	60	0.25	0.1	200	50	6	3	40	0.1884	0.002826	150	20.755	169.290
14	65	0.25	0.1	200	50	6	3	40	0.1884	0.002826	150	20.960	168.640
15	70	0.25	0.1	200	50	6	3	40	0.1884	0.002826	150	21.159	169.610
16	75	0.25	0.1	200	50	6	3	40	0.1884	0.002826	150	21.329	171.780
17	80	0.25	0.1	200	50	6	3	40	0.1884	0.002826	150	21.505	174.830
18	85	0.25	0.1	200	50	6	3	40	0.1884	0.002826	150	21.710	177.420
19	90	0.25	0.1	200	50	6	3	40	0.1884	0.002826	150	21.883	179.930
20	95	0.25	0.1	200	50	6	3	40	0.1884	0.002826	150	22.072	182.280
21	100	0.25	0.1	200	50	6	3	40	0.1884	0.002826	150	22.210	184.510
22	35	0.15	0.1	200	50	6	3	40	0.1884	0.002826	150	19.500	174.660
23	35	0.16	0.1	200	50	6	3	40	0.1884	0.002826	150	19.507	174.660
24	35	0.17	0.1	200	50	6	3	40	0.1884	0.002826	150	19.512	174.600
25	35	0.18	0.1	200	50	6	3	40	0.1884	0.002826	150	19.514	174.550
26	35	0.19	0.1	200	50	6	3	40	0.1884	0.002826	150	19.529	174.580
27	35	0.2	0.1	200	50	6	3	40	0.1884	0.002826	150	19.530	174.510
28	35	0.21	0.1	200	50	6	3	40	0.1884	0.002826	150	19.531	174.450
29	35	0.22	0.1	200	50	6	3	40	0.1884	0.002826	150	19.542	174.420
30	35	0.23	0.1	200	50	6	3	40	0.1884	0.002826	150	19.557	174.400
31	35	0.24	0.1	200	50	6	3	40	0.1884	0.002826	150	19.568	174.290
32	35	0.25	0.1	200	50	6	3	40	0.1884	0.002826	150	19.570	174.300
33	35	0.26	0.1	200	50	6	3	40	0.1884	0.002826	150	19.587	174.220
34	35	0.27	0.1	200	50	6	3	40	0.1884	0.002826	150	19.599	174.170
35	35	0.28	0.1	200	50	6	3	40	0.1884	0.002826	150	19.612	174.150
36	35	0.29	0.1	200	50	6	3	40	0.1884	0.002826	150	19.637	174.080
37	35	0.3	0.1	200	50	6	3	40	0.1884	0.002826	150	19.640	174.050
38	35	0.31	0.1	200	50	6	3	40	0.1884	0.002826	150	19.657	173.920
39	35	0.32	0.1	200	50	6	3	40	0.1884	0.002826	150	19.654	173.890
40	35	0.33	0.1	200	50	6	3	40	0.1884	0.002826	150	19.683	173.810
41	35	0.34	0.1	200	50	6	3	40	0.1884	0.002826	150	19.686	173.740
42	35	0.35	0.1	200	50	6	3	40	0.1884	0.002826	150	19.700	173.690
43	35	0.25	0.01	200	50	6	3	40	0.1884	0.002826	150	18.090	185.220
44	35	0.25	0.02	200	50	6	3	40	0.1884	0.002826	150	17.796	186.220
45	35	0.25	0.03	200	50	6	3	40	0.1884	0.002826	150	17.993	184.860
46	35	0.25	0.04	200	50	6	3	40	0.1884	0.002826	150	18.271	183.120
47	35	0.25	0.05	200	50	6	3	40	0.1884	0.002826	150	18.470	181.350
48	35	0.25	0.06	200	50	6	3	40	0.1884	0.002826	150	18.718	179.850
49	35	0.25	0.07	200	50	6	3	40	0.1884	0.002826	150	18.952	178.350
50	35	0.25	0.08	200	50	6	3	40	0.1884	0.002826	150	19.160	176.990
51	35	0.25	0.09	200	50	6	3	40	0.1884	0.002826	150	19.384	175.590

Option	A1	A2	A3	B1	B2	B3	B4	B5	B6	B7	B8	PDE/KJ	ESE/KJ
52	35	0.25	0.1	200	50	6	3	40	0.1884	0.002826	150	19.570	174.300
53	35	0.25	0.11	200	50	6	3	40	0.1884	0.002826	150	19.767	173.050
54	35	0.25	0.12	200	50	6	3	40	0.1884	0.002826	150	19.942	171.860
55	35	0.25	0.13	200	50	6	3	40	0.1884	0.002826	150	20.126	170.810
56	35	0.25	0.14	200	50	6	3	40	0.1884	0.002826	150	20.315	169.650
57	35	0.25	0.15	200	50	6	3	40	0.1884	0.002826	150	20.470	168.560
58	35	0.25	0.16	200	50	6	3	40	0.1884	0.002826	150	20.650	167.470
59	35	0.25	0.17	200	50	6	3	40	0.1884	0.002826	150	20.834	166.320
60	35	0.25	0.18	200	50	6	3	40	0.1884	0.002826	150	20.996	166.340
61	35	0.25	0.19	200	50	6	3	40	0.1884	0.002826	150	21.163	167.600
62	35	0.25	0.2	200	50	6	3	40	0.1884	0.002826	150	21.290	168.820
63	35	0.25	0.1	100	50	6	3	40	0.1884	0.002826	150	19.525	182.05
64	35	0.25	0.1	110	50	6	3	40	0.1884	0.002826	150	19.537	181.19
65	35	0.25	0.1	120	50	6	3	40	0.1884	0.002826	150	19.542	180.37
66	35	0.25	0.1	130	50	6	3	40	0.1884	0.002826	150	19.542	179.52
67	35	0.25	0.1	140	50	6	3	40	0.1884	0.002826	150	19.547	178.73
68	35	0.25	0.1	150	50	6	3	40	0.1884	0.002826	150	19.565	177.86
69	35	0.25	0.1	160	50	6	3	40	0.1884	0.002826	150	19.571	177.13
70	35	0.25	0.1	170	50	6	3	40	0.1884	0.002826	150	19.572	176.32
71	35	0.25	0.1	180	50	6	3	40	0.1884	0.002826	150	19.584	175.57
72	35	0.25	0.1	190	50	6	3	40	0.1884	0.002826	150	19.572	174.88
73	35	0.25	0.1	200	50	6	3	40	0.1884	0.002826	150	19.57	174.3
74	35	0.25	0.1	210	50	6	3	40	0.1884	0.002826	150	19.573	173.68
75	35	0.25	0.1	220	50	6	3	40	0.1884	0.002826	150	19.582	173.15
76	35	0.25	0.1	230	50	6	3	40	0.1884	0.002826	150	19.587	172.57
77	35	0.25	0.1	240	50	6	3	40	0.1884	0.002826	150	19.575	172.02
78	35	0.25	0.1	250	50	6	3	40	0.1884	0.002826	150	19.573	171.5
79	35	0.25	0.1	260	50	6	3	40	0.1884	0.002826	150	19.582	171
80	35	0.25	0.1	270	50	6	3	40	0.1884	0.002826	150	19.573	170.52
81	35	0.25	0.1	280	50	6	3	40	0.1884	0.002826	150	19.585	170.02
82	35	0.25	0.1	290	50	6	3	40	0.1884	0.002826	150	19.583	169.66
83	35	0.25	0.1	300	50	6	3	40	0.1884	0.002826	150	19.581	169.19
84	35	0.25	0.1	200	10	6	3	40	0.1884	0.002826	150	19.57	174.3
85	35	0.25	0.1	200	15	6	3	40	0.1884	0.002826	150	19.57	174.3
86	35	0.25	0.1	200	20	6	3	40	0.1884	0.002826	150	19.57	174.3
87	35	0.25	0.1	200	25	6	3	40	0.1884	0.002826	150	19.57	174.3
88	35	0.25	0.1	200	30	6	3	40	0.1884	0.002826	150	19.57	174.3
89	35	0.25	0.1	200	35	6	3	40	0.1884	0.002826	150	19.57	174.3
90	35	0.25	0.1	200	40	6	3	40	0.1884	0.002826	150	19.57	174.3
91	35	0.25	0.1	200	45	6	3	40	0.1884	0.002826	150	19.57	174.3
92	35	0.25	0.1	200	50	6	3	40	0.1884	0.002826	150	19.57	174.3
93	35	0.25	0.1	200	55	6	3	40	0.1884	0.002826	150	19.57	174.3
94	35	0.25	0.1	200	60	6	3	40	0.1884	0.002826	150	19.57	174.3
95	35	0.25	0.1	200	65	6	3	40	0.1884	0.002826	150	19.57	174.3
96	35	0.25	0.1	200	70	6	3	40	0.1884	0.002826	150	19.57	174.3
97	35	0.25	0.1	200	75	6	3	40	0.1884	0.002826	150	19.57	174.3
98	35	0.25	0.1	200	80	6	3	40	0.1884	0.002826	150	19.57	174.3
99	35	0.25	0.1	200	85	6	3	40	0.1884	0.002826	150	19.57	174.3
100	35	0.25	0.1	200	90	6	3	40	0.1884	0.002826	150	19.57	174.3
101	35	0.25	0.1	200	50	2	3	40	0.1884	0.002826	150	19.46	176.92
102	35	0.25	0.1	200	50	2.5	3	40	0.1884	0.002826	150	19.491	176.29
103	35	0.25	0.1	200	50	3	3	40	0.1884	0.002826	150	19.499	175.8
104	35	0.25	0.1	200	50	3.5	3	40	0.1884	0.002826	150	19.525	175.41
105	35	0.25	0.1	200	50	4	3	40	0.1884	0.002826	150	19.52	175.04
106	35	0.25	0.1	200	50	4.5	3	40	0.1884	0.002826	150	19.543	174.69
107	35	0.25	0.1	200	50	5	3	40	0.1884	0.002826	150	19.55	174.52
108	35	0.25	0.1	200	50	5.5	3	40	0.1884	0.002826	150	19.562	174.43

Option	A1	A2	A3	B1	B2	B3	B4	B5	B6	B7	B8	PDE/KJ	ESE/KJ
109	35	0.25	0.1	200	50	6	3	40	0.1884	0.002826	150	19.57	174.3
110	35	0.25	0.1	200	50	6.5	3	40	0.1884	0.002826	150	19.575	174.25
111	35	0.25	0.1	200	50	7	3	40	0.1884	0.002826	150	19.577	174.25
112	35	0.25	0.1	200	50	7.5	3	40	0.1884	0.002826	150	19.577	174.25
113	35	0.25	0.1	200	50	8	3	40	0.1884	0.002826	150	19.577	174.25
114	35	0.25	0.1	200	50	8.5	3	40	0.1884	0.002826	150	19.577	174.25
115	35	0.25	0.1	200	50	9	3	40	0.1884	0.002826	150	19.577	174.25
116	35	0.25	0.1	200	50	9.5	3	40	0.1884	0.002826	150	19.577	174.25
117	35	0.25	0.1	200	50	10	3	40	0.1884	0.002826	150	19.577	174.25
118	35	0.25	0.1	200	50	6	1	40	0.1884	0.002826	150	19.44	181.84
119	35	0.25	0.1	200	50	6	1.2	40	0.1884	0.002826	150	19.46	180.62
120	35	0.25	0.1	200	50	6	1.4	40	0.1884	0.002826	150	19.49	179.53
121	35	0.25	0.1	200	50	6	1.6	40	0.1884	0.002826	150	19.511	178.5
122	35	0.25	0.1	200	50	6	1.8	40	0.1884	0.002826	150	19.521	177.72
123	35	0.25	0.1	200	50	6	2	40	0.1884	0.002826	150	19.53	176.94
124	35	0.25	0.1	200	50	6	2.2	40	0.1884	0.002826	150	19.541	176.27
125	35	0.25	0.1	200	50	6	2.4	40	0.1884	0.002826	150	19.537	175.68
126	35	0.25	0.1	200	50	6	2.6	40	0.1884	0.002826	150	19.578	175.14
127	35	0.25	0.1	200	50	6	2.8	40	0.1884	0.002826	150	19.565	174.69
128	35	0.25	0.1	200	50	6	3	40	0.1884	0.002826	150	19.57	174.3
129	35	0.25	0.1	200	50	6	3.2	40	0.1884	0.002826	150	19.592	173.95
130	35	0.25	0.1	200	50	6	3.4	40	0.1884	0.002826	150	19.607	173.65
131	35	0.25	0.1	200	50	6	3.6	40	0.1884	0.002826	150	19.6	173.39
132	35	0.25	0.1	200	50	6	3.8	40	0.1884	0.002826	150	19.606	173.13
133	35	0.25	0.1	200	50	6	4	40	0.1884	0.002826	150	19.6	172.87
134	35	0.25	0.1	200	50	6	4.2	40	0.1884	0.002826	150	19.608	172.58
135	35	0.25	0.1	200	50	6	4.4	40	0.1884	0.002826	150	19.603	172.43
136	35	0.25	0.1	200	50	6	4.6	40	0.1884	0.002826	150	19.619	172.22
137	35	0.25	0.1	200	50	6	4.8	40	0.1884	0.002826	150	19.608	172.07
138	35	0.25	0.1	200	50	6	5	40	0.1884	0.002826	150	19.62	171.87
139	35	0.25	0.1	200	50	6	5	10	0.1884	0.002826	150	19.53	175.24
140	35	0.25	0.1	200	50	6	5	13	0.1884	0.002826	150	19.542	175.11
141	35	0.25	0.1	200	50	6	5	16	0.1884	0.002826	150	19.536	174.95
142	35	0.25	0.1	200	50	6	5	19	0.1884	0.002826	150	19.548	174.83
143	35	0.25	0.1	200	50	6	5	22	0.1884	0.002826	150	19.551	174.69
144	35	0.25	0.1	200	50	6	5	25	0.1884	0.002826	150	19.55	174.59
145	35	0.25	0.1	200	50	6	5	28	0.1884	0.002826	150	19.552	174.55
146	35	0.25	0.1	200	50	6	5	31	0.1884	0.002826	150	19.554	174.47
147	35	0.25	0.1	200	50	6	5	34	0.1884	0.002826	150	19.559	174.43
148	35	0.25	0.1	200	50	6	5	37	0.1884	0.002826	150	19.565	174.36
149	35	0.25	0.1	200	50	6	5	40	0.1884	0.002826	150	19.57	174.3
150	35	0.25	0.1	200	50	6	5	43	0.1884	0.002826	150	19.573	174.26
151	35	0.25	0.1	200	50	6	5	46	0.1884	0.002826	150	19.577	174.25
152	35	0.25	0.1	200	50	6	5	49	0.1884	0.002826	150	19.577	174.25
153	35	0.25	0.1	200	50	6	5	52	0.1884	0.002826	150	19.577	174.25
154	35	0.25	0.1	200	50	6	5	55	0.1884	0.002826	150	19.577	174.2
155	35	0.25	0.1	200	50	6	5	58	0.1884	0.002826	150	19.577	174.25
156	35	0.25	0.1	200	50	6	5	61	0.1884	0.002826	150	19.577	174.25
157	35	0.25	0.1	200	50	6	5	64	0.1884	0.002826	150	19.577	174.25
158	35	0.25	0.1	200	50	6	5	67	0.1884	0.002826	150	19.577	174.25
159	35	0.25	0.1	200	50	6	5	70	0.1884	0.002826	150	19.577	174.25
160	35	0.25	0.1	200	50	6	5	40	0.0628	0.002826	150	19.53	175.01
161	35	0.25	0.1	200	50	6	5	40	0.0785	0.002826	150	19.544	174.84
162	35	0.25	0.1	200	50	6	5	40	0.0942	0.002826	150	19.55	174.7
163	35	0.25	0.1	200	50	6	5	40	0.1099	0.002826	150	19.55	174.58
164	35	0.25	0.1	200	50	6	5	40	0.1256	0.002826	150	19.55	174.56
165	35	0.25	0.1	200	50	6	5	40	0.1413	0.002826	150	19.559	174.46

Option	A1	A2	A3	B1	B2	B3	B4	B5	B6	B7	B8	PDE/KJ	ESE/KJ
166	35	0.25	0.1	200	50	6	5	40	0.157	0.002826	150	19.569	174.41
167	35	0.25	0.1	200	50	6	5	40	0.1727	0.002826	150	19.57	174.35
168	35	0.25	0.1	200	50	6	5	40	0.1884	0.002826	150	19.57	174.3
169	35	0.25	0.1	200	50	6	5	40	0.2041	0.002826	150	19.576	174.25
170	35	0.25	0.1	200	50	6	5	40	0.2198	0.002826	150	19.579	174.25
171	35	0.25	0.1	200	50	6	5	40	0.2355	0.002826	150	19.584	174.24
172	35	0.25	0.1	200	50	6	5	40	0.2512	0.002826	150	19.58	174.28
173	35	0.25	0.1	200	50	6	5	40	0.2669	0.002826	150	19.579	174.23
174	35	0.25	0.1	200	50	6	5	40	0.2826	0.002826	150	19.582	174.24
175	35	0.25	0.1	200	50	6	5	40	0.2983	0.002826	150	19.604	174.18
176	35	0.25	0.1	200	50	6	5	40	0.314	0.002826	150	19.58	174.26
177	35	0.25	0.1	200	50	6	5	40	0.1884	0.000314	150	19.43	183.58
178	35	0.25	0.1	200	50	6	5	40	0.1884	0.00045216	150	19.447	183.8
179	35	0.25	0.1	200	50	6	5	40	0.1884	0.00061544	150	19.467	183.85
180	35	0.25	0.1	200	50	6	5	40	0.1884	0.00080384	150	19.493	183.8
181	35	0.25	0.1	200	50	6	5	40	0.1884	0.00101736	150	19.506	183.74
182	35	0.25	0.1	200	50	6	5	40	0.1884	0.001256	150	19.526	182.91
183	35	0.25	0.1	200	50	6	5	40	0.1884	0.00151976	150	19.522	181.39
184	35	0.25	0.1	200	50	6	5	40	0.1884	0.00180864	150	19.559	179.73
185	35	0.25	0.1	200	50	6	5	40	0.1884	0.00212264	150	19.574	177.89
186	35	0.25	0.1	200	50	6	5	40	0.1884	0.00246176	150	19.577	176.08
187	35	0.25	0.1	200	50	6	5	40	0.1884	0.002826	150	19.57	174.3
188	35	0.25	0.1	200	50	6	5	40	0.1884	0.00321536	150	19.578	172.72
189	35	0.25	0.1	200	50	6	5	40	0.1884	0.00362984	150	19.584	171.18
190	35	0.25	0.1	200	50	6	5	40	0.1884	0.00406944	150	19.586	169.74
191	35	0.25	0.1	200	50	6	5	40	0.1884	0.00453416	150	19.589	168.31
192	35	0.25	0.1	200	50	6	5	40	0.1884	0.005024	150	19.59	167.07
193	35	0.25	0.1	200	50	6	5	40	0.1884	0.00553896	150	19.571	166.68
194	35	0.25	0.1	200	50	6	5	40	0.1884	0.00607904	150	19.579	166.64
195	35	0.25	0.1	200	50	6	5	40	0.1884	0.00664424	150	19.58	167.18
196	35	0.25	0.1	200	50	6	5	40	0.1884	0.00723456	150	19.587	167.61
197	35	0.25	0.1	200	50	6	5	40	0.1884	0.00785	150	19.62	168.02
198	35	0.25	0.1	200	50	6	5	40	0.1884	0.002826	50	19.58	174.3
199	35	0.25	0.1	200	50	6	5	40	0.1884	0.002826	100	19.57	174.27
200	35	0.25	0.1	200	50	6	5	40	0.1884	0.002826	150	19.57	174.3
201	35	0.25	0.1	200	50	6	5	40	0.1884	0.002826	200	19.57	174.32
202	35	0.25	0.1	200	50	6	5	40	0.1884	0.002826	250	19.578	174.28
203	35	0.25	0.1	200	50	6	5	40	0.1884	0.002826	300	19.588	174.29
204	35	0.25	0.1	200	50	6	5	40	0.1884	0.002826	350	19.588	174.26
205	35	0.25	0.1	200	50	6	5	40	0.1884	0.002826	400	19.579	174.26
206	35	0.25	0.1	200	50	6	5	40	0.1884	0.002826	450	19.589	174.28
207	35	0.25	0.1	200	50	6	5	40	0.1884	0.002826	500	19.59	174.28
208	35	0.25	0.1	200	50	6	5	40	0.1884	0.002826	550	19.59	174.28
209	35	0.25	0.1	200	50	6	5	40	0.1884	0.002826	600	19.591	174.28
210	35	0.25	0.1	200	50	6	5	40	0.1884	0.002826	650	19.595	174.25
211	35	0.25	0.1	200	50	6	5	40	0.1884	0.002826	700	19.597	174.3
212	35	0.25	0.1	200	50	6	5	40	0.1884	0.002826	750	19.596	174.26
213	35	0.25	0.1	200	50	6	5	40	0.1884	0.002826	800	19.587	174.31
214	35	0.25	0.1	200	50	6	5	40	0.1884	0.002826	850	19.594	174.27
215	35	0.25	0.1	200	50	6	5	40	0.1884	0.002826	900	19.592	174.22
216	35	0.25	0.1	200	50	6	5	40	0.1884	0.002826	950	19.59	174.29
217	35	0.25	0.1	200	50	6	5	40	0.1884	0.002826	1000	19.59	174.26

References

1. He, M.; Xie, H.; Peng, S.; Jiang, Y.-D. Study on Rock Mechanics in Deep Mining Engineering. *Yanshilixue Yu Gongcheng Xuebao Chin. J. Rock Mech. Eng.* **2005**, *24*, 2803–2813.
2. Xie, H.-P.; Zhou, H.; Xue, D.; Wang, H.-W.; Zhang, R.; Gao, F. Research and Consideration on Deep Coal Mining and Critical Mining Depth. *J. China Coal Soc.* **2012**, *37*, 535–542.
3. Ranjith, P.G.; Zhao, J.; Ju, M.; De Silva, R.V.S.; Rathnaweera, T.D.; Bandara, A.K.M.S. Opportunities and Challenges in Deep Mining: A Brief Review. *Engineering* **2017**, *3*, 546–551. [CrossRef]
4. Xie, H.P. Research Review of the State Key Research Development Program of China: Deep Rock Mechanics and Mining Theory. *J. China Coal Soc.* **2019**, *44*, 1283–1305.
5. Shen, B.; King, A.; Guo, H. Displacement, Stress and Seismicity in Roadway Roofs during Mining-Induced Failure. *Int. J. Rock Mech. Min. Sci.* **2008**, *45*, 672–688. [CrossRef]
6. Prusek, S.; Rajwa, S.; Wrana, A.; Krzemień, A. Assessment of Roof Fall Risk in Longwall Coal Mines. *Int. J. Min. Reclam. Environ.* **2017**, *31*, 558–574. [CrossRef]
7. Wang, J.-C. Mechanism of the Rib Spalling and the Controlling in the Very Soft Coal Seam. *Mei Tan. Hsueh Pao J. China Coal Soc.* **2007**, *32*, 785–788.
8. Huang, S.; Zhao, G.; Meng, X.; Cheng, X.; Xu, W.; Liu, G.; Zhu, S. Study of Prevention and Control Technology for Roadway Excavation under the Soft and Extra-Thick Coal Roof in Luling Coal Mine. *Processes* **2022**, *10*, 1835. [CrossRef]
9. Cheng, L.; Ge, Z.; Chen, J.; Ding, H.; Zou, L.; Li, K. A Sequential Approach for Integrated Coal and Gas Mining of Closely-Spaced Outburst Coal Seams: Results from a Case Study Including Mine Safety Improvements and Greenhouse Gas Reductions. *Energies* **2018**, *11*, 3023. [CrossRef]
10. Li, H.; Zhang, Y.; Wu, J.; Zhang, X.; Zhang, L.; Li, Z. Grouting Sealing Mechanism of Water Gushing in Karst Pipelines and Engineering Application. *Constr. Build. Mater.* **2020**, *254*, 119250. [CrossRef]
11. Si, X.; Gong, F.; Luo, Y.; Li, X. Experimental Simulation on Rockburst Process of Deep Three-Dimensional Circular Cavern. *Rock Soil Mech.* **2018**, *39*, 621–634.
12. Liu, D.; Ling, K.; Guo, C.; He, P.; He, M.; Sun, J.; Yan, X. Experimental Simulation Study of Rockburst Characteristics of Sichuan–Tibet Granite: A Case Study of the Zheduoshan Tunnel. *Eng. Geol.* **2022**, *305*, 106701. [CrossRef]
13. Zhu, Z.; Du, M.; Xi, C.; Yuan, H.; He, W. Mechanics Principle and Implementation Technology of Surrounding Rock Pressure Release in Gob-Side Entry Retaining by Roof Cutting. *Processes* **2022**, *10*, 2629. [CrossRef]
14. Keneti, A.; Sainsbury, B.-A. Review of Published Rockburst Events and Their Contributing Factors. *Eng. Geol.* **2018**, *246*, 361–373. [CrossRef]
15. He, M.; Ren, F.; Liu, D. Rockburst Mechanism Research and Its Control. *Int. J. Min. Sci. Technol.* **2018**, *28*, 829–837. [CrossRef]
16. Skrzypkowski, K.; Korzeniowski, W.; Zagórski, K.; Zagórska, A. Modified Rock Bolt Support for Mining Method with Controlled Roof Bending. *Energies* **2020**, *13*, 1868. [CrossRef]
17. Zuo, J.; Wang, J.; Jiang, Y. Macro/Meso Failure Behavior of Surrounding Rock in Deep Roadway and Its Control Technology. *Int. J. Coal Sci. Technol.* **2019**, *6*, 301–319. [CrossRef]
18. Weishen, Z.; Feng, C. Constitutive Model of Energy Dissipation and Its Application to Stability Analysis of Ship-Lock Slope in Three Gorges Project. *Chin. J. Rock Mech. Eng.* **2000**, *19*, 261–264.
19. Xie, H.P.; Peng, R.D.; Ju, Y. Energy Dissipation of Rock Deformation and Fracture. *Chin. J. Rock Mech. Eng.* **2004**, *23*, 3565–3570.
20. Liang, P.; Zhang, Y.B.; Tian, B.Z.; Yao, X.L.; Sun, L.; Liu, X.X. Experimental Study on Energy Evolution Characteristics in the Process of Tunnel Rockbursts. *Chin. J. Rock Mech. Eng.* **2019**, *38*, 736–746.
21. Yu, H.; Zhang, X.; Li, B.; Chu, F. Macro-Micro Mechanical Response and Energy Mechanism of Surrounding Rock under Excavation Disturbance. *J. China Coal Soc.* **2020**, *45*, 60–69.
22. Dai, B.; Zhao, G.; Zhang, L.; Liu, Y.; Zhang, Z.; Luo, X.; Chen, Y. Energy Dissipation of Rock with Different Parallel Flaw Inclinations under Dynamic and Static Combined Loading. *Mathematics* **2022**, *10*, 4082. [CrossRef]
23. Dai, B.; Shan, Q.; Chen, Y.; Luo, X. Mechanical and Energy Dissipation Characteristics of Granite under Cyclic Impact Loading. *J. Cent. South Univ.* **2022**, *29*, 116–128. [CrossRef]
24. Zhang, D.; Guo, W.; Zhao, T.; Zhao, Y.; Chen, Y.; Zhang, X. Energy Evolution Law during Failure Process of Coal-Rock Combination and Roadway Surrounding Rock. *Minerals* **2022**, *12*, 1535. [CrossRef]
25. Yi, K.; Liu, Z.; Lu, Z.; Zhang, J.; Sun, Z. Transfer and Dissipation of Strain Energy in Surrounding Rock of Deep Roadway Considering Strain Softening and Dilatancy. *Energy Sci. Eng.* **2021**, *9*, 27–39. [CrossRef]
26. He, M.; Ding, M.; Yuan, Z.; Zhao, J.; Luo, B.; Ma, X. Numerical Simulation of Rock Bursts Triggered by Blasting Disturbance for Deep-Buried Tunnels in Jointed Rock Masses. *Comput. Geotech.* **2023**, *161*, 105609. [CrossRef]
27. He, M.; Zhao, J.; Deng, B.; Zhang, Z. Effect of Layered Joints on Rockburst in Deep Tunnels. *Int. J. Coal Sci. Technol.* **2022**, *9*, 21. [CrossRef]
28. Wang, H.; Zhang, C.; Zhou, B.; Xue, S.; Jia, P.; Zhu, X. Prediction of Triaxial Mechanical Properties of Rocks Based on Mesoscopic Finite Element Numerical Simulation and Multi-Objective Machine Learning. *J. King Saud. Univ.-Sci.* **2023**, *35*, 102846. [CrossRef]
29. Li, X.; Nishio, M.; Sugawara, K.; Iwanaga, S.; Chun, P. Surrogate Model Development for Slope Stability Analysis Using Machine Learning. *Sustainability* **2023**, *15*, 10793. [CrossRef]

30. Hecht-Nielsen, R. Theory of the Back Propagation Neural Network. In *Neural Networks for Perception*; Academic Press: Cambridge, MA, USA, 1992; pp. 65–93. [CrossRef]
31. Yan-yan, S.; Ying, L. Decision Tree Methods Applications for Classification and Prediction. *Shanghai Arch. Psychiatry* **2015**, *27*, 130–135. [CrossRef]
32. Ding, S.; Xu, X.; Nie, R. Extreme Learning Machine and Its Applications. *Neural Comput. Appl.* **2014**, *25*, 549–556. [CrossRef]
33. Biau, G. Analysis of a Random Forests Model. *J. Mach. Learn. Res.* **2012**, *13*, 1063–1095.
34. Hearst, M.A.; Dumais, S.T.; Osuna, E.; Platt, J.; Scholkopf, B. Support Vector Machines. *IEEE Intell. Syst. Their Appl.* **1998**, *13*, 18–28. [CrossRef]
35. Rodriguez-Dono, A.; Alejano, L.R.; Veiga, M. Analysis Of Longitudinal Deformation Profiles Using Flac3D. *OnePetro*, 23 October 2010.
36. Zhu, Z.; Wang, J.; Wu, M. DEM Simulation of Particle Crushing in a Triaxial Test Considering the Influence of Particle Morphology and Coordination Number. *Comput. Geotech.* **2022**, *148*, 104769. [CrossRef]
37. Wu, Q.; Liu, H.; Dai, B.; Cheng, L.; Li, D.; Qin, P. Influence of Base-Angle Bolt Support Parameters and Different Sections on Overall Stability of a Roadway under a Deeply Buried High Stress Environment Based on Numerical Simulation. *Sustainability* **2023**, *15*, 2496. [CrossRef]
38. Labuz, J.F.; Zang, A. Mohr–Coulomb Failure Criterion. *Rock Mech. Rock Eng.* **2012**, *45*, 975–979. [CrossRef]
39. Board, M. *FLAC (Fast Lagrangian Analysis of Continua)*, Version 2. 20; Itasca Consulting Group, Inc.: Minneapolis, MN, USA, 1989. [CrossRef]
40. Ma, T.; Zhang, Y.; Qiu, Y.; Liu, Y.; Li, Z. Effect of Parameter Correlation on Risk Analysis of Wellbore Instability in Deep Igneous Formations. *J. Pet. Sci. Eng.* **2022**, *208*, 109521. [CrossRef]
41. Kunpeng, Y.; Fengyu, R.; Ruslan, P. Optimization of Combined Support in Soft-Rock Roadway. *Tunn. Undergr. Space Technol.* **2020**, *103*, 103502.
42. Li, G.; Ma, W.; Yu, C. Optimization of anchorage support parameters for soft rock tunnel based on displacement control theory. *J. Mt. Sci.* **2023**, *20*, 2076–2092. [CrossRef]
43. Yuan, C.; Cao, L.; Wang, W.; Fan, L.; Huang, C. Case Study on Rock Support Technology for Roadways Based on Characteristics of Plastic Area. *KSCE J. Civ. Eng.* **2021**, *25*, 705–723. [CrossRef]

applied sciences

MDPI

Article

Numerical Simulation Analysis of the Influence of Interlayer Quantity on the Long-Term Stable Operation of Gas Storage Facilities

Lilong Li [1], Xin Jiang [2], Jiafeng Tan [3,*], Rong Liu [4,*], Xiaolinag Quan [1], Jinyang Fan [3], Cheng Qian [2] and Jinjie Suo [3]

[1] Sichuan Yanjiang Yijin Expressway Co., Ltd., Xichang 615099, China; zebra-2000@163.com (L.L.); quanxliang@163.com (X.Q.)
[2] Chongqing Chengtou Jiangchang Construction Co., Ltd., Chongqing 400023, China; jxeent@163.com (X.J.); qianchengqc@foxmail.com (C.Q.)
[3] School of Resources and Safety Engineering, Chongqing University, Chongqing 400044, China; jinyang.f@cqu.edu.cn (J.F.); 202120021018@stu.cap.edu.cn (J.S.)
[4] Institute of Future Civil Engineering Science and Technology, Chongqing Jiaotong University, Chongqing 400074, China
* Correspondence: 202320021058t@stu.cqu.edu.cn (J.T.); cqu_liurong@163.com (R.L.); Tel.: +86-13628232903 (J.T.); +86-17783002935 (R.L.)

Abstract: Salt rock is considered as an ideal energy storage medium, and compressed air energy storage by a salt cavern can improve the utilisation efficiency of renewable energy. Salt rock in China mostly contains different interlayers, among which mudstone interlayers are the most common. At present, there are relatively few studies on the influence of mudstone interlayers on the long-term stable operation of gas storage. FLAC3D software was used to simulate the long-term operation of salt rock gas storage with different numbers of interlayers in the Yexian area of Pingdingshan. The results show that with the passage of time, the vertical displacement of the surrounding rock of the vertical single-cavity gas storage tank increases gradually. The maximum settlement value at the top of the surrounding rock is always greater than the maximum uplift value at the bottom. The horizontal displacement shows obvious symmetry with the vertical displacement at the top and bottom of the surrounding rock. The effect of the cyclic pressure interval on horizontal displacement is the same as that of vertical displacement. With the increase in the number of interlayers, the volume of the plastic zone gradually increases with the increase in the running time, and the increasing speed shows a growing trend.

Keywords: salt cavern; mudstone interlayer; FLAC3D; simulation

1. Introduction

Salt rock is regarded as an optimal medium for energy storage, given its favourable rheological properties, low porosity, low permeability, and damage self-healing attributes. The utilisation of salt caverns for compressed air energy storage has emerged as an efficacious methodology for enhancing the utilisation efficiency of renewable energy. The characteristics of a foreign salt cavern's underground gas storage include a shallow burial, a large stratum thickness, high-grade salt rock, and a minimal number of interlayers. However, the buried salt mine in China is typically situated at greater depths, exceeding 2000 m in certain regions. The rock layer in these mines is relatively thin, ranging from 60 to 250 m in thickness. Additionally, the salt rock in Chinese mines is predominantly layered, comprising up to eight layers of thicker interlayers [1–3]. The mudstone interlayer is particularly prevalent, representing a significant portion of the rock composition.

In light of the increasingly severe global climate crisis, it is imperative that China prioritises the utilisation of renewable energy sources in order to achieve its dual-carbon objective. According to estimates, to reach carbon neutrality by 2060, clean energy should account for about 60 percent of the country's energy consumption [4]. The large-scale utilisation of salt cavern space gives full play to its advantages of a large capacity, cleanliness, environmental protection, safety, reliability, and economy, and provides important support for reaching national strategic objectives [5]. However, due to the layered structure of Chinese salt rocks, which is different from that of foreign countries, the presence of interlayers and impurities poses a serious threat to the safety of the reservoirs [6]. In order to guarantee the secure operation of salt cavern gas storage reservoirs, a considerable number of scholars have conducted research on the salt rocks. However, there has been comparatively little investigation into the mudstone interlayers present within the reservoirs.

Taheri et al. [7] found that creep of Gachsaran rock salt shows strong sensitivity to temperature changes; however, its sensitivity to different stress changes is low. Zhao K. et al. [8] investigated the creep fatigue characteristics of salt rock under different loading and unloading paths, and explained the deformation behaviour from the perspective of microscopic mechanisms. Malinjian et al. [9] employed a nonlinear cumulative damage rule to predict the cyclic life of rock salt under creep-fatigue loading. Hakan et al. [10] studied the creep behaviour of salt rocks and proposed a mathematical model that can explain their creep behaviour. Wang Junbao et al. [11] studied the creep characteristics of salt rock under low-frequency cyclic loading, and obtained the axial creep equation of Burgers' model under cyclic loading. By comparing the fatigue characteristics of salt rock in intermittent fatigue tests with those of salt rock without intermittent fatigue tests, Jiang Deyi et al. [12] found that the fatigue life of salt rock was shortened under intermittent action. Fan Jinyang et al. [13] proposed an intrinsic model of creep fatigue in salt rock based on hardening coefficients. Li Zongze et al. [14] investigated the effects of different circumferential pressures and stress levels on the interval fatigue of salt rock, and found that the presence of circumferential pressure enhanced the compressive strength of salt rock while increasing the fatigue life. Cui Yao et al. [15] investigated the acoustic emission characteristics of salt rock through intermittent fatigue tests, further revealing the effect of the time interval on the residual strain and fatigue life of salt rock. The evolution of rock salt deformation and elastic constants under the influence of time interval factors was investigated by Li Xin et al. [16]. By summarising a large amount of the literature on salt rock creep tests, Bingren Hou et al. [17] found that under constant axial pressure and cyclic peripheral pressure paths, the creep properties of salt rocks are significantly affected by the peripheral pressure and cyclic cycle.

In addition, a large number of scholars have studied the factors affecting the stability of salt cavern cavities for the safe and stable operation of salt cavern gas storage reservoirs. Yu Haibing et al. [18] studied the influence law of salt layer inclination and burial depth on the deformation and volume contraction of the salt cavity, and gave the safe mining depth of the proposed salt mine. Liang Wuxing et al. [19] combined Ansys and FLAC3D to carry out numerical simulation, and judged that the effect of three factors on the stability of the cavity based on the results of linear fitting was cavity burial depth > pressure interval > pressure difference. Fu Xing et al. [20] used FLAC3D 3.00 software to analyse the long-term operational stability of energy storage cavities in order to discuss the possibility of the safe operation of CAES reservoirs under Chinese domestic stratigraphic conditions. Liu Wei et al. [21] conducted permeability tests based on interface-bearing salt rocks from domestic salt mines, and numerical simulations based on the test results on the stability and confinement of natural gas reservoirs containing entrapped underground salt caverns. Zhixin Zhang et al. [22] conducted a series of indoor tests, theoretical and numerical simulations, and discussions to simulate hydrogen leakage from cave chambers under different conditions using COMSOL Multiphysics 6.0 software, and analysed the leakage range, pore pressure, and the amount of hydrogen leakage. Chunhe Yang et al. [23] combined the stability characteristics of salt caverns and gave five key scientific and technological issues facing the deep large-scale salt cavern energy storage technology in China. Tongtao Wang et al. [24] explored the effects of different gas frequencies on the stability of horizontal salt caverns under different geological conditions. They found that shallow salt caverns are suitable for compressed air energy storage, medium and deep salt caverns are suitable for natural gas energy storage, and deep salt caverns are suitable for helium energy storage. Leszek Lankof et al. [25] discussed the suitability of underground hydrogen storage facilities in layered salt caverns based on a geographic information system (GIS), taking into account the results of previous studies on the storage capacity of rock salt deposits.

This paper presents a study of the salt cavern gas storage reservoir in the Yexian area of the Pingdingshan Mountain. The research employs a comprehensive approach combining theoretical analysis and numerical simulations to investigate the impact of varying numbers of interlayers on the long-term stability of the gas storage reservoir. The objective is to address the research gap concerning mudstone interlayers and to provide theoretical support and numerical simulation analyses for the safe operation of gas storage reservoirs.

2. Materials and Methods

2.1. Introduction to FLAC3D Finite Difference Software

FLAC is an acronym that stands for "Fast Lagrangian Analysis of Continua". It is a computer program designed for the numerical modelling and simulation of geotechnical and rock mechanics problems. The FLAC program employs a finite difference method to simulate the mechanical behaviour of continuous media in a discrete form. This includes processes such as deformation, fracture, and stress transfer in rocks. The software is frequently employed to model a range of problems in diverse fields, including rock engineering, groundwater flow, earthquake geology, and underground storage. FLAC is a frequently utilised finite element software in the domain of geotechnical engineering, offering a diverse array of applications and a comprehensive range of functionalities.

The FLAC3D software offers a number of advantages, including multi-physics field coupling, dynamic equations of motion solving, and an explicit solution scheme. These features enable the simulation of complex physical processes, such as rock mechanics, heat conduction, and fluid flow, while reducing the required memory space. However, the software requires high input parameters, the solution time is affected by the proportion of natural cycles, and the pre-processing function is relatively limited. Furthermore, FLAC3D boasts a plethora of features, including exclusivity, a command-driven mode, five calculation modes, multiple boundary conditions and structural forms simulations, and openness. The solution process comprises the following steps: modelling, the verification of the static equilibrium, modification of conditions, recalculation until a new equilibrium is reached, the exporting of data, and the analysis of images.

Salt rocks have good creep properties, and the creep deformation of salt rocks is also one of the causes of reservoir instability. FLAC3D software has the function to realise the simulation calculation of five creep material calculation models, so it is necessary to introduce the rheological calculation models of FLAC3D software [26].

2.1.1. WIPP Model

It is essential to consider the effects of time and temperature on the behaviour of salt rock in thermodynamic analyses simulating underground isolation studies of nuclear waste in salt rock. One such model is the WIPP (Waste Isolation Pilot Plant) model. In the FLAC3D finite difference software, the bias strain rate tensor in the WIPP model is composed of a viscous component and an elastic component, which can be expressed as follows:

$$\begin{bmatrix} -d\dot{\varepsilon}_{ij} \\ -d\dot{e}_{ij} \\ -d\dot{\alpha}_{ij} \end{bmatrix} = \begin{bmatrix} \varepsilon_{ij} \\ e_{ij} \\ \alpha_{ij} \end{bmatrix} + \begin{bmatrix} \dot{\varepsilon}_{ij} \\ \dot{e}_{ij} \\ \dot{\alpha}_{ij} \end{bmatrix}, \quad (1)$$

where ε_{ij}, e_{ij}, and α_{ij} are the corresponding strain components; $\dot{\varepsilon}_{ij}$, $\dot{e}_{ij}$, and $\dot{\alpha}_{ij}$ are the corresponding strain rate components; and $\begin{bmatrix} \varepsilon_{ij} \\ e_{ij} \\ \alpha_{ij} \end{bmatrix}$ is the elastic part of the biased strain rate tensor.

The objective of this model is to simulate the time and temperature dependence of the subsurface isolation process of nuclear waste in a salt layer. The FLAC3D software is capable of more accurately describing the complex behaviour of salt rock, including the interaction of the viscous and elastic components, by employing this form of the biased strain rate tensor.

2.1.2. Viscoplastic Model

Viscoelastic models are typically a combination of viscoelastic and plastic models, employed to describe the behaviour of materials during deformation. The WIPP model is a representative viscoelastic model, whereas the Drucker–Prager model is a prevalent plastic model. The combination of these two models provides a more comprehensive description of the deformation behaviour of the material, whereby both viscous and plastic effects are taken into account, thus enhancing the accuracy of the simulation.

The shear yield function for the Drucker–Prager model is shown below:

$$f^s = \tau + q_\phi \sigma_o - k_\phi, \tag{2}$$

where q_ϕ and k_ϕ are the material parameters.

Concurrently, the distinctive attributes and operational procedures of FLAC3D software are integrated with the specific attributes of layered salt-rock underground gas storage reservoirs and the non-linear and creep characteristics of salt rocks. In the simulation, the viscoplastic model is employed for the salt rock layer, while the Moore-Cullen model is utilised for the remaining strata. The objective of employing a combination of models is to provide a more accurate description of the mechanical behaviour of the salt-rock reservoir system. The utilisation of FLAC3D finite-difference software in simulation studies facilitates a more comprehensive comprehension of the long-term stability of salt-rock reservoirs, alongside an evaluation of their viability and safety in practical applications.

2.2. *Saltstone Stratigraphic Conditions and Cavity Building Plan*

The Pingdingshan Salt Field is rich in mineral deposits, and after nearly 20 years of mining, a large number of abandoned mining cavities have been formed. The salt rocks are distributed in the Cenozoic Paleoproterozoic Loop strata, showing thinly bedded salt rocks and mudstones with alternating lamination, relatively simple tectonics, and horizontal production. The main salt rocks are buried at depths between −1400 and −1000 m [27].

The Pingdingshan salt deposit is a set of Mesozoic and Cenozoic clastic chemical rock series which is contained in the Wuyang Depression. From the Cretaceous to the Quaternary, the Hu Gang Formation (K_2h), Yuhuangding Formation (E_2y), Dacangfang Formation (E_2d), Walnut Garden Formation (E_2h), Liaozhuang Formation (E_3l), Shangshi Formation (N_sh), and Plain Formation (Q_p) were successively deposited. The salt-bearing strata are the nuclear section (top) of the Paleocene Oligocene Walnut Garden Formation. A set of mudstones with a stable distribution is developed at the top of the nuclear section [28]. The profile of the Mesozoic and Cenozoic strata in the Wuyang Depression is shown in Table 1.

Table 1. Sequence stratigraphy list of Wuyang Depression [28].

Stratigraphic System: Scope	Faculty	Unite	Group		Thickness/m	Lithologic Association Characteristics
Cenozoic K_z	Quaternary Q		Plains group Q_p		25~168	Mottled, light brownish-yellow, lower clay with gravel, middle and upper sandy clay, and clay with several layers of gravel. Unconsolidated contact with underlying strata.
	Neogene Z		Shangji Formation N_sh		300~800	Light yellow, grayish-yellow, and mottled fine conglomerate and conglomerate sandstone are interbedded with grayish-yellow, brownish-red, and grayish-green mudstone. Unconsolidated contact with underlying strata.
	Paleocene system E	Oligocene system E_3	Liaozhuang group E_3l		315~869	Variegated conglomerate sandstone and light gray conglomerate-bearing sandstone are interbedded with brown and grayish-yellow mudstone in consolidated contact with underlying strata.
			Walnut Orchard Group	First nuclear segment E_3h_1	200~1138	The lower part is mudstone, rock salt rock, and cream-bearing mudstone; the middle part is cream-bearing rock salt rock and mudstone; and the upper part is mudstone and cream-bearing mudstone interbedded, sandwiched with rock salt rock. Saltstone is gray and brownish-yellow; other mudstone is gray and brownish-red.
		Eocene system E_2		Second nuclear segment E_2h_2	489~1070	The lower part is dark gray paste-bearing mudstone and mudstone interbedded with fine sandstone, and the bottom is mottled conglomerate sandstone interbedded with purplish-red mudstone; the upper part is gray mudstone, siltstone, oil shale, and grayish-white and brownish-yellow-lithium salts and paste-bearing mudstone.
				Tertiary nuclear segment E_2h_3	784~986	The lower part is grayish-red conglomerate-bearing sandstone, conglomerate sand-stone, light grayish-red fine sandstone, brownish-red, brownish-purple mudstone, locally interspersed with dolomitic mud-stone; the upper part is grayish-red, brownish mudstone, mottled conglomerate sand-stone, and grayish-white greystone.
			Dachangfang Group E_2d		>1200	Variegated sandstone, conglomerate, locally interbedded with brownish-red and purple sandy mudstone and pink-fine sandstone. Consolidated contact with underlying strata.
Cenozoic K_z	Paleocene system E	Eocene system E_2	Yuhuangding Group E_2y		1943	The lower part is brown and purple sand-stone and mudstone interbedded with colourful conglomerate and gray siltstone, etc.; the middle part is purple sandstone, mudstone, and dolomite and marl; and the top part is light gray mudstone, siltstone, and gray–white dolomite.
Mesosphere M_z	Cretaceous K	Unification K_2	Hugang Group K_2h		300	The lower part is yellow–green conglomerate with black coal second layer at the bottom; the upper part is mottled conglomerate in unconformable contact with underlying Carboniferous, Cambrian, or Aurignacian strata.

The sedimentary characteristics and stratigraphic assemblage of the Wulipu salt mine section in Yexian Zhongyan Haolongmazhuang are highly complex. The overall stratigraphy demonstrates a typical lithological sequence comprising alternating salt rocks and mudstones. These are intermixed and distributed, with a large number of intercalated layers, a thin thickness of the single layer, and a deeper burial depth. The geographic location of Ye County is shown in Figure 1.

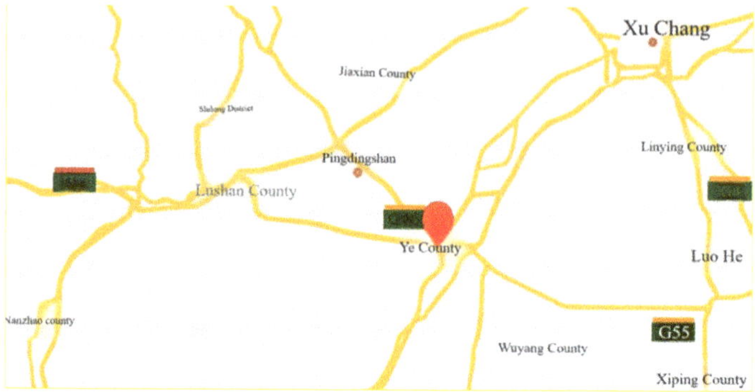

Figure 1. Geographic location map of Ye County, Pingdingshan.

Further investigation reveals that the overall thickness of the salt layers in the Pingdingshan Salt Field region ranges from 489.0 to 1070.0 m, which can be categorised into 22 distinct salt group assemblages, with a maximum of 61 layers reachable. The thickness of a single layer of salt is up to 27.7 m, with the thinnest layers measuring less than 1.0 m, and typically within 10.0 m. The thickness of the mudstone interlayer is variable, with an average of approximately 2.8 m and a maximum of 12.3 m. The majority of the interlayers are within 5 m, representing approximately 30% of the overall thickness.

From a macroscopic perspective, a stable distribution of interlayers is observed at the base of each salt group. In the 0–20 salt group, for instance, the interlayer thickness ranges from 2 to 13.2 m, with an average of 4.1 m. Notably, the interlayer thickness is minimal, reaching a minimum of 0 m. The interlayer thickness at the bottom of the 7, 10, and 13 salt groups is greater than 10 m, with a maximum of 13.2 m. The interlayer thickness in the 14–20 salt groups is 2–8 m, with an average of 5.1 m, and contains eight interlayers in total. In total, there are eight sandwiches.

In order to establish the mechanical model of mudstone entrapment, it is necessary to understand the basic characteristics of the subsurface cavities in detail and obtain relevant data to support the modelling. The consistency of the laboratory simulation studies with the material and joint properties of the field-scale salt cavern conditions helps to improve the relevance and applicability of the studies to real-world engineering, thus providing a reliable guide to real-world engineering applications [29]. Jie Chen and Yichao Rui et al. [30,31]. proposed an acoustic emission positioning technique to effectively reduce the TDOA error by combining the microseismic (MS) source localisation technique and the refraction path study, respectively, which is of good significance for the detection of underground salt cavern gas reservoirs. Wang, Z. R. et al. explored the underground salt cavern gas storage reservoir in Ye County, Pingdingshan, by using the Controlled Source Audio Geomagnetic Method (CSAMT) and the Geodetic Electric Field Petrographic Detection (CYT) method, and obtained the basic features such as the burial depth, dimensions, and spatial location of the underground caverns [32]. According to the summary of Wang, Z. R. et al.'s study, the exploration results show that there are more cavities in the underground of the exploration area, with depths ranging from −1148.0 to −1424.0 m, and with moderate spacing between each other of about 90.0 to 450.0 m. The results are summarised in the following table. The cavity shows a narrow top and wide bottom, with a moderate volume of individual cavities, a range of cavity heights from

153.0 to 233.0 m, averaging about 199.4 m, and a cavity radius between 8.0 and 60.0 m. A schematic of the cavity profile is shown in Figure 2 [33].

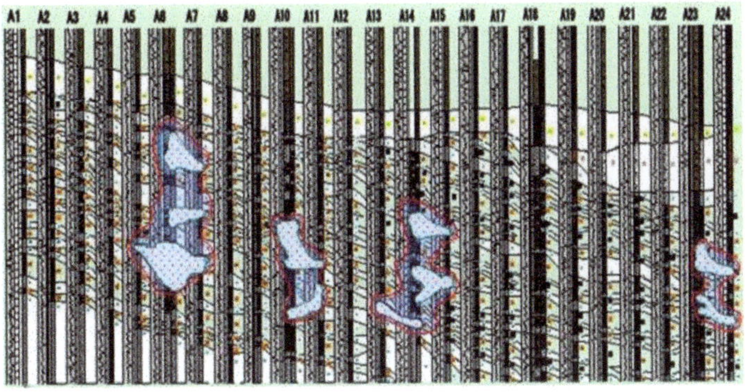

Figure 2. Schematic of underground cavity profile [33].

This paper presents the establishment of a single-cavity model of the salt cavern, created using Ansys 2020R1 and FLAC3D 6.00 software in accordance with the geological conditions observed in the salt cavern gas storage reservoir in Ye County, Pingdingshan. The study concentrated on the impact of varying numbers of mudstone interlayers on displacement changes at the top of the cavity and the volumetric shrinkage rate of the cavity, and presented a summary of the influencing laws. In regard to the long-term safety of salt cavern gas storage reservoirs, it is of paramount importance to select the optimal perimeter rock permeability and operating pressure to effectively prevent deformation resulting from the combined effects of creep and seepage, as well as potential incidents such as cavity instability and gas leakage. In order to facilitate the analysis, an ideal ellipsoidal cavity was constructed. In order to minimise the impact of boundary effects, the model length was set to be between six and eight times the diameter of the cavity. The specific model parameters are set as follows:

All lines (NL1) have a cell boundary size (SIZE) of 2 and a segmentation ratio of 1 (SPACE).

Number 1 (NL1) has a cell boundary size (SIZE) of 2 and is divided into 10 parts (NDIV) with a segmentation ratio of 5 (SPACE).

Control over the cell shape: (1,3D) tetrahedral cell division mesh, (1,2D) triangular division; (0,2D) quadrilateral division, (0,3D) hexahedral division. Tetrahedral cell division mesh is selected.

The grid type for mesh division: 0 free mesh division, 1 mapping mesh division, 2 first choice with mapping, if not, then free mesh division. Free mesh division was chosen.

In this paper, an ellipsoidal cavity will be set up with a long axis of 200 m, a short axis of 100 m, and an aspect ratio of 2. The total length of the model is 1000 m, the width is 500 m, the thickness is 400 m, and the top of the cavity is 1100 m above the surface. Considering the geological data, the 14–20 salt group interlayers were studied and the thickness of the mudstone interlayer was set at 5 m. In order to deeply study the influence law of the number of mudstone interlayers on the long-term operation of the gas storage reservoir, two groups of control experiments will be set up: one group is 6 layers of mudstone interlayers, and the other group is 10 layers of mudstone interlayers.

This design will help to understand the impact of the amount of mudstone entrapment on the long-term operation of a gas storage reservoir. The comparative analysis of the two sets of experiments can reveal more clearly the influence of different numbers of mudstone interlayers on the stability of gas storage reservoirs.

As shown in Figure 3, the storage model schematic shows, in detail, the structure and location distribution of the storage reservoir. In this model, we choose the following key locations as observation points for displacement changes.

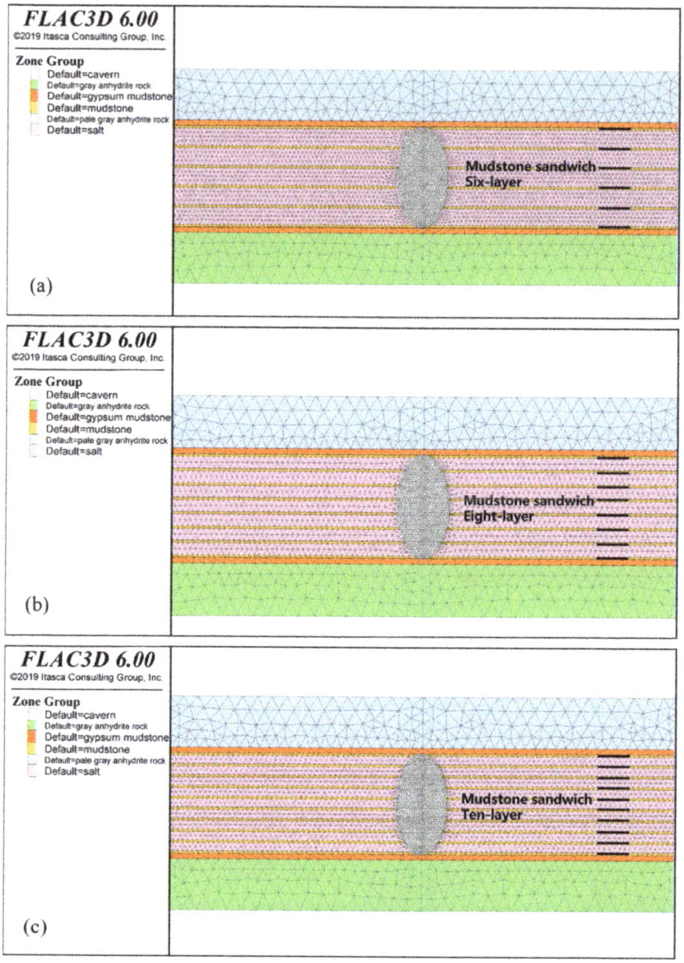

Figure 3. Schematic diagram of the gas storage model: (**a**) Six-layer sandwich; (**b**) Eight-layer sandwich; (**c**) Ten-layer sandwich.

Top Plate Position (T): This is the top position of the model, above the cavity. Monitoring displacement changes at this point can help assess the stability of the roof and potential risk of collapse.

Left position of the cavity side gang (L): this is the position on the left side of the cavity, and monitoring displacement changes at this point can provide information on lateral stresses and deformations.

Right position of the cavity side gang (R): this is the position on the right side of the cavity, corresponding to the left position, and it is equally important to monitor displacement changes at this point to get a full picture of lateral stresses and deformations.

Baseplate position (B): This is the bottom position of the model, located below the cavity. Monitoring displacement changes at this point helps to assess the stability of the footing and potential subsidence or uplift.

The data from these observation points can provide a comprehensive understanding of the displacement changes in different parts of the gas storage reservoir during long-term operation, thus providing an important reference for the safe operation and structural optimisation of the gas storage reservoir. Firstly, the construction of the cavity model was carried out, and subsequently, the mechanical parameters were set in FLAC3D to determine the computational parameters for the creep properties of the rock [34,35], as shown in Table 2. (In the context of salt mines, the aforementioned data are considered standard and are subject to testing at the mine level. The data presented in Table 2 have been sourced from the mine's operational documentation).

Table 2. Physical parameters of rock layer.

Rock Mass	Bulk Modulus/GPa	Shear Modulus/GPa	Cohesive Force/MPa	Tensile Strength/MPa	Internal Friction Angle/°
salt rock	8.67	3.95	1.93	1.20	37.5
sandstone	13.04	6.81	2.9	1.6	36
mud rock	13.04	6.81	2.9	1.60	36
Calcareous	8.45	3.85	0.50	0.50	30.0
gypsum rock	10.92	5.54	2.2	1.4	35
Light gray	13.04	6.81	2.9	1.6	36
anhydrite rock	13.04	6.81	2.9	1.60	36

The initial and boundary conditions of the reservoir cavity have an important influence on the mechanical characterisation of salt cavern reservoirs. According to the research of related scholars, a certain consensus exists on the value of the air pressure cycle in the cavity of the gas storage reservoir, i.e., it is believed that the value of the pressure during the air pressure cycle should be limited to 0.3–0.8 times that of the geopathic stress in the vertical direction. Studies have shown [36] that the main factor affecting the long-term stable operation of gas storage reservoirs is low pressure, so related scholars have proposed that the minimum air pressure is set at 33% or more. Considering multiple factors, the range of circulating gas pressure in the reservoir was finally determined to be 0.3–0.8 times the vertical geopathic stress (10–22 MPa) [37–39]. In this paper, the range of cyclic air pressure is set to be 12–18 MPa, and the main manifestations of single salt cavern cavity destabilisation include roof failure and cavity deformation, while the stability of the sidewalls needs to be considered as well [40–42]. In this study, three sets of tests were conducted: a 6-layer mudstone entrapment, an 8-layer mudstone entrapment, and a 10-layer mudstone entrapment. By comparing the displacement changes of the top and bottom plates and sidewalls, as well as the volume shrinkage rate, in these three sets of tests, it is possible to discuss the influence of different numbers of mudstone interlayers on the long-term operation of gas storage reservoirs. The aforementioned test setup can facilitate an understanding of the stability performance of salt caverns under different numbers of entrapment conditions, thereby providing a valuable reference point for the design and operation of gas storage facilities. By monitoring the alterations to the top and sidewalls of the salt caverns, it is possible to ascertain the degree to which varying quantities of mudstone entrapment influence the stability of the salt caverns. This enables the formulation of more logical operational and management strategies.

3. Results

After simulating the 30-year operation of a vertical single-cavity gas storage reservoir in FLAC3D, and after the calculation by FLAC3D software, we obtained the displacement maps of the salt cavities under cyclic gas pressure for different numbers of mudstone interlayers. As the operating time increases, the displacement distribution of the surrounding rock changes, especially the vertical and horizontal displacements. Figure 3 illustrates the cloudy distribution of the vertical displacement of the surrounding rock after 1, 5, 10, 15, 20, and 30 years of operation for a single-cavity vertical gas storage reservoir with a number of eight interlayers, while Figure 4 illustrates the cloudy distribution of the horizontal displacement of the surrounding rock for the same reservoir over the same period of time.

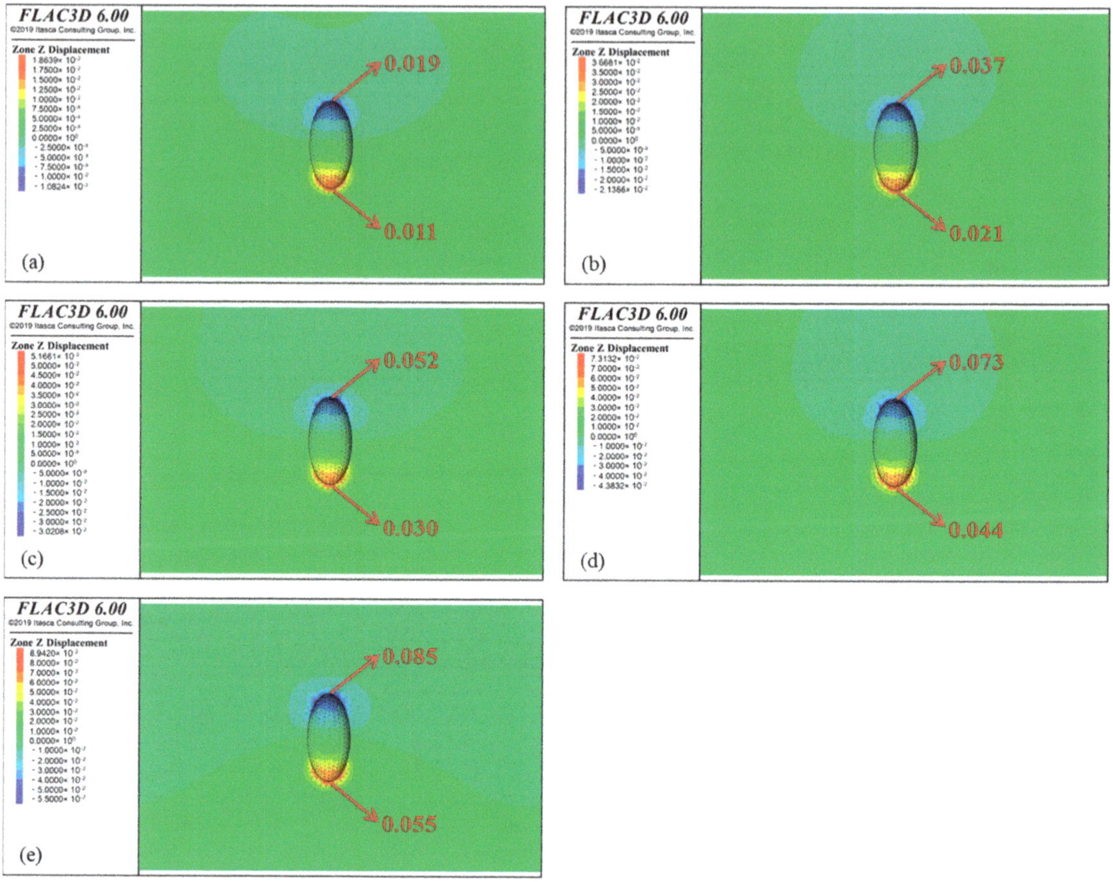

Figure 4. Vertical displacement of the surrounding rock of a single-cavity vertical gas storage reservoir with time: (**a**) One year; (**b**) Five years; (**c**) Ten years; (**d**) Twenty years; (**e**) Thirty years.

The plots offer valuable insights into the temporal displacement of the gas storage enclosure. The observation of displacement cloud maps at varying time points allows for the comprehension of the deformation of the surrounding rock during the operation of the gas storage reservoir.

A detailed examination of the test results reveals that the cavities under different numbers of mudstone interlayers all exhibit a contraction towards the interior. In contrast, the deformation of the upper surface of the cavity is consistently greater than that of the lower surface, and a symmetrical tendency is observed on the lateral walls. This phenomenon can be attributed to the fact that the cavity is situated in a gradient distribution of low stress states, with high-pressure gases filling the cavity and exhibiting an isobaric distribution. Consequently, the top and bottom of the force are not uniform.

The gradient distribution of ground stress causes the ground stress on the top of the cavity to potentially be greater than that on the bottom, so the deformation of the top will be relatively larger. And the high-pressure gas is distributed as isobaric inside the cavity, which means that the stress state inside the cavity is uniform. However, the ends of the cavity are in a state of uneven stress and may have been subjected to greater stresses, so the deformation at the ends of the cavity will be greater.

In addition, the further away the surrounding rock is from the cavity, the less it is affected, but by and large, it shows a symmetrical distribution up and down along the short axis of the cavity. This is because the surrounding rock away from the cavity is less affected, but the deformation of the cavity will affect the surrounding rock, so that the deformation along the short axis of the cavity presents a symmetrical distribution on the whole. Thus, these factors combine to characterise the deformation of the cavity under different conditions.

As illustrated in Figure 4, the vertical displacement of the surrounding rock of the vertical single-cavity gas storage reservoir is increasing over time. To illustrate, the settlement value at the upper limit of the reservoir increased from 0.019 m to 0.089 m over the course of 30 years, while the maximum uplift value at the lower limit rose from 0.011 m to 0.055 m over the same period.

By observing the vertical displacement cloud diagram of the surrounding rock of the gas storage reservoir in Figure 4, it is found that with the increase in time, the maximum settlement value of the surrounding rock at the top of the reservoir is always larger than the maximum uplift value at the bottom of the reservoir, e.g., the maximum settlement value of the top of the single-cavity reservoir with eight layers of mudstone interlayers after 30 years of operation is 0.089 m, and the maximum uplift value of the bottom of the reservoir is 0.055 m. The asymmetry in the vertical displacement of the gas reservoir envelope is due to the different effects of creep action on the top and bottom of the reservoir.

During the operation of a gas storage reservoir, there may be differences in the stresses and temperatures to which the top and bottom are subjected, which could result in disparate creep effects between the two. The upper portion of the reservoir may experience greater stress levels than the lower portion. Furthermore, given that temperature typically increases gradually with depth, the temperature at the bottom may be higher than at the top, resulting in disparate creep effects.

Thus, the fact that the maximum settlement value at the top of the gas storage enclosure is consistently greater than the maximum bulge value at the bottom over time is due to the different effects of creep on the top and bottom of the storage reservoir.

The horizontal displacement cloud map of the surrounding rock after 30 years of operation was obtained by numerical simulations for the vertical single-cavity salt rock reservoir with eight layers of mudstone interlayers, as illustrated in Figure 5. The remaining models exhibited comparable patterns.

Figure 5 illustrates that the horizontal displacement of the surrounding rock in the single-cavity vertical gas storage reservoir is influenced by the surrounding pressure, with a tendency for expansion into the cavity. This suggests that the surrounding rock has undergone horizontal deformation in the direction of inward contraction, which has resulted in a reduction in the effective volume of the gas storage reservoir.

At the same time, the horizontal displacement of the enclosing rock shows a clear symmetry with respect to the vertical displacement of the top and bottom of the enclosing rock. This means that the maximum value of the displacement change towards the interior of the cavity at the horizontal ends is approximately equal. For example, the horizontal displacement values at the horizontal ends of the reservoir in year 30 are 0.1149 m and 0.1227 m. This symmetry may be attributed to the deformation of the surrounding rocks occurring horizontally under the influence of similar geologic stresses and confining pressures, which results in the deformation of the surrounding rocks occurring horizontally showing a similar tendency at the two ends.

A summary the variation of perimeter rock displacement with time for a single-cavity vertical gas storage reservoir is shown in the Figure 6. Overall, these observations help to understand the deformation characteristics of the gas storage enclosure rock when subjected to subsurface pressure, which is of great engineering significance for the design and management of gas storage reservoirs.

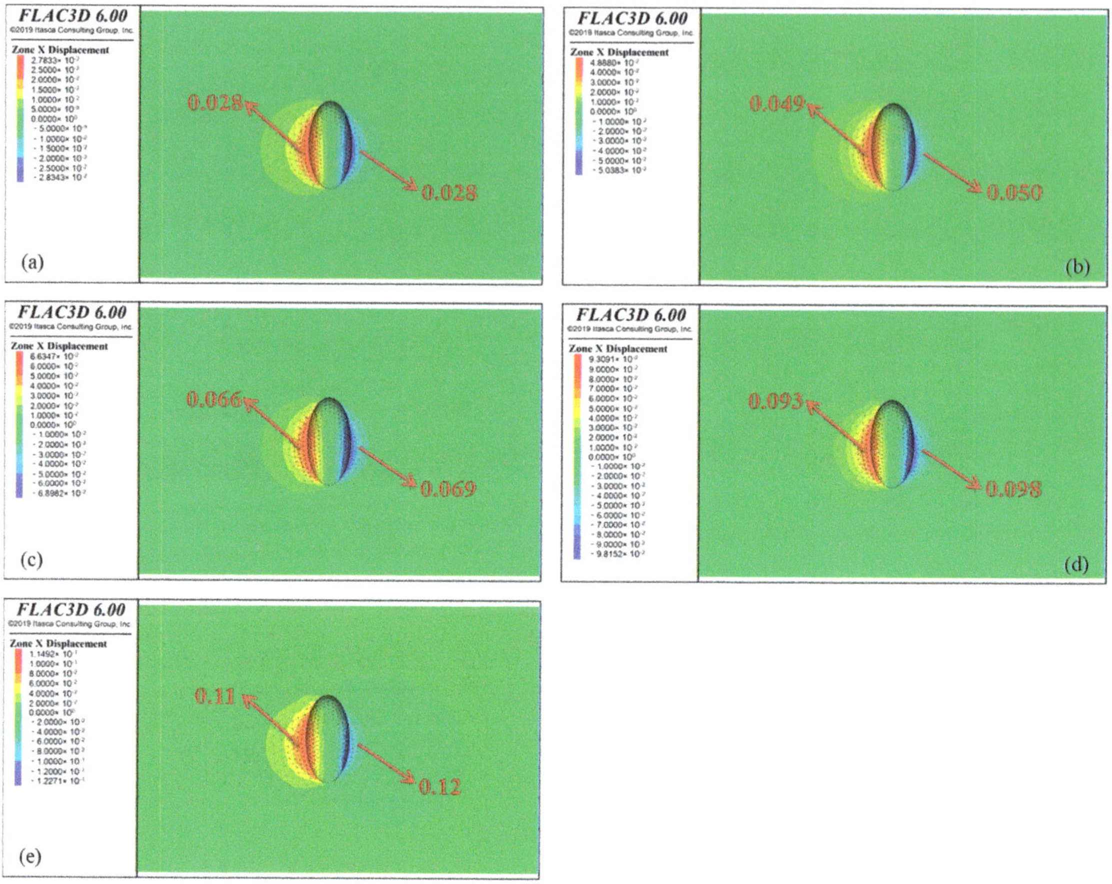

Figure 5. Horizontal displacement of the surrounding rock of a single-cavity vertical gas storage reservoir with time: (**a**) One year; (**b**) Five years; (**c**) Ten years; (**d**) Twenty years; (**e**) Thirty years.

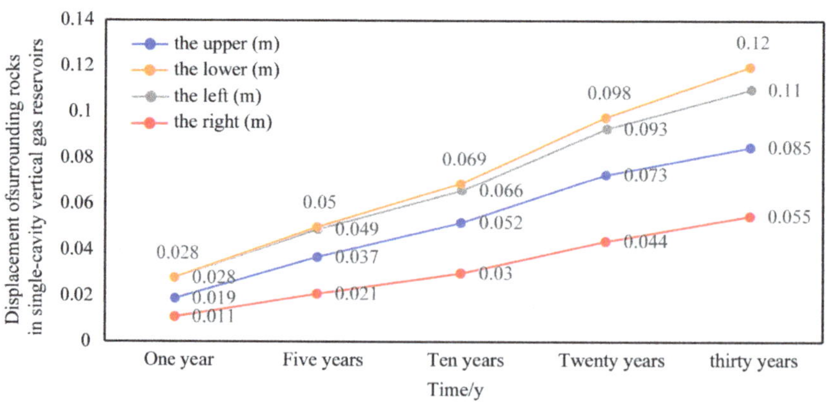

Figure 6. Plot of perimeter rock displacement versus time for a single-cavity vertical gas reservoir.

4. Discussion

The data pertaining to displacement, obtained from the numerical simulation of single-cavity vertical gas storage reservoir monitoring points, were subjected to processing using Origin 2018 software, with the objective of obtaining the rule of change.

The changes in vertical displacement at monitoring points T and B and the top and bottom plates of the cavity are shown in Figure 5, respectively.

From Figure 7, a trend can be clearly observed: as the number of mudstone interlayers increases, the displacement of the top and bottom plates of the reservoir increases. This trend reflects the relatively weaker nature of the mudstone interlayer in the surrounding rock. The increase in the number of interlayers under the same cyclic air pressure directly leads to an increase in displacement deformation.

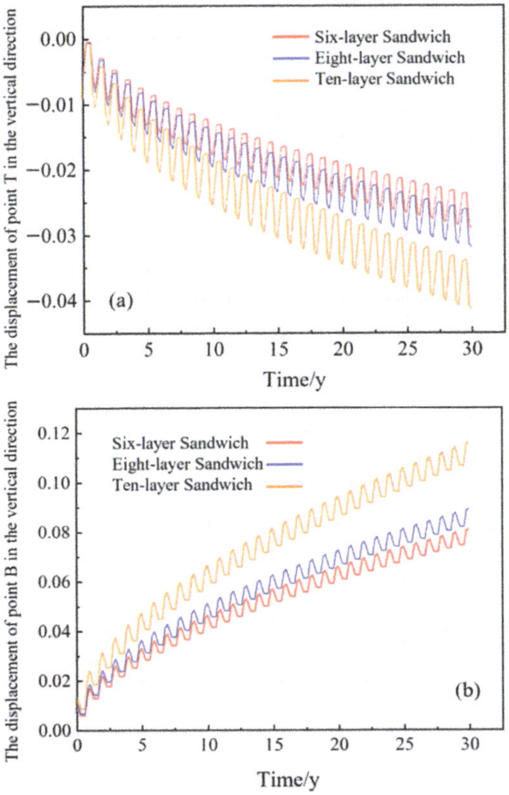

Figure 7. Vertical Displacement of Monitoring Points T and B: (**a**) The top plate; (**b**) The base plate slab.

The explanation for this phenomenon can be attributed to the role of mudstone interlayers in the reservoir structure. Mudstone interlayers typically have lower strength and higher deformability and are, therefore, more susceptible to displacement and deformation when subjected to external pressures. When the number of mudstone entrapments increases, the relative proportion of mudstone entrapment in the gas storage structure increases, and the effects of their displacement and deformation increase accordingly.

It is, therefore, imperative that the impact of the quantity of mudstone inclusions on the stability of the surrounding rock is given due consideration at all stages of the design and operational phases of a gas storage reservoir. Appropriate support and reinforcement measures are implemented to mitigate the adverse effects of mudstone interlayers on the structural stability of gas storage reservoirs, thereby ensuring the safe operation of the reservoirs. These findings have significant implications for the optimisation of reservoir operation strategies, the reduction of surrounding rock deformation, and the improvement of long-term reservoir stability. The deformation of the surrounding rock can be effectively mitigated through the implementation of a rationalised air pressure peaking strategy and the regulation of the air pressure action time, thereby enhancing the performance and safety of the gas storage reservoir.

From Figure 8, the changes in horizontal displacement show a similar pattern to the vertical displacement, which is due to the contraction of the cavity under the influence of the circulating air pressure. It is worth noting that the displacement changes on the left and right sides are nearly symmetrical, which means that the trend of changes on the left and right sides is similar when the cavity is affected. Similar to the vertical displacement, the effect of the number of interlayers on the horizontal displacement change is also consistent. When the number of mudstone inclusions is increased, it leads to greater horizontal displacement changes, which suggests a positive correlation between the number of inclusions and displacement changes.

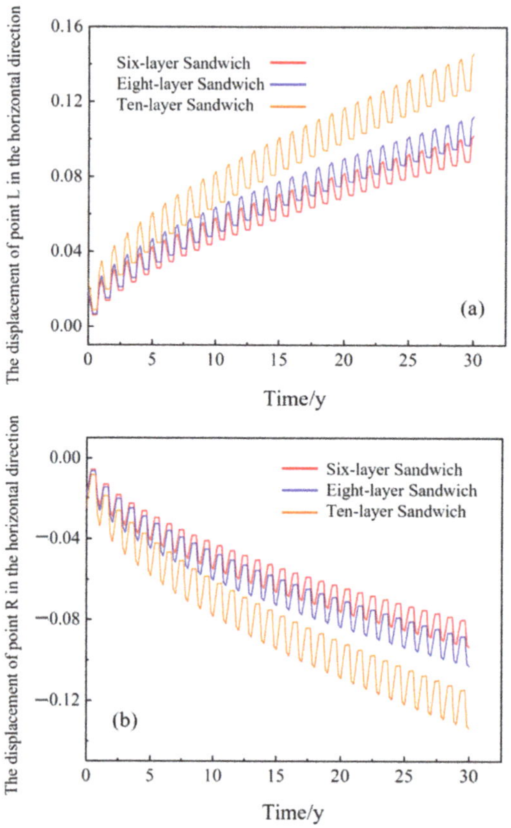

Figure 8. Horizontal Displacement of Monitoring Points L and R: (**a**) The left plate; (**b**) The right plate slab.

The volumetric shrinkage of a cavity is defined as the extent to which its volume decreases due to deformation of the surrounding rock under specific conditions. The volume change of the cavity is calculated and monitored using FLAC3D, and the resulting data are instrumental in comprehending the deformation of the cavity under disparate operational conditions, as well as in appraising the stability and performance of the gas storage reservoir. Figure 9 illustrates the variation in cavity volume shrinkage at different air pressure cycling frequencies.

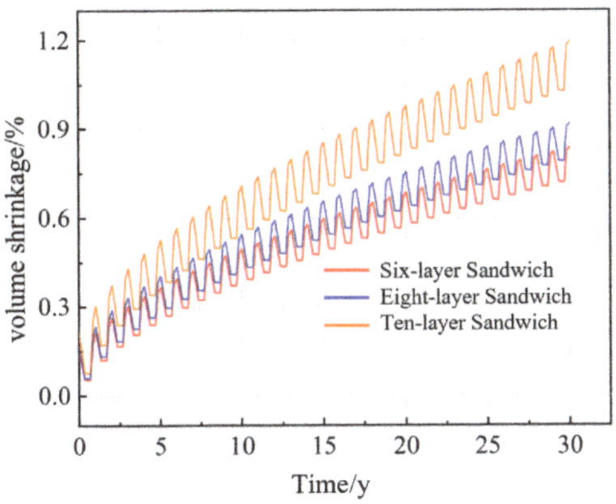

Figure 9. Volume shrinkage with time at different intervals.

As can be seen from the above figure, the volumetric shrinkage of the gas storage reservoirs all show a nearly exponential and steady increase for different numbers of mudstone interlayers. This trend suggests that the rate of the volumetric shrinkage of the reservoir does not change linearly over time and with the continued action of the circulating air pressure, but rather increases gradually at a faster rate, especially if the number of interlayers is high.

The exponential increase in volumetric shrinkage indicates that the mudstone interlayer plays a pivotal role in the long-term operation of the reservoir. The increase in the number of mudstone interlayers not only results in an increase in the displacement and deformation of the surrounding rocks, but also presents the gas storage reservoir face with greater challenges in terms of its overall stability. In particular, as the number of mudstone interlayers rises, the intricacy of the reservoir's internal configuration intensifies, and the mechanical interconnections between the layers become more pronounced. This renders the reservoir more vulnerable to deformation when subjected to pressure, and consequently, volumetric shrinkage is accelerated.

This phenomenon places higher demands on the design and management of gas storage reservoirs. In order to slow down the exponential growth of volumetric shrinkage, it is necessary to carry out fine geological investigations and mechanical analyses at the early stage of gas storage reservoir construction. Meanwhile, during the operation of the gas storage reservoir, it is necessary to regularly monitor the volume changes and perimeter rock deformation, and take timely remedial measures, such as adjusting the circulating air pressure and strengthening the support structure, in order to maintain the long-term stability and safety of the gas storage reservoir.

Figure 10 shows the plastic zone volume for a single cavity vertical gas storage. In this figure, it can be observed that the plastic zone volume is, overall, small with a maximum value of only 1 m^3. This smaller plastic zone volume may be due to the fact that the reservoirs are designed for a long service life, usually more than 30 years, and also have long injection and extraction cycles with very stable pressures in the chamber.

Furthermore, the volume of the plastic zone demonstrates a persistent augmentation with time for varying numbers of interlayers, exhibiting a consistently accelerating growth rate. This indicates that the volume of the plastic zone will increase gradually and at an accelerated rate as the reservoir is operated over time.

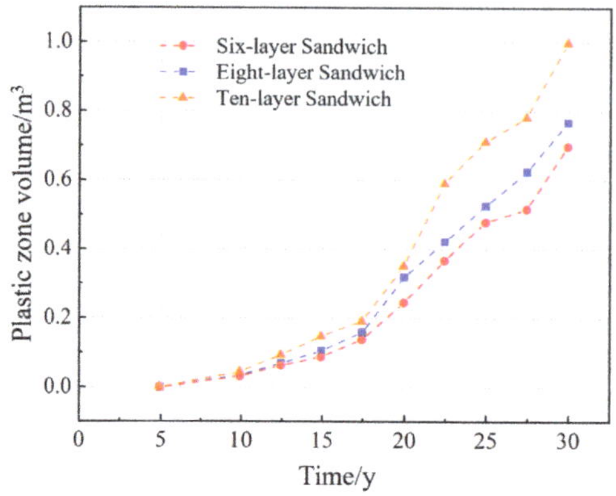

Figure 10. Plastic zone volume variation with interval time.

5. Conclusions

In view of the special geological conditions of layered salt rocks in China, and in order to deeply investigate the influence of the number of mudstone interlayers on the long-term safe and stable operation of gas storage reservoirs, this study conducted numerical simulation analyses of gas storage reservoirs containing 6, 8, and 10 layers of mudstone interlayers, respectively, and came up with the following main conclusions.

(1) As the number of mudstone interlayers increases, the vertical displacement of the surrounding rocks of the vertical single-cavity gas storage reservoir also increases. This is evidenced by the observation that the maximum settlement value at the upper boundary of the gas storage enclosure is consistently less than the maximum uplift value at the lower boundary. This asymmetry in vertical displacement is primarily attributable to discrepancies in ground stresses stemming from the disparate depths at which the top and bottom slabs are situated. The top is subjected to less geopathic stress, whereas the bottom is subjected to higher geopathic stress, which results in a greater tendency for the surrounding rock at the bottom to undergo uplift, given that it bears the weight of the reservoir. This asymmetry is a reflection of the impact of a stratigraphic depth on the distribution of ground stress, and is a principal factor contributing to the observed differences in the vertical displacement of the surrounding rocks of the gas storage reservoir.

(2) The horizontal displacement exhibits a clear symmetry. In comparison with the vertical displacement observed at the top and bottom of the surrounding rock formation, the horizontal displacement of the surrounding rock formation demonstrates a distinct symmetry. This symmetry may be attributed to the resemblance in geological stress and confining pressure within the gas reservoir, which gives rise to the horizontal deformation of the surrounding rock exhibiting analogous trends at both ends. This finding demonstrates that, when subjected to horizontal stress, an increase in the interlayer does not result in the destruction of the symmetry of horizontal displacement.

(3) The influence of circulating air pressure on the horizontal displacement of surrounding rock is consistent with its influence on vertical displacement. An increase in the number of mudstone entrapments leads to greater horizontal displacement changes. This implies that an increase in the number of interlayers will further amplify the amplitude of the horizontal displacement of the surrounding rock under cyclic gas pressure, indicating that the number of interlayers has an important influence on the stability of the gas reservoir.

(4) A change in the volume of the plastic zone was observed. It was determined that the volume of the plastic zone within the gas storage reservoir is relatively limited in scope. However, the rate of contraction for the reservoir demonstrates a notable exponential and consistent increase over time. The volume of the plastic zone continues to increase with time for different numbers of interlayers, and the rate of increase continues to accelerate.

This trend indicates that the number of interlayers has a considerable impact on the plastic deformation of the surrounding rocks of the reservoir. Consequently, an increase in the number of interlayers results in a proportional acceleration in the expansion rate of the plastic zone volume, which in turn exerts a more pronounced influence on the long-term stability of the reservoir.

Author Contributions: Conceptualization, L.L.; Methodology, X.J.; Software, J.T. and J.S.; Formal analysis, R.L.; Resources, X.Q.; Project administration, J.F. and C.Q. All authors have read and agreed to the published version of the manuscript.

Funding: This work was supported by Chongqing Science and Technology Bureau (CSTB2022NSCQ-MSX1595).

Institutional Review Board Statement: Not applicable.

Informed Consent Statement: Not applicable.

Data Availability Statement: The data presented in this study are available on request from the corresponding author. The data is not publicly available due to privacy.

Conflicts of Interest: Authors Lilong Li and Xiaolinag Quan were employed by the company Sichuan Yanjiang Yijin Expressway Co., Ltd.; Authors Xin Jiang and Cheng Qian were employed by the company Chongqing Chengtou Jiangchang Construction Co., Ltd. The remaining authors declare that the research was conducted in the absence of any commercial or financial relationships that could be construed as a potential conflict of interest.

References

1. Wanyan, Q. *Evaluation of Geological Conditions and Countermeasures for Salt Cavern Underground Gas Storage in China*; Southwest Petroleum University: Chengdu, China, 2015.
2. Guo, B.; Fang, D.; Wang, X.; Zhang, S. Development of technology of salt cavern underground natural gas storage in foreign countries. *Fault Block Oil Field* **2002**, *1*, 78–80+86.
3. Wang, Z.; Yuan, G.; Ban, F. Experimental study on the utilization of sediment space in salt cavern gas storage in butt well. *Chin. Mine Salt* **2020**, *51*, 27–30.
4. Yang, C.; Wang, T. Progress in deep earth energy storage research. *J. Rock Mech. Eng.* **2022**, *41*, 1729–1759.
5. Ren, P.; Qi, L.; Wang, W.; Jiao, Y.; Zhou, Z.; Wang, W. Current status and development trend of space utilisation in salt caverns. *Oil Gas Field Surf. Eng.* **2023**, *42*, 1–8.
6. Chen, J.; Jiang, D.; Liu, W.; Huang, G.; Fan, J. Progress of water-soluble cavity creation and comprehensive utilisation of cavities in salt mines. *China Sci. Found.* **2021**, *35*, 911–916.
7. Taheri, S.; Pak, A.; Shad, S.; Mehrgini, B.; Razifar, M. Investigation of rock salt layer creep and its effects on casing collapse. *Int. J. Min. Sci. Technol.* **2020**, *30*, 357–365. [CrossRef]
8. Zhao, K.; Li, H.; Ma, H.; Yu, H.; Li, H.; Zeng, Z.; Zheng, Z.; Chen, X.; Yang, C. Creep-fatigue characteristics of rock salt under different loading paths. *J. Pet. Sci. Eng.* **2022**, *218*, 111036. [CrossRef]
9. Ma, L.; Wang, Y.; Wang, M.; Xue, B.; Duan, L. Mechanical properties of rock salt under combined creep and fatigue. *Int. J. Rock Mech. Min. Sci.* **2021**, *141*, 104654. [CrossRef]
10. Ozsen, H.; Ozkan, I.; Sensogut, C. Measurement and mathematical modelling of the creep behaviour of Tuzköy rock salt. *Int. J. Rock Mech. Min. Sci.* **2014**, *66*, 128–135. [CrossRef]
11. Wang, J.; Liu, X.; Huang, M.; Yang, X. Analysis of axial creep of salt rock under low frequency cyclic loading by Burgers model. *Rock Soil Mech.* **2014**, *35*, 933–942.
12. Jiang, D.; Fan, J.; Chen, J.; Yang, C.; Chui, Y. Effect of intermittent fatigue test on fatigue characteristics of salt rock. *J. Geotech. Eng.* **2016**, *38*, 1181–1186.
13. Fan, J.; Tang, L.; Chen, J.; Yang, Z.; Jiang, D. Creep fatigue constitutive model of salt rock based on hardening parameters. *Rock Soil Mech.* **2023**, *44*, 1271–1283.
14. Li, Z.; Jiang, D.; Fan, J.; Chen, J.; Liu, W.; Wu, F.; Du, C.; Kang, Y. Experimental study on fatigue characteristics of salt rock at triaxial interval. *Rock Soil Mech.* **2020**, *41*, 1305–1312.
15. Chui, Y.; Jiang, D.; Du, F.; Chen, J.; Ren, S.; Fan, J. Experimental study on acoustic emission characteristics of salt rock interval fatigue. *J. Cent. South Univ. (Nat. Sci. Ed.)* **2017**, *48*, 1875–1882.
16. Li, X.; Sun, X.; Xu, Q.; Liu, B. Experimental study on fatigue characteristics of rock salt under the influence of time interval. *J. Undergr. Space Eng.* **2017**, *13*, 606–611.
17. Hou, B.; Shangguan, S.; Niu, Y.; Su, Y.; Yu., C.; Liu, X.; Li, Z.; Li, J.; Liu, X.; Zhao, K. Unique properties of rock salt and application of salt caverns on underground energy storage: A mini review. *Energy Sources Part A Recovery Util. Environ. Eff.* **2024**, *46*, 621–635. [CrossRef]

18. Yu, H.; Ji, W.; Hou, J.; Shang, H.; Wan, J.; Ma, H. Stability study of a compressed air energy storage power station with shallow buried salt caverns. *Electr. Surv. Des.* **2024**, *3*, 69–75.
19. Liang, W.; Tang, L.; Li, X.; Chen, G.; Jiang, D. Stability analysis of near-field perimeter rock for salt cavern pressurised gas storage reservoirs. *J. Undergr. Space Eng.* **2023**, *19*, 698–706.
20. Fu, X.; Xia, Y.; Xu, Z.; Yuan, G.; Ban, F. Numerical simulation of the stability of compressed air energy storage cavities in salt caverns. *J. Undergr. Space Eng.* **2018**, *14*, 930–934.
21. Liu, W.; Zhang, Z.; Chen, J.; Jiang, D.; Wei, F.; Fan, J.; Li, Y. Feasibility evaluation of large-scale underground hydrogen storage in bedded salt rocks of China: A case study in Jiangsu province. *Energy* **2020**, *198*, 117348. [CrossRef]
22. Zhang, Z.; Liu, W.; Guo, Q.; Duan, X.; Li, Y.; Wang, T. Tightness evaluation and countermeasures for hydrogen storage salt cavern contains various lithological interlayers. *J. Energy Storage* **2022**, *50*, 104454. [CrossRef]
23. Yang, C.; Wang, T.; Chen, H. Theoretical and Technological Challenges of Deep Underground Energy Storage in China. *Engineering* **2023**, *25*, 168–181. [CrossRef]
24. Wang, T.; Ding, Z.; He, T.; Xie, D.; Liao, Y.; Chen, J.; Zhu, K. Stability of the horizontal salt cavern used for different energy storage under varying geological conditions. *J. Energy Storage* **2024**, *84*, 110817. [CrossRef]
25. Lankof, L.; Tarkowski, R. GIS-based analysis of rock salt deposits' suitability for underground hydrogen storage. *Int. J. Hydrogen. Energy* **2023**, *48*, 27748–27765. [CrossRef]
26. Bai, X. *Stability Analysis of Horizontal Oil Storage Groups in Stratified Salt Rock*; Chongqing University: Chongqing, China, 2022.
27. Wang, Z.; Li, Y.; Zhang, L.; Chen, L. Engineering geological conditions and feasibility analysis of thin bedded salt rock underground gas storage. *J. Eng. Geol.* **2015**, *23*, 148–154.
28. Chen, B.; Shen, X.; Wanyan, Q.; Sun, L. Geological condition evaluation of Pingdingshan salt cavern gas storage. *Sci. Technol. Guide* **2016**, *34*, 135–141.
29. Chen, J.; Tong, J.; Rui, Y.; Cui, Z.; Pu, Y.; Du, J.; Derek, B.A. Step-path failure mechanism and stability analysis of water-bearing rock slopes based on particle flow simulation. *Theor. Appl. Fract. Mech.* **2024**, *131*, 104370. [CrossRef]
30. Chen, J.; Chen, J.; Rui, Y.; Pu, Y. Joint Inversion of AE/MS Sources and Velocity with Full Measurements and Residual Estimation. *Rock Mech. Rock Eng.* **2024**, *57*, 7371–7386. [CrossRef]
31. Rui, Y.; Chen, J.; Chen, J.; Qiu, J.; Zhou, Z.; Wang, W.; Fan, J. A robust triaxial localization method of AE source using refraction path. *Int. J. Min. Sci. Technol.* **2024**, *34*, 521–530. [CrossRef]
32. Wang, Z.; Wang, Q.; Zhang, L. Application of CSAMT and CYT combined detection method in the construction of deep salt cavern gas storage. *J. Eng. Geol.* **2016**, *24*, 142–149. [CrossRef]
33. Wang, Y. *Study on the Stability and Collapse Disaster of Interlayer of Pingdingshan Salt Cavern Gas Storage*; Zhengzhou University: Zhengzhou, China, 2019.
34. Xiao, Y.; Xu, M. FLAC(3D) and ANSYS modeling characteristics and their applications. *Hydropower Technol. Gansu* **2011**, *47*, 11–12.
35. Zhang, G.; Wang, L.; Wu, Y.; Li, Y.; Yu, S. Failure mechanism of bedded salt formations surrounding salt caverns for underground gas storage. *Bull. Eng. Geol. Environ.* **2017**, *76*, 1609–1625. [CrossRef]
36. Jiang, D.; Wang, Y.; Liu, W.; Li, L.; Qiao, W.; Chen, J.; Li, D.; Li, Z.; Fan, J. Construction simulation of large-spacing-two-well salt cavern with gas blanket and stability evaluation of cavern for gas storage. *J. Energy Storage* **2022**, *48*, 103932. [CrossRef]
37. Liu, W.; Chen, J.; Jiang, D.; Shi, X.; Li, Y.; Daemen, J.J.K.; Yang, C. Tightness and suitability evaluation of abandoned salt caverns served as hydrocarbon energies storage under adverse geological conditions (AGC). *Appl. Energy* **2016**, *178*, 703–720.
38. Yang, C.; Zhou, H.; Li, Y. *Disaster Mechanism and Protection of Large Salt Cavern Gas Storage Group*, 3rd ed.; Science Press: Beijing, China, 2014.
39. Zhang, G.; Wang, Z.; Liu, W.; Luo, L.; Dong, J. Study on stability of key roof of compressed air energy storage in horizontal salt cavern. *Rock Soil Mech.* **2021**, *42*, 800–812.
40. Xu, S.; Zhang, Q.; Li, X.; Gai, Z.; Wen, P.; Wei, Q. Research on optimization design of injection and production operation parameters of Jintan Salt Cave gas storage. *Oil Drill. Prod. Technol.* **2020**, *42*, 490–496.
41. Jing, W.; Yang, C.; Li, L.; Yuan, Z.; Shi, X.; Li, Y. Sensitivity analysis of factors affecting shrinkage risk of salt cavern gas storage. *J. Rock Mech. Eng.* **2012**, *31*, 1804–1812.
42. Tang, Z.; Jiang, D.; Chen, J.; Fan, J.; Liu, W. Comprehensive evaluation on stability of salt cavern storage with small well spacing and double well spacing. *J. Undergr. Space Eng.* **2021**, *17*, 1997–2006.

applied sciences

MDPI

Article

Rock Slope Instability Mechanism Induced by Repeated Mining in Mountain Mining Areas

Rong Luo [1,2,3], Guangyue Li [1], Lu Chen [2,*], Ling Zeng [2], Ke Pei [2] and Xiangxi Yu [2]

1 School of Resource Environment and Safety Engineering, University of South China, Hengyang 421001, China; luorong-lr@stu.usc.edu.cn (R.L.)
2 School of Civil Engineering, Changsha University of Science & Technology, Changsha 410110, China; peike@stu.csust.edu.cn (K.P.); 17740683572@163.com (X.Y.)
3 Periodical Press, Changsha University of Science & Technology, Changsha 410110, China
* Correspondence: chenlu@csust.edu.cn

Abstract: When mineral resources are extracted using underground mining methods in hilly regions, landslides or slope failures can be induced frequently. In this study, slope collapse disasters in mountain mining areas were analyzed. The model test and numerical simulation of the slope impacted by repeated mining were carried out. The crack evolution and failure process were analyzed to reveal the instability mechanism. The results show that the rock mass would topple to the inside of the slope first, when the subsidence of overlying rock was induced by the mining of the upper coal seam. When repeated mining was performed in the lower coal seam, the mining induced macro-cracks that could connect with natural fissures, inducing the outward displacement of the slope. Then, the rock mass at the foot of the slope has to bear the upper load, which is also squeezed out by the collapsed rock mass, forming the potential slip zone. Finally, the instability is caused by the shear slip of the slope toe rock mass. Therefore, the instability evolution of the slope under underground repeated mining disturbance can be divided into four stages as follows: roof caving and overlaying rock subsidence, joint rock toppling, fracture penetration, and slope toe shearing and slope slipping.

Keywords: repeated mining; crack evolution; shear slip; slope instability

1. Introduction

In mining areas, subsidence or landslides will be caused by underground mining activities [1,2]. In particular, many mineral resources are buried in mountainous areas in the southwest of China. When the underground mining method is used to excavate the mineral resource, mountain cracking and collapse disasters of the slope can be induced easily [3,4]. Therefore, the characteristics of deformation and crack evolution are analyzed, and the influence factors of underground mining on mountain stability are investigated [5–7].

Usually, scholars focus on the investigation of the engineering conditions of the mining-induced landslide disasters [8–10]. And the model test or numerical simulation is performed to analyze the causes of instability [11–13]. Salmi et al. [14] established a discontinuous numerical model to investigate the mechanisms of the underground-mining-induced landslide that occurred in Nattai North, Australia. The results show the shear and compressive failure will be induced at the escarpment base, then the forward sliding and toppling of the rock column can be initiated. Arca et al. [15] analyzed the landslide-related factors to identify landslide susceptibility of mined disturbed areas in Kozlu, Turkey. Then, the landslide susceptibility map was established. Cui et al. [16] identified the key features and contributing factors for mining-induced landslides, and the dynamic mechanism was analyzed using geological field surveys, numerical simulation and theoretical analysis. Yang et al. [17] used the interferometric synthetic aperture radar technique and landslide and fissure spatiotemporal statistics to investigate the spreading process of the slow subsidence caused by underground coal mining. Then, the mining impact on the occurrence of shallow loess landslides was revealed. Lai et al. [18] considered the soft–hard interlayer structure of the slope disturbed by underground mining, and centrifuge model tests and 3D discrete element mode were performed to investigate the Madaling landslide. The displacement of the deep rock mass is proposed to serve as the early-warning criterion for mining landslides. Liu et al. [19] designed a two-stage monitoring system to identify the potential landslide area and record multiple real-time data regarding an impending landslide. Zhong et al. [20] investigated the formation mechanism of a landslide occurring in coal-mining subsidence area at the Tanshan Coal Mine in Ningxia. The stress and deformation characteristic during the failure process were discussed, which was divided into four stages: slope creep, slope deformation, landslide movement and landslide accumulation.

The previous studies show that the collapse of slopes located in hilly regions is easily induced by underground mining. Researchers focus on the description of the collapse phenomenon, offering critical insight into the mechanism of the landslide [21,22]. However, the key factor of repeated mining has inadvertently been ignored. Some mining landslides revealed that multiple coal seams were mined in the collapse area [23]. To reveal the instability mechanism induced by the repeated mining, the mined induced collapse disasters that occurred in the Pusa Village were analyzed. The model test and numerical simulation of slope disturbed by underground repeated mining were established. The failure process was analyzed to investigate the characteristics of deformation, crack evolution and sliding.

2. The Background of the Landslide

2.1. Overview of the Landslide

Abundant coal resources are buried in the Laoyingyan Mountain, Pusa Village, Zhangjiawan Town, Nayong County, Guizhou Province, which have been extracted using underground mining methods for many years, forming a complex goaf. As shown in Figure 1, due to the combined effect of the primary structural plane, mining and weathering, deep cracks occurred on the mountaintop. The stability of the mountain slope decreases significantly. At about 10:40 PM on 28 August 2017, a large slope collapse disaster occurred. The length and width of the collapse were about 840 m and 410 m, respectively. The volume was about 6×10^5 m^3. The surrounding woodland and farmland were destroyed. Some houses located in the lower part of the village were buried, resulting in a total of 35 deaths [8].

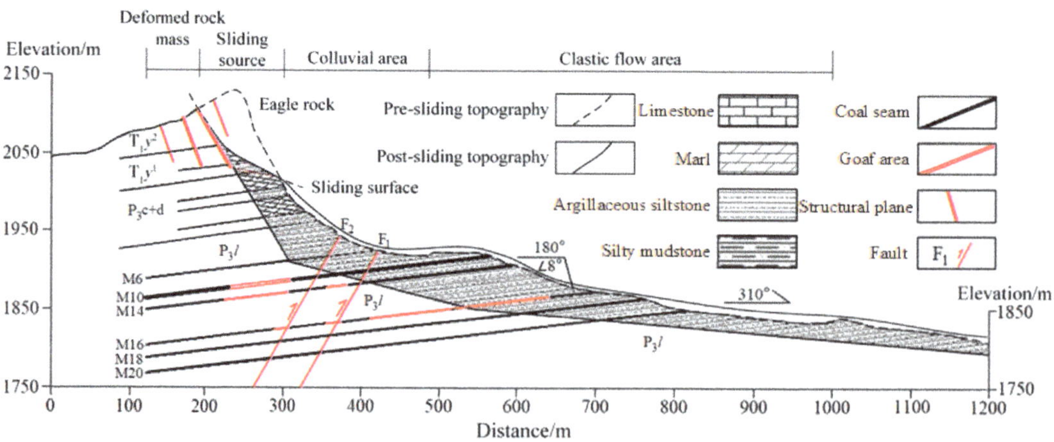

Figure 1. Geological profile of the collapse area.

2.2. Engineering Geological Condition

2.2.1. Stratigraphic and Lithologic Structure

The strata in the collapse area include the following: clay and sandy clay in the Quaternary; in the Triassic Yelang Formation (T1y), the upper part is gray-brown medium-thick limestone with marl, and the lower part is gray sandy mudstone with siltstone; in the Triassic Changxing–Dalong Formation (P_2c + d), the upper part is marl with shale, and the lower part is gray medium-thick-layered and thin-layered limestone; the Permian Longtan Formation (P_3l), which is the coal measure strata, includes gray-green-deep gray silty mudstone, gray-deep gray argillaceous siltstone and coal seam. Some rock mechanics tests are performed; the mechanical parameters of rock are listed in Table 1.

Table 1. Mechanical properties of rock.

Lithology	Density/(g/cm^3)	Strength/MPa	Young's Modulus/GPa
Limestone	2.67~2.72	57.6~85.6	9.51~36.7
Marl	2.62~2.69	43.3~64	6.47~30
Silty mudstone	2.54~2.65	33.9~56.2	5.18~14
Coal	1.3~1.4	4.65~20	0.76~3.5
Muddy siltstone	2.58~2.66	22.4~47.5	2.15~12

2.2.2. Geological Structure

The strata in the collapse region has a monoclinic structure. The dip direction is 138–187°, and the dip angle is 7–10°. The occurrence of a coal seam is consistent with that of the strata. Affected by lithology composition, the limestone and marl located at the upper of slope are relatively hard in strength, and the lower mudstone and siltstone are low, forming the upper hard and lower soft structure. Therefore, the slope rock mass is characterized as soft and hard interbedded. Furthermore, the mining area is disturbed by three faults, F1, F2 and F3 [24]. F1 is a normal fault, which has an NE dip direction with an angle of 155–167° and a dip angle of 63–70°. F2 is a reverse fault, dipping to southeast, with an angle of 70–75°. F3 is a normal fault, dipping to the northeast, with an angle of 75–80°. Among them, F1 and F2 are developed in the middle and front of the slope at the lower part of the cliff in the collapse area, cutting the coal seam in the mining area. F3 is located about 500 m to the east of the collapse area. The collapse area cannot be affected by those faults. However, the rock mass structure at the top of the eagle rock is broken, and the joint fissures are developed, forming the four groups of structural planes. The depth is 50–80 m, and the width is 2–3 m. Those are considered as the seepage channel. The surface water can flow into the underground goaf. The strength of the rock will deteriorate, impacting the stability of the slope.

2.3. The Underground Mine

Six minable coal seams are buried under slope, which are named M6, M10, M14, M16, M18 and M20, from top to bottom. The dip angle of the coal seams is between 7° and 12°. M10, M14 and M16 are mainly excavated in the collapse area, and the average thickness of the coal seams is 2.12, 1.23 and 1.48 m, respectively. The roof and floor of the coal seams are argillaceous siltstone. The mining area under the mountain is about 0.96 km^2. According to the characteristics of coal seams, the longwall retreating mining method is used. Before 2010, M16 was mainly excavated. The goaf of M16 was located at the front end of the slope, which has little effect on the deformation of the slope. The M10 and M14 coal seams were mainly excavated from 2011 to the time before the occurrence of collapse. The goafs of the M10 and M14 coal seams were located directly below the slope. When the M10 and M14 coal seams were excavated, a macro-fracture was caused at the top of the mountain. Then, the mountain began to deform violently.

3. The Model Test of the Collapse

3.1. Model Design

A frame and a non-contact displacement monitoring system are used in the model test. The maximum dimension of the model is 1300 mm × 1000 mm × 150 mm. The monitoring system is composed of an image acquisition and analysis system. A charge-coupled device camera was used to record the digital images of the model. And the deformation characteristic is analyzed using the digital speckle correlation method. By tracking the same pixels occurring in the digital images recorded before and after deformation of the model, the displacement and deformation of the monitored area can be calculated [25]. In this study, the charge-coupled device camera (CCD) and the image storage device were used. The maximum resolution and speed of the CCD is 2456 pixels × 2058 pixels and 75 M/s, respectively. During the mining process, a resolution of 1280 pixels × 960 pixels and a 5 Hz acquisition rate were set to capture the deformation digital images.

As shown in Figure 2, considering the similarity ratio and experimental feasibility, the simplified similarity model was established based on the engineering and mining conditions of the Laoyingyan Mountain. The mechanical and geometrical similarity ratios of the similarity model were designed as follows: C_L = 1:300, C_S = 1:330 and C_ρ = 1:1.1 ($C_{L,S}$ and C_ρ are the similarity constant for geometry, stress and density, respectively). Details of the material similarity for the model test and similarity criteria can be found in our previous publications [26,27]. The mechanical parameters of stratum are shown in Table 2. Two mineable coal seams were designed. The length and height mined of the upper coal seam were 60 cm and 2.5 cm, respectively. The length mined of the lower coal seam was also 60 cm, and the mining height was 1.5 cm. A speckled pattern was sprayed on surface of the model. The coal mining was carried out from the inside of the slope to the free face continually. In this study, a positive value indicates that the horizontal displacement of the slope is towards the free surface, and the vertical displacement is downward. The six characteristic states of the model are analyzed. Both the upper and lower coal seams are divided into three stages. The sequence is S1 → S2 → S3 → X1 → X2 → X3, and the interval length was 20 cm.

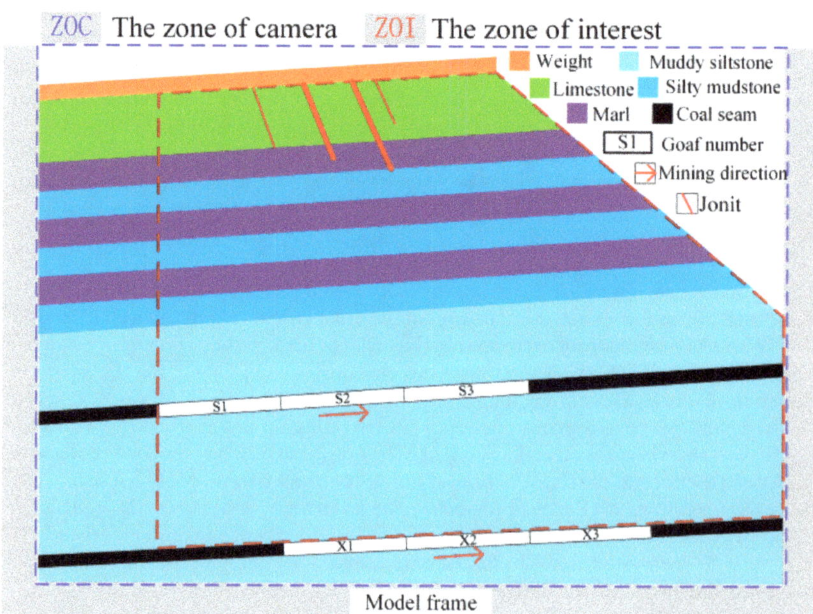

Figure 2. The similarity model.

Table 2. Mixture ratio and mechanical parameters of similarity material.

Lithology	Mixture Composition					Strength (MPa)	Young's Modulus (MPa)
	Sand (kg)	Barite Powder (kg)	$CaCO_3$ (kg)	Gypsum (kg)	Water (kg)		
Limestone	22.91	9.82	0.16	1.47	3.82	0.227	29.6
Marl	11.50	4.93	0.33	0.49	1.92	0.165	20
Silty mudstone	12.81	3.20	0.40	0.40	1.33	0.157	15.4
Muddy siltstone	54.36	6.04	1.81	1.21	3.02	0.115	6.6
Coal seam	2.82	0.31	3.14	0.00	1.15	0.014	2.4

3.2. The Displacement Induced by the Mining

3.2.1. The Displacement Induced by the Mining of the Upper Coal Seam

The displacement of the model induced by the mining of S1 and S2 in the upper coal seam is shown in Figure 3. When S1 and S1 were mined, a slight displacement was caused. The horizontal displacement mainly occurred in the roof caving zone and slope toe. The vertical displacement mainly occurred at the middle of the roof of the working face, and the maximum value was 5.5 mm, which indicates that the mining of S1 and S2 had a great influence on the deformation of the immediate roof and had little influence on the slope.

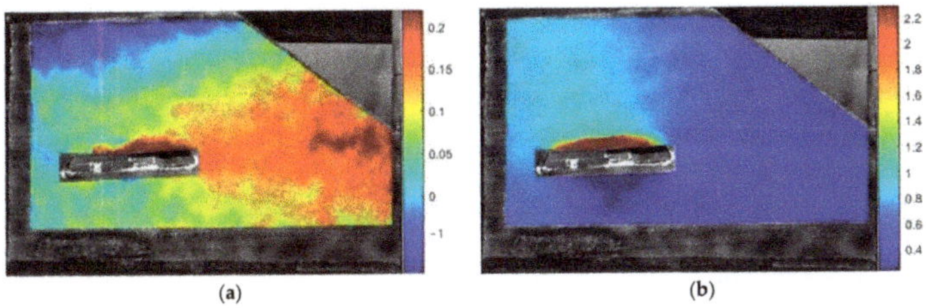

Figure 3. The displacement induced by the mining of S1 and S2. (**a**) Horizontal displacement; (**b**) vertical displacement.

The displacement of the model induced by the mining of S3 is shown in Figure 4. The significant displacement was caused in the upper rock mass of the slope. The boundary of mining-induced fracture could be clearly identified. The vertical displacement mainly occurred above the goaf. The closer to the goaf, the greater the displacement. The settlement value of the model top was between 5 mm and 7 mm, which indicates that the mining influence had been transmitted from the roof to the top of the model through the overburden rock. However, the vertical displacement of the slope surface was still small. The horizontal displacement mainly occurred in the upper part of the goaf and the slope surface. The deformation boundary was generated at the middle of the goaf, and the displacement direction was different in that area. This indicates that the roof caved into the goaf, and the collapsed rock mass squeezed the slope, which moved towards the free surface. Therefore, the stability of the mountain was significantly reduced.

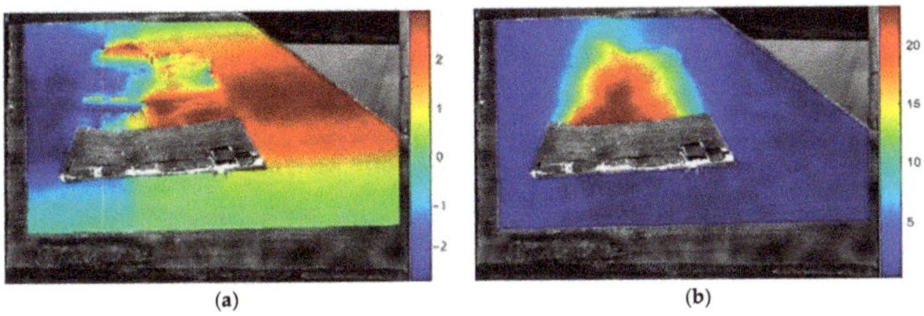

Figure 4. The displacement induced by the mining of S3. (**a**) Horizontal displacement; (**b**) vertical displacement.

3.2.2. The Displacement Induced by the Mining of the Lower Coal Seam

The displacement of the model induced by the mining of X1 is shown in Figure 5. As the model in this stage is very different from the initial state, the initial state of mining of the lower coal seam is taken as the reference frame in the digital speckle analysis. When X1 was mined, the displacement of the roof of the lower coal seam was relatively small. However, the obvious stratification was caused in the overlying strata. The impact zone was larger than the mined range. This indicates that the smaller subsidence of the roof of the lower coal seam may lead to a larger displacement of the upper rock layer.

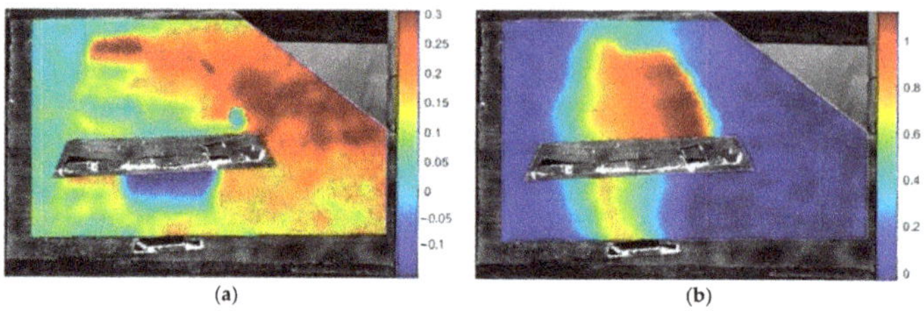

Figure 5. The displacement induced by the mining of X1. (**a**) Horizontal displacement; (**b**) vertical displacement.

The displacement of the model induced by the mining of X2 is shown in Figure 6. Affected by the large-area caving of the lower coal seam roof, the deformation of the slope increased obviously, which resulted from the mining of X2. In the horizontal direction, significant displacement occurred at the position on the right side of the upper roof and the middle of the lower coal seam, and the maximum value was 3.7 mm. This indicates that the broken rock layer was continuously toppled outward with the excavation of the lower coal seam, and the slope surface rock mass was squeezed at the same time. In the vertical direction, significant displacement occurred at the position on the middle of the roof of the lower coal seam. Obvious displacement was also caused at the position on the right side of the upper roof. At this time, the influence of underground excavation on the top of the slope was slight, and the settlement value was less than 2 mm.

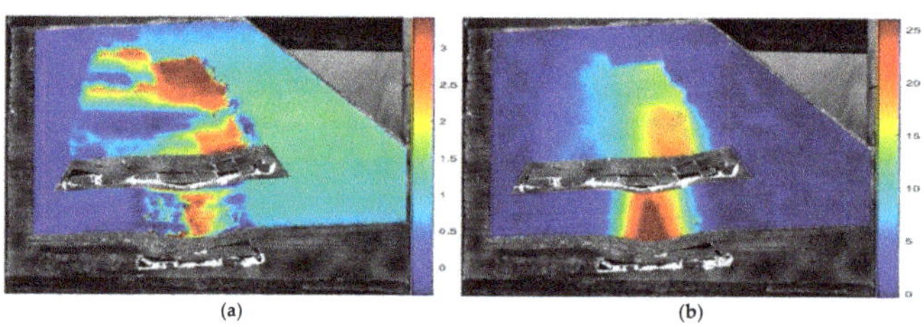

Figure 6. The displacement induced by the mining of X2. (**a**) Horizontal displacement; (**b**) vertical displacement.

The displacement of the model induced by the mining of X3 is shown in Figure 7. When the lower coal seam was completely excavated, part of the slope displacement value of the model failed to display, as the rock had been seriously broken, and the speckle pattern on the surface was damaged. Hence, the displacement cannot be identified using the digital speckle correlation method. It can be seen from the existing data that the significant horizontal displacement was caused in the middle of the broken rock mass, and the maximum value reached 19.8 mm. The slope toppled to the free surface, and the potential sliding surface was penetrated. Significant vertical displacement was also induced at the right side of the broken rock mass. The slope had been in a critical state. Further underground mining or other disturbances may induce landslides.

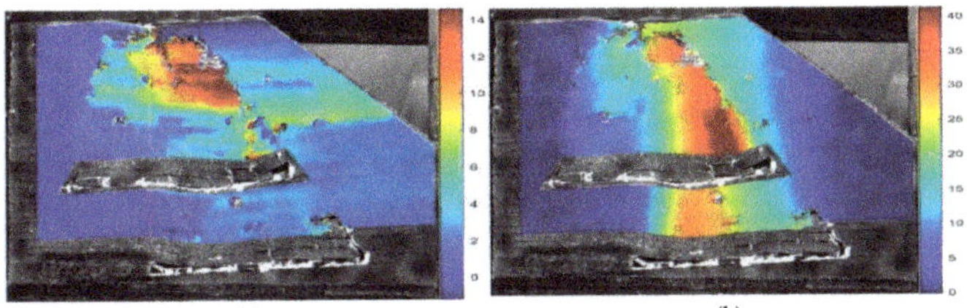

Figure 7. The displacement induced by the mining of X3. (a) Horizontal displacement; (b) vertical displacement.

4. Numerical Simulation Analysis of the Collapse

4.1. Scheme of Numerical Simulation

4.1.1. The Numerical Model

Based on the engineering geological condition, the repeated-mining-disturbed numerical model was established using the PFC 2D (Particle Flow Code 5.0) discrete element software, as shown in Figure 8. The complete process of crack initiation and propagation induced by the excavation disturbance can be simulated [28,29]. The length of the model was 500 m, and the width was 345 m. The geological characteristics of the slope were designed basically. Two minable coal seams buried under the slope were established. The thickness of the upper and lower coal seams was 7 m and 4 m, respectively. The limestone, marl silty mudstone and muddy siltstones were designed as the overburden. The dip angle of the rock stratum in the numerical model was 10°. Four joint fissures were set at the top of the model to simulate the structural plane of the mountain prototype. The fissure depth changed from 50 m to 80 m, and the dip angle was about 70°. The horizontal displacement of the left and right boundaries of the model was limited; the vertical displacement of model at the potion of the bottom boundary was also limited. The parallel bond model was used for simulating the mechanical behavior of the coal and rock mass. If the tensile-strength limit is exceeded, then the bond in tension is broken. If the bond has not broken in tension, then the shear-strength limit is enforced, and the stiffness will reduce with the breaking of the bond. Hence, this model is considered as a suitable bond model for the rock materials [30]. The parameters in the numerical model were listed in Table 3, which were determined by referring to the previous research and parameter calibration test [24,31]. In the numerical model, the unconfined compression strength of limestone, marl, silty mudstone, coal and muddy siltstone is about 69 MPa, 54 MPa, 51 MPa, 4.7 MPa and 38 MPa, respectively, according to the parameters listed in Table 3.

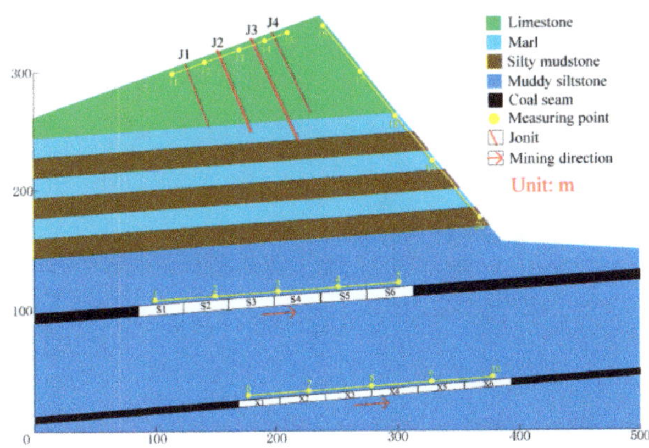

Figure 8. The numerical model.

Table 3. The parameters in the numerical model.

Parameter	Limestone	Marl	Silty Mudstone	Coal Seam	Muddy Siltstone
Porosity (%)	0.1	0.1	0.1	0.1	0.1
Particle density (kg/m^3)	2700	2650	2600	1350	2560
Effective modulus/GPa	10	8	7	1.84	5.6
Bond normal-to-shear stiffness Ratio ($\bar{k}^*$)	1	1	1	1	1
Bond effective modulus/GPa	10	7	5	1.84	4.6
Bond tensile strength/MPa	15	12.5	10.2	3.15	8
Bond cohesion/MPa	25	18.1	15.4	5.53	14.45
Bond friction angle/°	38	35	33.8	18	30.1

4.1.2. Excavation and Monitoring

The upper and lower coal seams were excavated, respectively. Each coal seam was mined from the inside of the slope to the outside of the slope. The length of the work face was 240 m. In order to investigate the deformation response law of the layered rock slope under excavation disturbance, a total of 20 monitoring points were laid out as shown in Figure 8. On the roof of the lower and upper coal seams and the top and surface of slope were set five monitoring points, respectively. When the coal seam was mined for 40 m, the crack and displacement were recorded and analyzed. The sequence was marked as S1 → S2 → S3 → S4 → S5 → S6 → X1 → X2 → X3 → X4 → X5 → X6.

4.2. Analysis of the Failure Process

4.2.1. The Mining of the Upper Coal Seam

Figure 9 exhibits the fracture evolution of the slope during the excavation of the upper coal seam. When the upper coal seam was mined for 40 m, the separation was caused in the immediate roof. Due to the large span of the working face, the immediate roof first collapsed after the mining length of 80 m. Some separations were caused in the overlying strata. After the mining of S3, the overlying strata above the working face collapsed as an inverted trapezoid, and some separation fissures were caused between the marl and mudstone. At this time, the rock mass cut by the deep structural plane at the top of the slope toppled towards the inside slope, resulting in the closure of J3. When S4 was mined, the caving area continued to increase, and the mining-induced cracks on the left boundary also expanded upward. Tensile cracks appeared at the bottom of the rock bridge on the top of the slope, and J2 gradually closed. When S5 was mined, a large displacement had been caused in the overlying strata, and the marl and mudstone above the working face collapsed completely. The mining-induced cracks had been transmitted upward to the surface, and the joint rock toppled to the trailing edge of the mountain. Some cracks were induced at the foot of the slope. After the mining of S6, the range of the fracture zone above the goaf was further expanded. The slope stability was significantly reduced due to the oblique tension crack in the middle of the slope.

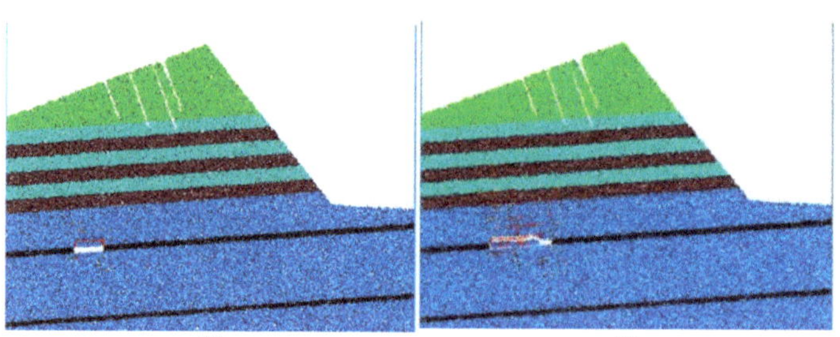

(a) (b)

Figure 9. *Cont.*

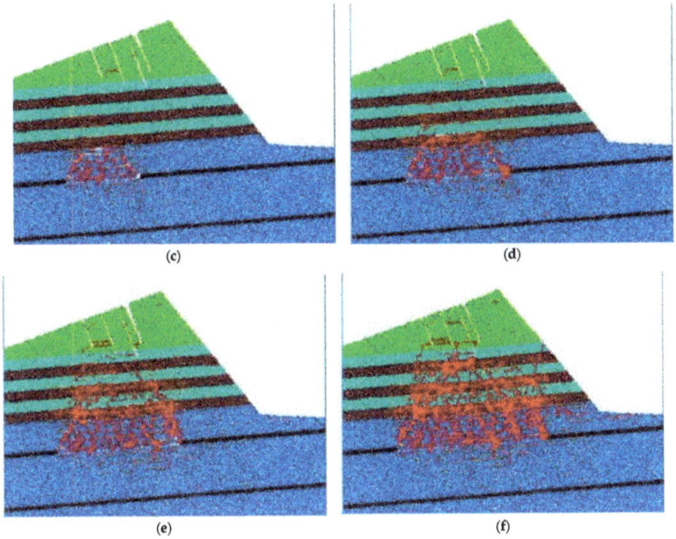

Figure 9. The fracture evolution during the mining process of the upper coal seam. (**a**) S1; (**b**) S2; (**c**) S3; (**d**) S4; (**e**) S5; (**f**) S6.

4.2.2. The Mining of the Lower Coal Seam

Figure 10 exhibits the fracture evolution of the slope during the excavation of the lower coal seam. After the excavation of X1, the separation was caused in the roof. However, the mining has a slight effect on the roof of the upper coal seam. When X2 was mined, the immediate roof collapsed, and the main roof subsided. The mining fracture zone expanded upward, resulting in a small settlement of the mining zone of the upper coal seam. After the mining of X3, a large-area roof of the working face caved, and the fracture zone had expanded to the floor of the upper coal seam. After the mining of X4, the overlying strata of the working face sank, which led to obvious subsidence in the upper mining zone, and the joint rock on the top continued to move into the slope. At this time, the surface rock mass at the foot of the slope was uplifted. After the mining of X5, the influence range of the goaf was further expanded, and the rock mass at the bottom of J1 was cracked. Furthermore, the rock mass at the upper part of the slope had a tendency to slide downward. The potential sliding surface was formed. However, part of roof at the right boundary of the goaf had not collapsed completely, which could support the overlying rock temporarily. After the mining of X6, the rock mass above the working face collapsed completely, the rock mass at the inside of the slope was broken extremely and step-shaped subsidence pits were caused on the joint surface of the slope top. The rock mass at the foot of the slope was squeezed out, and the slope body slipped slightly in the direction of the free surface; the potential sliding surface had essentially penetrated.

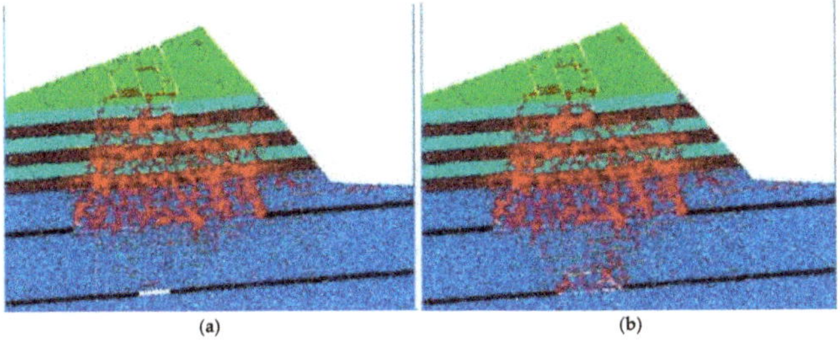

Figure 10. *Cont.*

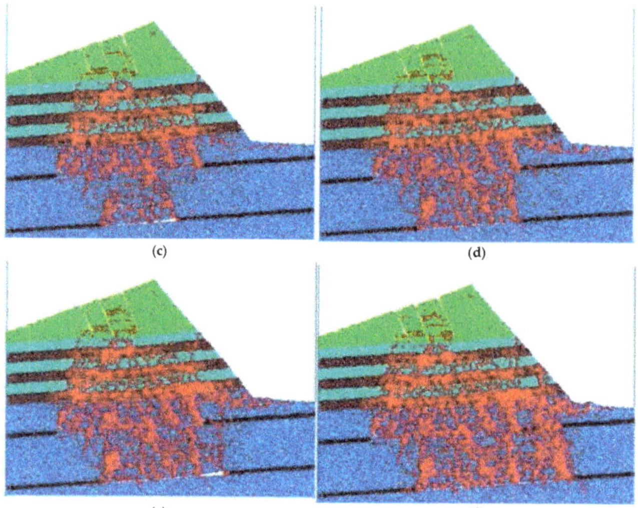

(c) (d) (e) (f)

Figure 10. The fracture evolution during the mining process of the lower coal seam. (**a**) X1; (**b**) X2; (**c**) X3; (**d**) X4; (**e**) X5; (**f**) X6.

4.3. Slope Deformation Induced by Mining

4.3.1. The Displacement Characteristic of the Slope

The displacement characteristic of the slope during the mining process of the upper coal seam is shown in Figure 11. After the mining of S3, the overlying rock of the working face caved, and the settlement was characterized as an arch shape. The rock above J3 was toppled into the slope. After the mining of S6, a huge arch-shaped settlement area was formed in the middle of the slope. At this time, the vault settlement area had expanded upward to the bottom of the joint rock, resulting in the rock at the top of the slope dumping into the slope. When X3 was mined, the bending deformation was induced in overlying rock above the working face. The collapsed rock mass induced by the upper coal seam mining area continued to sink. At this time, the joint rock continued to pour into the slope. The strata of the slope above the goaf were borne by the rock mass at the toe of the slope. After the mining of X6, the overlying strata above the lower coal seam completely collapsed, and the impacted area had expanded to the foot of the slope. The support capacity of the slope toe was weakened, which failed to bear the load of the upper rock mass. The rock mass at the foot of the slope was cut off, forming a potential slip surface, and the upper rock mass began to slip towards the free surface.

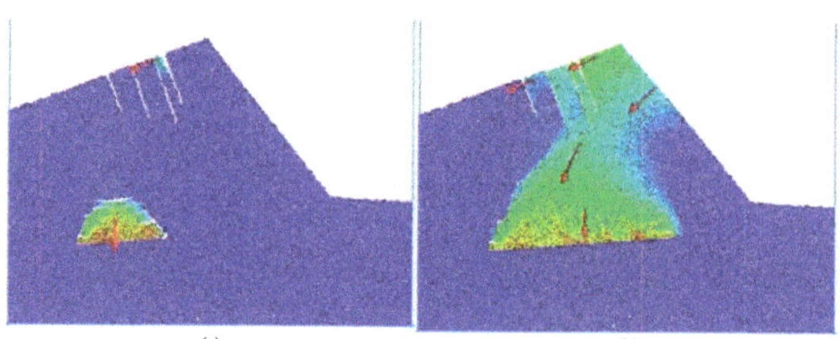

(a) (b)

Figure 11. *Cont.*

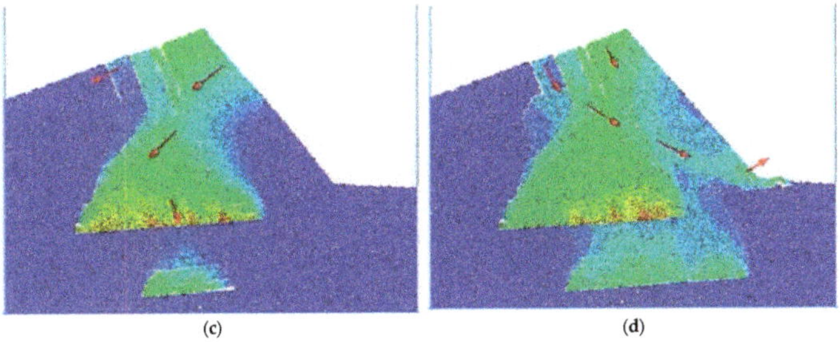

Figure 11. The displacement characteristic of the slope during the mining process. (**a**) S3; (**b**) S6; (**c**) X3; (**d**) X6.

4.3.2. The Displacement Characteristic of the Measuring Point

The displacement of measuring points in the roof of the upper coal seam is shown in Figure 12. When the upper coal seam was mined, the measuring points moved rapidly, as the immediate roof collapsed with mining. Then, the displacement stopped increasing, resulting from the end of the roof collapse. With the excavation of the lower coal seam, the displacement of the measuring points in the lower coal seam roof increased accordingly, as shown in Figure 13. When X4 was mined, significant displacement was induced again in the measuring points No. 6–No. 10. This indicates that the immediate roof collapsed gradually, and the deformation of the upper coal seam floor was then activated, causing the secondary subsidence of collapsed strata of the upper coal seam. Hence, the displacement of the measuring points started to increase. During the mining process of the lower coal seam, No. 6 and No. 7 first moved to the outside of the slope, and a sudden turning change was then caused. This may be caused by the bending deformation and collapse of the immediate and main roof.

The displacement of the measuring points at the top of the slope is shown in Figure 14. The horizontal displacement of the No. 11 measuring point was different from that of the other measuring points. When the upper coal seam was mined, the displacement value of the No. 22–No. 15 measuring points reduced, indicating the rock moved to the inside of the slope. When the influence range of the mined-out area had extended to the rock above J2, the displacement of No. 14 at the position above J2 increased rapidly, indicating the rock began to fall towards the slope. During the mining of S5, the displacement of No. 12 and No. 13 increased and then decreased. This indicates the joint rock was broken first, due the subsidence of the strata, which toppled to the inside. With the increase in the subsidence range, the measured points also moved to the outside of the slope. When the lower coal seam was mined, with the increase in the excavation distance, the displacement of most measuring points gradually moved to the outside. However, the displacement of the No. 15 measuring point continued to increase until X5 was mined. This indicates the rock mass above J4 also moved to the outside. Then, the rate of change increased suddenly, indicating the slope had begun to accelerate to slip at this time. During the mining process, the displacement of No. 11 located on the left side of J1 was lower than that of the other measuring points laid out at the top of the slope. This indicates that, under the disturbance of mining, differential deformation was induced on both sides of the joint rock mass at the top of the slope. The broken joint rock would lose its support, toppling into the slope. Then, the collapsed rock had a tendency to slip outward, which contributed to form the potential sliding of the slope.

The displacement of the measuring points at the surface of the slope is shown in Figure 15. When the upper coal seam was mined, bending deformation of the overburden rock was induced. Hence, the No. 16–18 measuring points mainly moved towards the inside of the slope first, while a tiny upward movement was induced in the No. 19 and No. 20 measuring points. During the mining process of the lower coal seam, as the cracks were induced at the foot of the slope, the No. 16–18 measuring points started to move towards the outside of the slope. The horizontal displacement value of the No. 20 measuring point was much larger than the vertical displacement value, indicating that the slope toe was mainly squeezed out in the direction of the free surface.

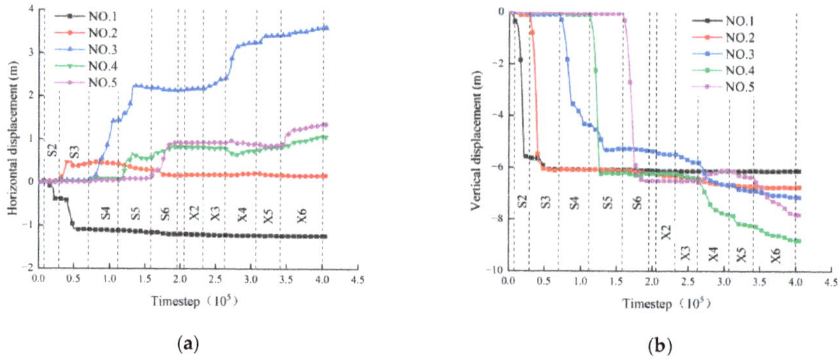

Figure 12. The displacement of measuring points in the roof of the upper coal seam. (**a**) Horizontal displacement; (**b**) vertical displacement.

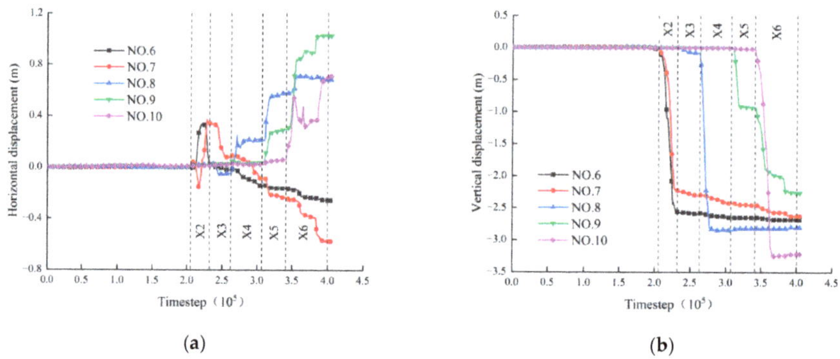

Figure 13. The displacement of measuring points in the lower coal seam roof. (**a**) Horizontal displacement; (**b**) vertical displacement.

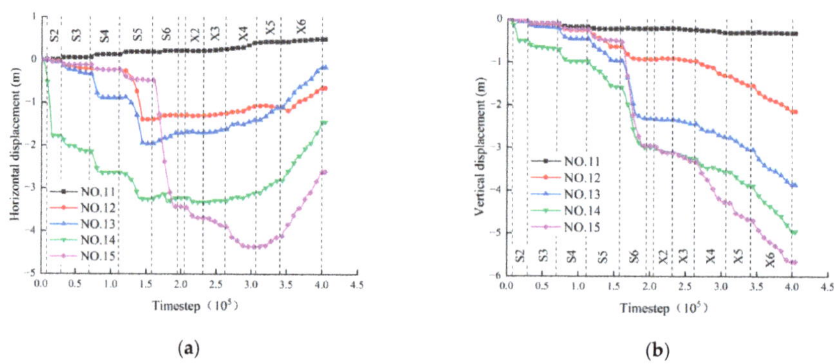

Figure 14. The displacement of measuring points at the top of the slope. (**a**) Horizontal displacement; (**b**) vertical displacement.

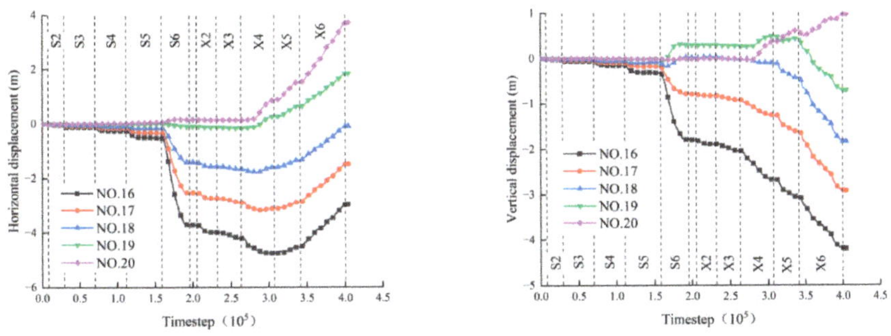

(**a**) (**b**)

Figure 15. The displacement of measuring points at the surface of the slope. (**a**) Horizontal displacement; (**b**) vertical displacement.

5. Slope Instability Slip Mechanism

As shown in Figure 16, combined with the results of model test and numerical simulation analysis, the instability evolution of the slope under underground repeated mining disturbance can be divided into four stages as follows: roof caving and overlaying rock subsidence, joint rock toppling, fracture penetration, and slope toe shearing and slope slipping. The failure mechanism of slope instability can be summarized as strata collapse–joint rock toppling–shear slip. The main reasons for the instability and slip of the slope will be divided into geological factors and non-geological factors. In the collapse area, the geological factors were characterized as the typical upper hard and lower soft, upper steep and lower gentle slope. During the long period of weathering and karstification, many deep and large joints are developed at the top of the slope, and the stability of the mountain will be impacted. Non-geological factors include external disturbances such as underground excavation, rainfall and earthquakes.

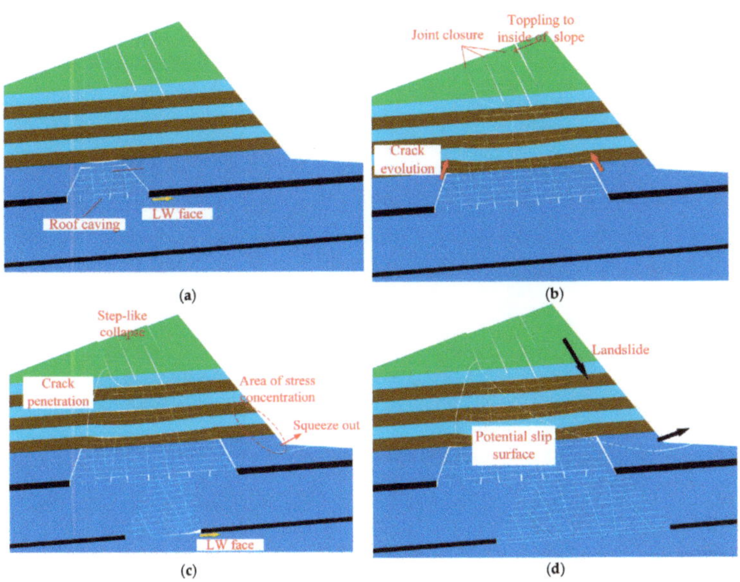

Figure 16. Slope instability process. (**a**) Roof caving and overlaying rock subsidence; (**b**) joint rock toppling; (**c**) crack penetration; (**d**) slope toe shearing and slope slipping.

When the underground resource is mined repeatedly, the joint rock can be broken during the mining process of the upper coal seam. The deep and large fissures of the slope continuously developed, and the mining-induced cracks will connect with natural fissures. In addition, water can penetrate into the slope along the cracks, and the conditions of the slope are further disturbed and deteriorated. Finally, the potential slip surface will be formed. When the rock mass at the foot of the slope is sheared and squeezed out, the rock of the slope loses support, forming a potential sliding zone and eventually leading to the slip of the slope.

6. Conclusions

In this study, the landslide disaster in Pusa Village is taken as the engineering background; the model test and numerical simulation were performed to investigate crack propagation and the mechanism of a landslide induced by repeated mining. The following specific conclusions can be made:

(1) During the repeated mining process, the joint rock mass at the top of the mountain will be broken by the mining of the upper coal seam. The rock mass topples to the inside of slope first. Then, the outward displacement of the slope will be caused by the mining of the lower coal seam. The potential slip zone will be formed.
(2) The immediate rock caving and strata subsidence will be caused by the mining. When the mining-induced crack connects with natural fissures, the rock mass at the foot of the slope has to support the overlying rock mass, which is also squeezed by the collapsed rock mass. Hence, shear cracking can be caused at the foot of the slope.
(3) The instability evolution of the slope under underground repeated mining disturbance can be divided into four stages as follows: roof caving and overlaying rock subsidence, joint rock toppling, fracture penetration, and slope toe shearing and slope slipping.

Author Contributions: Conceptualization, R.L.; formal analysis, G.L.; funding acquisition, L.C.; investigation, R.L.; methodology, L.Z.; project administration, L.C. and R.L.; testing, X.Y. and R.L.; software, X.Y. and K.P.; writing—original draft, R.L. and G.L.; validation, L.C. and L.Z. All authors have read and agreed to the published version of the manuscript.

Funding: This research is funded by the National Natural Science Foundation of China (52004036) and the Key Laboratory of Advanced Engineering Materials and Structural Mechanical Behavior and Intelligent Control for Universities in Hunan Province (19ka01).

Institutional Review Board Statement: Not applicable.

Informed Consent Statement: Not applicable.

Data Availability Statement: The data and materials are contained within the article.

Acknowledgments: We would like to acknowledge the editors and reviewers for their invaluable comments.

Conflicts of Interest: The authors declare no conflicts of interest.

References

1. Takashi, S.; Hiroshi, T.; Hideki, S.; Jiro, O.; Akihiro, H.; Kikuo, M. Surface subsidence due to underground mining operation under weak geological condition in Indonesia. *J. Rock Mech. Geotech. Eng.* **2015**, *7*, 337–344.
2. Chi, S.; Yu, X.; Wang, L. Method of predicting dynamic deformation of mining areas based on synthetic aperture radar I interferometry (InSAR) time series Boltzmann function. *Appl. Sci.* **2024**, *14*, 7917. [CrossRef]
3. Chen, L.; Yu, X.; Luo, R.; Zeng, L.; Cao, H. High steep rock slope instability mechanism induced by the pillar deterioration in the mountain mining area. *Mathematics* **2023**, *11*, 1889. [CrossRef]
4. Zhong, Z.; Xu, Y.; Wang, N.; Liu, X.; Gao, G. Environmental characteristics and unified failure mode classification system for mining landslides in the Karst mountainous areas of southwestern China. *Carbonates Evaporites* **2023**, *38*, 2. [CrossRef]
5. Qiu, Y.; Shang, Y. Analysis of progressive failure mechanism of loose slope under seismic load. *J. China Foreign Highw.* **2022**, *42*, 15–21.
6. Fernández, P.; Granda, G.; Krzemień, A.; Cortés, S.; Valverde, G. Subsidence versus natural landslides when dealing with property damage liabilities in underground coal mines. *Int. J. Rock Mech. Min. Sci.* **2020**, *126*, 104175.

7. Chen, J.; Tong, J.; Rui, Y.; Cui, Y.; Pu, Y.; Du, J.; Apel, D.B. Step-path failure mechanism and stability analysis of water-bearing rock slopes based on particle flow simulation. *Theor. Appl. Fract. Mech.* **2024**, *131*, 104370.
8. Fan, X.; Xu, Q.; Scaringi, G.; Zheng, G.; Huang, R.; Dai, L.; Ju, Y. The "long" runout rock avalanche in Pusa, China, on august 28, 2017: A preliminary report. *Landslides* **2019**, *16*, 139–154. [CrossRef]
9. Esterhuizen, G.S.; Tyrna, P.L.; Murphy, M.M. A case study of the collapse of slender pillars affected by through-going discontinuities at a limestone mine in Pennsylvania. *Rock Mech. Rock Eng.* **2019**, *52*, 4941–4952.
10. Reed, M.; Steve, K. Peripheral gully and landslide erosion on an extreme anthropogenic landscape produced by mountaintop removal coal mining. *Earth Surf. Process. Landf.* **2020**, *45*, 2078–2090.
11. Luo, R.; Li, G.Y.; Chen, L.; Yang, Q.Y.; Zang, C.W.; Cao, W.Z. Ground subsidence induced by pillar deterioration in abandoned mine districts. *J. Cent. South Univ.* **2020**, *27*, 2160–2172.
12. Yang, C.; Shi, W.; Peng, X.; Zhang, S.; Wang, X. Numerical simulation of layered anti-inclined mining slopes based on different free face characteristics. *Bull. Eng. Geol. Environ.* **2022**, *81*, 359.
13. Li, Z.F.; Zhang, Y.J.; Huang, J.X.; Luo, Z.M. Stability analysis of karst subgrade roof considering spatial form of the Karst cave. *Highw. Automot. Appl.* **2023**, *216*, 66–72.
14. Salmi, E.; Nazem, M.; Karakus, M. Numerical analysis of a large landslide induced by coal mining subsidence. *Eng. Geol.* **2017**, *217*, 141–152.
15. Arca, D.; Kutoğlu, H.; Becek, K. Landslide susceptibility mapping in an area of underground mining using the multicriteria decision analysis method. *Environ. Monit. Assess.* **2018**, *190*, 725.
16. Cui, F.; Xiong, C.; Wu, Q.; Li, B.; Luo, W.; Pei, Y.Q. Dynamic mechanism triggering an underground-mining induced landslide in karst mountains, Southwest China. *Bull. Eng. Geol. Environ.* **2023**, *82*, 449.
17. Yang, D.; Qiu, H.; Ma, S.; Liu, Z.; Du, C.; Zhu, Y.; Cao, M. Slow surface subsidence and its impact on shallow loess landslides in a coal mining area. *Catena* **2022**, *209*, 0341–8162.
18. Lai, Q.; Zhao, J.; Shi, B.; Liu, H.; Ji, L.; Li, Q.; Huang, R. Deformation evolution of landslides induced by coal mining in mountainous areas case study of the Madaling landslide, Guizhou, China. *Landslides* **2023**, *20*, 2003–2016.
19. Liu, F.; Yang, Z.; Deng, W.; Yang, T.; Zhou, J.; Yu, Q.; Mao, Y. Rock landslide early warning system combining slope stability analysis, two-stage monitoring, and case-based reasoning a case study. *Bull. Eng. Geol. Environ.* **2021**, *80*, 8433–8451.
20. Zhong, J.; Mao, Z.; Ni, W.; Zhang, J.; Liu, G.; Zhang, J.; Geng, M. Analysis of formation mechanism of slightly inclined bedding mudstone landslide in coal mining subsidence area based on finite–discrete element method. *Mathematics* **2022**, *10*, 3995. [CrossRef]
21. Dhahri, F.; Benassi, R.; Mhamdi, A.; Zeyeni, K.; Boukadi, N. Structural and geomorphological controls of the present-day landslide in the Moulares phosphate mines (western-central Tunisia). *Bull. Eng. Geol. Environ.* **2016**, *75*, 1459–1468. [CrossRef]
22. Rui, Y.; Chen, J.; Chen, J.; Qiu, J.; Zhou, Z.; Wang, W.; Fan, J. A robust triaxial localization method of AE source using refraction path. *Int. J. Min. Sci. Technol.* **2024**, *34*, 521–530. [CrossRef]
23. Li, H.J.; Dong, J.H.; Zhu, Y.Q.; Zou, Y.X.; Ding, H. Characteristics and genesis mechanism of Jianshanying landslide in Faer coal mine, Guizhou province. *Sci. Technol. Eng.* **2019**, *19*, 345–351.
24. Zhao, Q.; Yang, Z.; Jiang, Y.; Liu, X.; Cui, F.; Li, B. Discrete element analysis of deformation features of slope controlled by Karst fissures under the mining effect: A case study of Pusa landslide, China. *Geomat. Nat. Hazards Risk* **2023**, *14*, 1–35.
25. Stephen, H.; Valery, V.; Richard, G.; Philippe, C. Behavior in mixed-mode of desiccation cracks on a clayey rock front gallery. *Int. J. Rock Mech. Min. Sci.* **2022**, *154*, 105104.
26. Chen, L.; Zhou, Z.; Zang, C.; Zeng, L.; Zhao, Y. Failure pattern of large-scale goaf collapse and a controlled roof caving method used in gypsum mine. *Geomech. Eng.* **2019**, *18*, 449–457.
27. Zhu, W.; Chen, L.; Zhou, Z.; Shen, B.; Xu, Y. Failure propagation of pillars and roof in a room and pillar mine induced by longwall mining in the lower seam. *Rock Mech. Rock Eng.* **2019**, *52*, 1193–1209.
28. Zhao, Y.; Zhao, G.; Zhou, J.; Ma, J.; Cai, X. Failure mechanism analysis of rock in particle discrete element method simulation based on moment tensors. *Comput. Geotech.* **2021**, *136*, 104215.
29. Zhong, H.; Liang, C.; Chen, L.; Gao, S. Characteristics of large deformation and prevention methods for shallow buried large span tunnels. *J. Transp. Sci. Eng.* **2024**, *40*, 71–78.
30. Potyondya, D.; Cundall, P. A bonded-particle model for rock. *Int. J. Rock Mech. Min. Sci.* **2004**, *41*, 1329–1364. [CrossRef]
31. Zhou, Z.; Chen, L.; Zhao, Y.; Zhao, T.; Cai, X.; Du, X. Experimental and numerical investigation on the bearing and failure mechanism of multiple pillars under overburden. *Rock Mech. Rock Eng.* **2017**, *50*, 995–1010. [CrossRef]

applied sciences

MDPI

Article

Analysis of Granite Deformation and Rupture Law and Evolution of Grain-Based Model Force Chain Network under Anchor Reinforcement

Jiangfeng Guo [1,2], Doudou Fan [1,3], Liyuan Yu [1,3,*], Meixia Shi [1], Haijian Su [1], Tao Zhang [1] and Bowen Hu [1]

1 State Key Laboratory of Intelligent Construction and Healthy Operation and Maintenance of Deep Underground Engineering, China University of Mining and Technology, Xuzhou 221116, China; ts23030054a31tm@cumt.edu.cn (M.S.); hjsu@cumt.edu.cn (H.S.); bwhu@cumt.edu.cn (B.H.)

2 Shanxi Huaning Coking Coal Co., Ltd., China Coal Group Huajin Co., Ltd., Linfen 042100, China

3 State Key Laboratory of Disaster Prevention and Mitigation of Explosion and Impact, Army Engineering University of PLA, Nanjing 210007, China

* Correspondence: yuliyuancumt@163.com

Abstract: In actual underground rock engineering, to prevent the deformation and damage of the rock mass, rock bolt reinforcement technology is commonly employed to maintain the stability of the surrounding rock. Therefore, studying the anchoring and crack-stopping effect of rock bolts on fractured granite rock mass is essential. It can provide significant reference and support for the design of underground engineering, engineering safety assessment, the theory of rock mechanics, and resource development. In this study, indoor experiments are combined with numerical simulations to explore the impact of fracture dip angles on the mechanical behavior of unanchored and anchored granite samples from both macroscopic and microscopic perspectives. It also investigates the evolution of the anchoring and crack-stopping effect of rock bolts on granite containing fractures with different dip angles. The results show that the load-displacement trends, displacement fields, and debris fields from indoor experiments and numerical simulations are highly similar. Additionally, it was discovered that, in comparison to the unanchored samples, the anchored samples with fractures at various angles all exhibited a higher degree of tensile failure rather than shear failure that propagates diagonally across the samples from the regions around the fracture tips. This finding verifies the effectiveness of the numerical model parameter calibration. At the same time, it was observed that the internal force chain value level in the anchored samples is higher than in the unanchored samples, indicating that the anchored samples possess greater load-bearing capacity. Furthermore, as the angle α_s increases, the reinforcing and crack-stopping effects of the rock bolts become increasingly less pronounced.

Keywords: granite; crystalline structure; deformation and fracture process; 3D equivalent mineral crystal model; force chain network analysis

1. Introduction

In the realm of underground rock engineering, anchor reinforcement technology is commonly employed to avert deformation and damage to the rock mass, thereby ensuring the stability of the surrounding rock [1–3]. This technology, by offering proactive resistance, markedly enhances the shear strength of the structural surfaces and fortifies the constraints on ground deformation. It also optimizes the stress conditions within the excavated rock and soil mass [4]. Additionally, anchor reinforcement effectively harnesses the geotechnical body's intrinsic strength and load-bearing capacity [5], unleashing its latent strength potential. This approach not only reduces project costs but also expedites construction timelines and supports the project's sustainable long-term operation. In summary, the crack-stopping effect exerted by anchors on the fissured granite rock mass is of pivotal importance in underground rock reinforcement engineering. It offers substantial guidance and support for the design of underground engineering projects, safety assessments, the advancement of rock mechanics theory, and the exploitation of resources. The fracture stopping effect of anchors on the interior of the rock body cannot be solved by formula derivation of its rupture mechanism using theoretical methods, so indoor tests have become one of the main methods of studying the anchoring mechanism of anchors on fissured rock bodies, and many scholars and experts at home and abroad have carried out a lot of work on the mechanical behavior of the anchored fissured rock bodies.

In 1974, Bjurstrom [6] conducted a series of shear tests, revealing that anchors are capable of effectively containing deformation damage in jointed rock bodies and enhancing the shear resistance of joint surfaces. In 1975, Littlejohn and Bruce [7] categorized the damage modes of rock anchors into the following three distinct types: damage to the perimeter rock, interfacial failure, and damage to the anchor body itself. In 1987, Egger and Spang [8] carried out a series of large-scale in situ shear tests and similarly found that the anchorage angle of the anchor affects the anchorage effect of the anchor on the rock body. In 1990, Spang and Egger [9] investigated the anchorage effect of anchors in different rock materials by dividing their shear stress–displacement curves into a linear-elastic deformation, a yielding stage, and a plastic deformation stage. In 2010, Bezuijen [10] demonstrated that even with anchorage, jointed rock masses can still exhibit shear displacement when subjected to external loads. As the shear displacement across the joint surfaces accumulates, the anchor is likely to form a plastic hinge, at which point the maximum shear stress is reached.

On the domestic front, studies on the mechanical behavior of anchored fractured rock bodies have also focused on uniaxial compression and shear, which are the two most common and simplest stress states. Chen [11] conducted an experimental investigation on bolted rock samples to assess the anchoring efficacy of threaded steel bolts and D-bolts. The study incorporated a range of variables, including various angles of displacement, different joint spacings, and three distinct rock materials, namely, weak concrete, strong concrete, and concrete–granite. For anchored rock with perforated weak interlayers, Zong et al. [12] found that the anchoring effect could change the damage mode of the rock samples from brittle to ductile, with tensile damage being replaced by more shear damage. These above conclusions are consistent with the studies of Ren et al. [13] and Ding et al. [14]. In 2016, Li et al. [15] and Yang et al. [16] conducted laboratory tests on the effect of prestressing anchors in fractured rock bodies, which indicated the optimal anchorage angle and location. They also found that the anchors acted to inhibit fracture creation and expansion, and to improve the integrity of the fractured rock body. Under shear conditions, many scholars focused on the effects of anchors on shear force, displacement, and deformation, and achieved good results. In 2020, Yang et al. [17] explored in detail the fracture and anchorage mechanisms of jointed rock bodies on a microscopic scale by using the digital speckle correlation method, acoustic emission technology, and X-ray CT observation. In 2021, Li et al. [18] conducted anchorage shear tests. It was identified that the residual resistance contribution of full-length bond bolts was the largest at anchorage angles of 55° to 70°. The maximum contribution of the anchor bolts increased linearly with the increase in the joint extension angle. In addition, the effects of different anchorage types, anchorage patterns, joint roughness values, loading angles, and displacement angles were investigated in the anchorage shear tests.

With the rapid development of computer technology, numerical simulation has gradually become one of the mainstream research methods in the field of rock mechanics [19,20]. In the study of the mechanical behavior of anchored fractured rock, there are usually two methods, i.e., the finite element method (FEM) and discrete element method (DEM).

Han et al. [21] utilized the finite element analysis software ANSYS 2013R2 to construct a finite element model of fractured rock anchorage, investigating the impact of rock fractures on the load transfer mechanism of anchoring. Through numerical calculations, they analyzed the effects of fracture aperture, fracture length, and fracture position on the stress transfer mechanism in the anchored fractures. Kang et al. [22] analyzed the mechanism of "S"-type joint fractures in the fractured rock body and carried out numerical simulation research on the anchoring effect of this joint. They established the model of the fractured rock body under different anchoring methods using FEM software FLAC3D 5.0, and carried out simulation tests to analyze the mechanical properties of the fractured rock body under different anchoring methods. Comparison and analysis of the anchoring effect and the main damage mode under different anchoring methods. Yang et al. [16] used the FEM software ABAQUS (2018) to carry out a series of numerical simulation studies on the mechanical properties and anchoring effect of prestressed anchors reinforcing the fissured rock mass. In summary, the internal monitoring values of the samples, as well as the deformation and stress of the anchors can be easily obtained using FEM; however, the microscopic changes inside the rock mass are difficult to capture.

DEM can better capture some microscopic information inside the sample during the load damage process compared to FEM, and simulate the microcracks sprouting and expanding in the sample. Among them, many researchers have used bonded particles (BPs) in DEM to simulate anchors and rock masses [23]. Wu et al. [24] used three-dimensional particle flow code (PFC3D) to investigate the effect of different anchorage angles on crack extensions from a microscopic point of view, and used BPs in PFC to simulate both the rock body and the anchor system. Chen [25] conducted a series of numerical simulations using PFC3D to explore the mechanical impacts of anchoring on cracked rock. In these simulations, the rock mass was represented by the bonded particle model (BP) within PFC, while the prestressed anchor system was modeled using the Clump method.

It is not difficult to find that the internal monitoring values of the samples, as well as the deformation and stress of the anchors, can be easily obtained using FEM. However, the microscopic changes inside the rock mass are difficult to be captured [26–28]. In contrast, DEM offers a distinct advantage over FEM in capturing microscopic details within a sample during the loading and damage process, effectively simulating the initiation and propagation of microcracks within the material [29–31]. However, it falls short in accurately reflecting the elastic-plastic behavior of continuous media, such as the response of anchor rods to stress [25]. Therefore, many researchers have combined the advantages of these two methods by coupling FEM-DEM to provide a solution to the above mentioned problems. Luo et al. [32] established a numerical model of anchored fissured rock mass by coupling FEM and DEM, and investigated the internal stress evolution in the sample and the stress distribution in the anchor under the anchoring action.

However, for natural rock materials with a crystalline structure, such as granite, the complexity of their internal mineral composition and structure means that using a homogeneous model to simulate granite materials has certain limitations. Currently, for the numerical modeling of granite materials, the grain-based model (GBM) has become the choice of many scholars [33,34]. Potyondy et al. [35,36] were the first to establish a GBM based on the particle flow code (PFC). The PFC-GBM is capable not only of reproducing the interactions between crystals but also of simulating the internal mechanical behavior of mineral particles. In recent years, numerous scholars have improved upon the PFC-GBM and applied it to validate various indoor tests and to analyze the related mechanisms [37–39].

When rock materials are subjected to external loads, localized stress concentrations can lead to the destruction and fracturing of particles and bonds within the material, culminating in macroscopic failure. However, both the theoretical analysis and traditional indoor experimental methods fall short in capturing this macroscopic process. The force chain network (FCN) model offers a valuable perspective. It represents a network of interwoven force chains within granular materials. According to granular mechanics, under external loads, particles within the granular system are pressed against each other, creating a network known as the connection network of granular materials. These contact paths constitute the force chains and the overall force chain network. This network allows for the investigation of damage and fractures within the granular particles and their bonds in the rock material.

Currently, the force chain network is widely used in geotechnical discrete element numerical simulation studies and is considered as an important source of mechanical information in the microscopic perspective of geotechnics. In 2002, Zhang et al. [40] simulated a straight shear test in geotechnics using DEM to investigate the particle morphology and force chain distribution. In 2020, Leśniewska et al. [41] used DEM to study the quasi-static behavior under active earth pressure conditions, and compared with the results of photoelastic tests to successfully predict the overall structure of the force chain network and its characteristics. Zhang et al. [40] utilized the discrete element method (DEM) to simulate the direct shear test in soil mechanics, investigating the effects of particle shape and force chain distribution. In 2023, Wang et al. [42] investigated the evolution and distribution characteristics of the internal contact force in samples with different particle size gradations. In summary, the force chain network is a crucial indicator for elucidating the shifts in the mechanical behavior of rock materials. It serves as a vital research instrument for dissecting the deformation and rupture processes of rock materials under load from a granular perspective. Consequently, examining the mechanical response of the fractured and anchored rock mass through the lens of the force chain network is imperative.

To provide a scientific basis for the design and safety assessment of underground rock engineering, a combination of laboratory experiments and numerical simulations is employed, focusing on unanchored and anchored granite samples to capture their mechanical behavior during uniaxial compression tests. By comparing the macroscopic mechanical properties and the evolution of the force chain network, the research investigates the impact of fracture dip angles on the mechanical behavior of anchored granite samples from both macroscopic and microscopic viewpoints. It explores the evolution of the role of rock bolts in anchored granite samples containing fractures with varying dip angles, as well as the progression of the anchoring effect in granite with different fracture dips.

2. Research Methods

2.1. Uniaxial Compression Test

To ensure the tests closely reflect the actual project conditions, natural granite samples from Jining City, Shandong Province, are utilized. For the simulation of the anchor system, 8.8-grade low carbon alloy steel screws are selected, whose composition includes carbon (C), phosphorus (P), sulfur (S), and boron (B). The screws have an effective diameter of 3.88 mm, a length of 140 mm, and an elastic modulus of 121 GPa. The preparation of the granite-anchor composite samples involves several steps: initially, waterjet technology is employed to drill holes with a 6 mm diameter into the granite samples. The screws are then inserted into these holes, and a syringe is used to inject an anchoring filler. The filler is the anchoring agent, that fills the gap between the screw and the inner wall of the hole, ensuring the anchor adheres firmly to the hole. After allowing the anchors to cure for 1–2 days, shims, trays, and nuts are installed to complete the assembly. Figure 1 illustrates the schematic diagram of the final granite-anchor sample.

Considering the stress state of the granite samples and facilitating the recording of the deformation and rupture process of the samples, the MTS816 electro-hydraulic servo rock tester (MTS Systems Corporation, Eden Prairie, MN, USA) was used in this test. The maximum axial compression force of MTS816 is 1459 kN, which is used for uniaxial compression experiments on the granite samples, as shown in Figure 2. The test process was loaded by axial displacement control, with a displacement loading rate of 1.5×10^{-3} mm/s and a preset compression contact force of 1.0 kN.

The monitoring tools used in this study were digital image correlation (DIC), computerized tomography (CT), and polarized light microscopy.

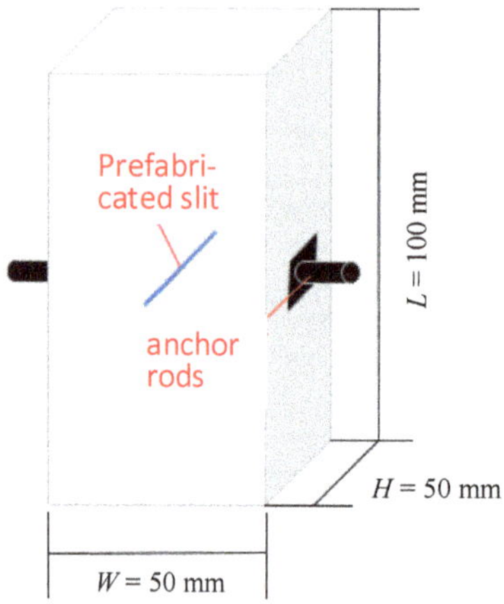

Figure 1. Diagram of anchored granite sample.

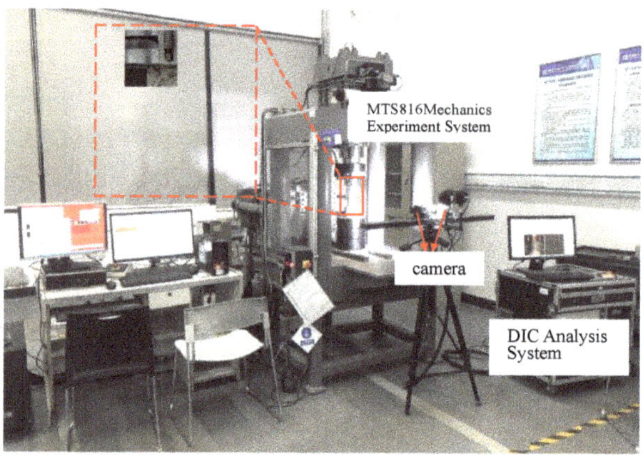

Figure 2. Test equipment and deformation failure monitoring device.

2.2. Construction of Continuous-Discrete Coupled Numerical Models

To incorporate the effects of non-homogeneity, mineral composition, and microstructure on the fracture behavior of granite, in this section, the development of a 3D numerical model is presented. This model is constructed using the three-dimensional equivalent mineral crystal model (3D-GBM) in particle flow code 3D (PFC3D). PFC3D 6.0 software is a powerful particle analysis program that simulates the dynamic motion and interaction of aggregates composed of disks or spherical particles of any size. The detailed procedure is depicted in Figure 3.

To advance the construction of the numerical model incorporating granite and anchor samples, it is imperative to identify an appropriate method for simulating the anchor system. Utilizing the finite element method (FEM) for simulation can more accurately capture the elastic–plastic mechanical behavior of the anchor during the stress process while significantly reducing computational expenses. Consequently, in this study, a continuous-discrete coupling method was employed to develop the numerical model for the granite-anchor composite sample.

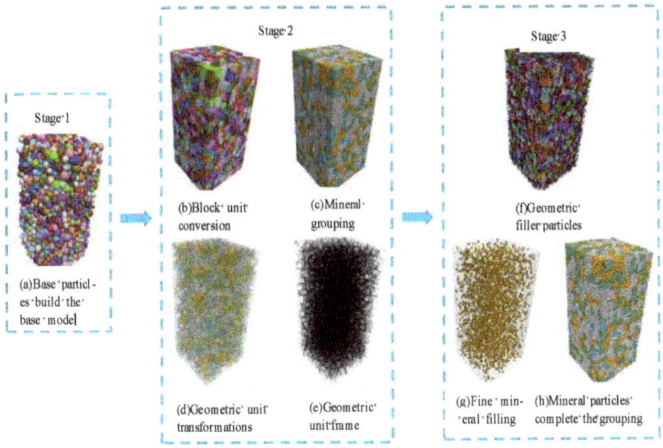

Figure 3. Process of constructing 3D GBM of granite.

2.3. *Quantitative Analysis and Representation of Force Chain Network*

The force chain network is a model formed by the interweaving of force chain structures within granular materials. According to particle material mechanics, particles in a granular system are squeezed together under external loads to form a network. This network, known as the connection network between particle materials, consists of contact paths that form force chains and force chain networks. Consequently, the failure and fracture of particles and bonds within rock materials can be explored through the analysis of force chain networks.

The built-in program of the PFC3D software can extract the values of the unit vectors of the internal force chain of the specimen in the x, y, and z directions in 3D space during the loading process. The center of the rose diagram is the coordinate origin, and the direction of the coordinate axis is consistent with the direction of the specimen placement, as shown in Figure 4. The 3D rose diagrams, also known as 3D group configuration diagrams, are widely used to show the distribution of force chains in 3D space [43,44]. The tendency in the rose diagram is the tendency of the force chain, and the height and color of individual columns can illustrate the number of force chains distributed on this tendency.

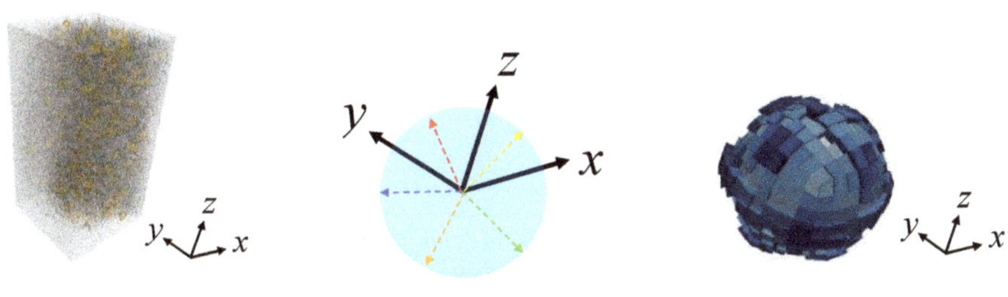

(**a**) Placement direction; (**b**) Orientation distribution; (**c**) 3D rose diagram.

Figure 4. Three-dimensional rose diagram form of force chain orientation distribution.

2.4. *Effectiveness Verification of Numerical Model*

From a microscopic perspective, the mechanical properties of the rock are primarily determined by the mineral grains themselves and the extent of their cementation. The extensive research by predecessors has demonstrated that the parallel bond model (PBM) within PFC3D is the most appropriate simulation method for capturing the mechanical behavior of rocks, offering superior simulation capabilities compared to other methods.

Consequently, the numerical simulation of natural granite rock materials presented in this paper will be grounded in the PBM framework. Furthermore, in this paper, the trial-and-error method is employed for calibrating the microscopic parameters of the numerical model. This approach has garnered widespread acceptance and recognition among scholars due to its proven effectiveness and reliability in various applications [44–46].

Through an iterative process of refining the micro-parameters and meticulously comparing the outcomes of the calibrated numerical simulations with those of the indoor tests, the set of micro-parameters that align closely with the experimental data is identified as the definitive configuration. The specific micro-parameters of the finalized numerical simulation samples, which correspond to the indoor test results, are presented in Table 1.

Table 1. Micro-parameters of the numerical sample.

Microscopic Parameter	Numerical Value				
Mineralogical composition	Sapphire	Plagioclase	Potassium feldspar	Micas	Fine mineral
Percentage by volume/%	29.4	42.2	21.9	2.6	4.0
Mineral Radius Minimum R_G/mm	1.8	1.4	1.6	1.2	0.65
Mineral radius maximum to minimum ratio r_G			2.0		
Basic unit	Sapphire	Plagioclase	Potassium feldspar	Micas	Fine mineral
Radius minimum R_p/mm			0.65		
Ratio of maximum to minimum radius r_p			2.0		
Density ρ_p/kg/m^3	2600	2650	3000	1700	2400
Modulus of elasticity E_p/GPa	62.0	47.0	32.0	27.0	22.0
Rigidity ratio $k_{n\text{-}p}/k_{s\text{-}p}$	1.4	1.6	1.8	2.0	2.2
Friction factor μ_p	0.25	0.30	0.40	0.50	0.60
Intracrystalline contact	Quartz inner	Plagioclase	Potassium feldspar	Mica	Inside the fine minerals
Modulus of elasticity $E_{c\text{-}tra}$/GPa	62.0	47.0	32.0	27.0	22.0
Rigidity ratio $k_{n\text{-}tra}/k_{s\text{-}tra}$	1.4	1.6	1.8	2.0	2.2
Angle of internal friction ϕ_{tra}/°	12	14	16	18	22
Bonding strength c_{tra}/MPa	366.0	306.0	286.0	246.0	106.0
Tensile strength σ_{tra}/MPa	183.0	153.0	143.0	123.0	83.0
Intergranular contact	Between identical minerals		Between different minerals		
Parallel stiffness ratio $k_{pb\text{-}n\text{-}ter\text{-}s}/k_{pb\text{-}s\text{-}ter\text{-}s}$	2.6		2.8		
Linear stiffness ratio $k_{c\text{-}n\text{-}ter\text{-}s}/k_{c\text{-}s\text{-}ter\text{-}s}$	2.6		2.8		
Friction factor $\mu_{ter\text{-}s}$	0.7		0.8		
Angle of internal friction $\phi_{ter\text{-}s}$/°	26.0		28.0		
Parallel modulus of elasticity $E_{pb\text{-}ter\text{-}s}$/GPa	2.5		2.2		
Linear modulus of elasticity $E_{c\text{-}ter\text{-}s}$/GPa	2.5		2.2		
Bonding strength $c_{ter\text{-}s}$/MPa	54.0		44.0		
Tensile strength $\sigma_{ter\text{-}s}$/MPa	27.0		22.0		

Subsequently, aligning with the dimensions and positions of the anchor holes in the indoor tests, the particles at corresponding locations within the 3D GBM model were selectively removed. The anchor system was then integrated into the model at the post-removal locations. The numerical simulation introduces a finite element-based anchor with isotropic elastic properties, mirroring the characteristics of actual anchors. The model anchor is defined with a density of 7850 kg/m^3, a Young's modulus of 9.7 GPa, and a Poisson's ratio of 0.3. The interaction between the anchors and the granite rock mass is governed by a parallel bond contact model, with detailed fine-grained parameters provided in Table 2.

Table 2. Mesoscopic parameters of the interface contact bonding.

Basic Particle Properties	Effective modulus of elasticity E_c (GPa)	20.0
	Normal to tangential stiffness ratio k_n/k_s	2.8
	Interparticle friction coefficient μ	0.65
Parallel bonding model	Parallel bonding tensile strength σ (MPa)	12.0
	Parallel bonding cohesion c (MPa)	24.0
	Bonding activation gap g_c (mm)	0.0001
	Parallel bonded effective modulus of elasticity E_c (GPa)	20.0
	Normal to tangential stiffness ratio k_n/k_s	2.8
	Friction angle φ (°)	45.0

By comparing and analyzing the uniaxial compression test results of the sample under laboratory testing and numerical simulation, the effectiveness of the micro-parameter calibration of the numerical model is verified. Figure 5a shows a comparison of the load–displacement curves from the uniaxial compression tests. The comparison indicates a strong consistency between the load–displacement results of the laboratory tests and the numerical simulations, demonstrating the reliability of the numerical model's micro-parameter calibration.

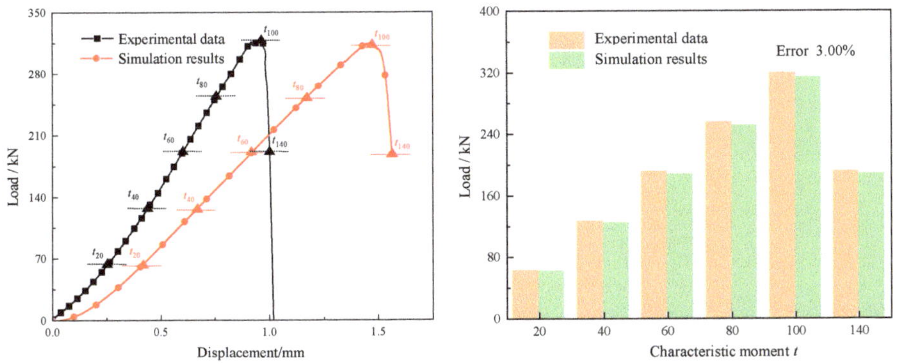

(**a**) Load–displacement curve (**b**) Load

Figure 5. Comparison of uniaxial compression test results.

In order to investigate the fracture process of the specimen in depth, in this study, the following six characteristic moments were selected during the loading process: the peak load moments of 20%, 40%, 60%, 80%, and 100% before the peak, and the peak load moments of 40% after the peak. These moments are denoted as t_{20}, t_{40}, t_{60}, t_{80}, t_{100}, and t_{140}, respectively. The load values at these six characteristic moments are extracted for comparison, as shown in Figure 5b. The figure demonstrates that the load levels of the experimental and simulation results are almost identical at the same characteristic time. Specifically, at peak load time t_{100}, the load values for the experimental and simulation results are 320.55 kN and 315.12 kN, respectively, with an error of only 3.00%.

3. Macroscopic Mechanical Properties

3.1. Mechanical Parameters

Figure 6 shows the load–displacement curves of the unanchored and anchored samples of granite with four different inclination fissures containing α_s = 30°, 45°, 60°, and 75°. Among them, the blue curves are the load–displacement curves of each inclined fissure sample under the unanchored condition, while the black color is the result of the sample under anchoring.

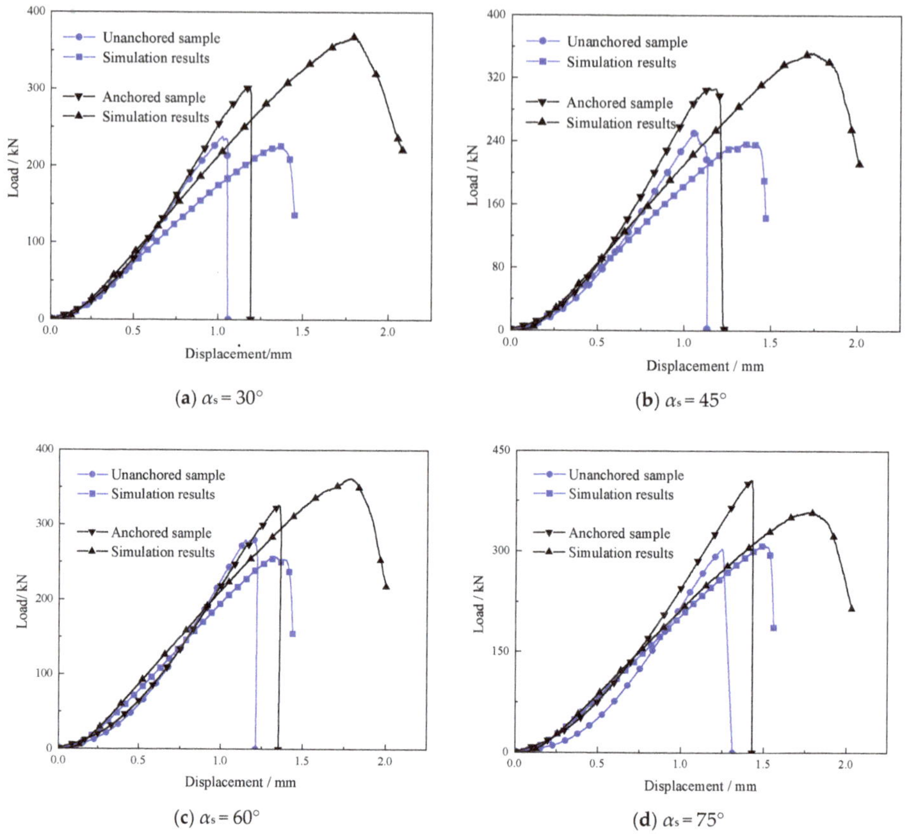

Figure 6. Load–displacement curves.

Through the comparison of the sample compression process, the results show a strong consistency in the trend of the load–displacement results of the indoor tests and numerical simulation samples, which further validate the numerical model. It can be clearly observed that the ground curves of the anchored samples of each inclined fissure are above the unanchored samples. The curves of each inclined fissure sample under both indoor test conditions and numerical simulation conditions mainly go through the following stages: initial compaction stage, linear elasticity growth stage, nonlinear deformation stage, and post-peak damage stage. The experimental results show that as α_s increases from 30° to 75°, the peak load of the anchored sample rises from 301.07 kN to 402.12 kN, representing a 33.5% increase. As α_s increases, the peak load levels of anchored samples are increased by 62.72%, 47.97%, 40.99%, and 15.58% compared to unanchored samples.

3.2. Macroscopic Rupture Models

Then, we analyzed the displacement field, rupture surface, and fragmentation field by comparing the displacement field, rupture surface, and fragmentation field of anchored samples containing different α_s fissures based on indoor tests and numerical simulation conditions.

(1) Comparison of displacement field features

Figure 7 illustrates the displacement field results of anchored samples containing different α_s fractures based on both indoor tests and numerical simulation conditions. It can be observed that the displacement field results of the test and simulated samples are highly similar. Specifically, the cleft of the sample becomes the demarcation of the displacement field for both the test and simulation results, which means that the tip region of the cleft undergoes a large displacement. The presence of cracks induces an uneven stress distribution within the sample, resulting in localized displacement and deformation. Specifically, stress concentrations form at the crack tips, which in turn, drive crack propagation in these regions. This process cumulatively leads to the overall damage of the sample.

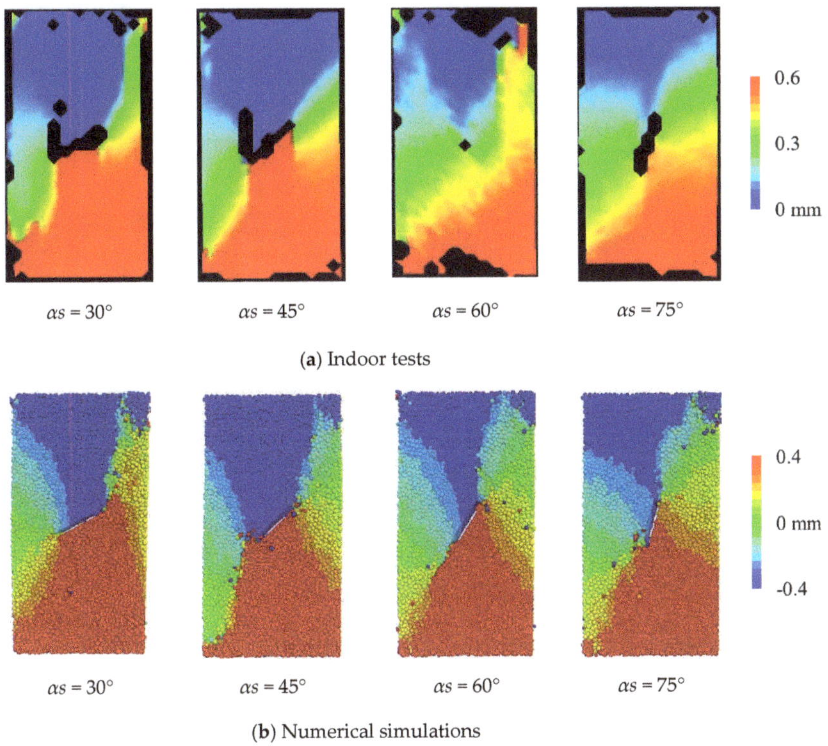

Figure 7. Variation in displacement field for anchored samples with different α_s.

(2) Comparison of fragmentation fields

Figure 8 presents the macroscopic damage patterns of unanchored and anchored samples of the fissures containing different α_s. Figure 9 presents the comparative results of the fragmentation fields of the unanchored and anchored samples under numerical simulation conditions. The results are similar to those of the simulated samples; the fragmentation field of the unanchored sample is concentrated at the tip of the fissure and distributed along the diagonal of the sample, which shows an obvious shear damage pattern. In contrast, the fragment length of the anchored sample is more uniformly distributed, which implies that the anchor structure improves the stress distribution inside the granite sample. This results in a larger and more evenly distributed stress area within the samples, thereby enhancing their load-bearing capacity.

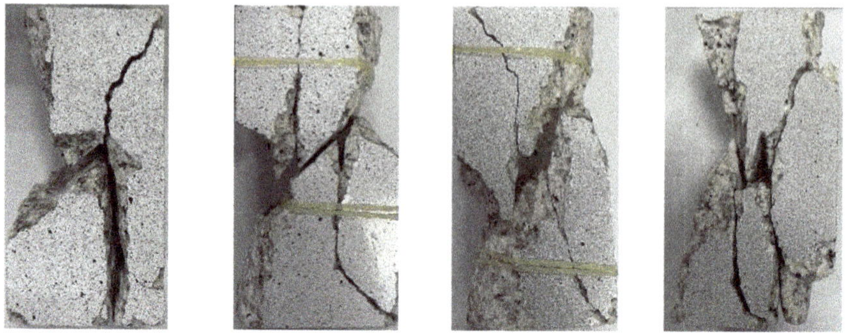

$\alpha s = 30°$ $\alpha s = 45°$ $\alpha s = 60°$ $\alpha s = 75°$

(**a**) Unanchored samples

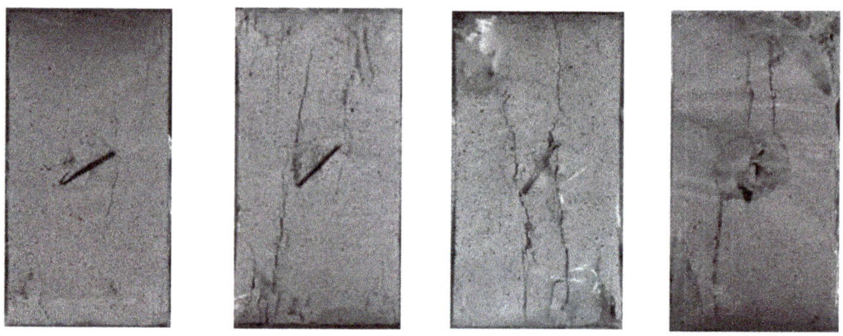

$\alpha s = 30°$ $\alpha s = 45°$ $\alpha s = 60°$ $\alpha s = 75°$

(**b**) Anchored samples

Figure 8. Comparison of failure modes for samples with different α_s.

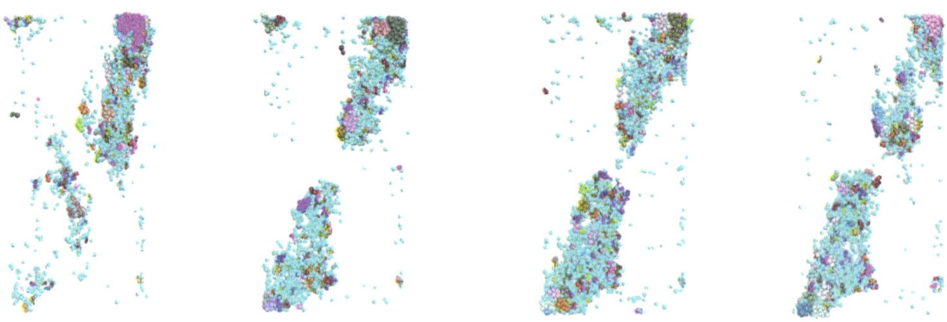

$\alpha s = 30°$ $\alpha s = 45°$ $\alpha s = 60°$ $\alpha s = 75°$

(**a**) Unanchored samples

Figure 9. *Cont.*

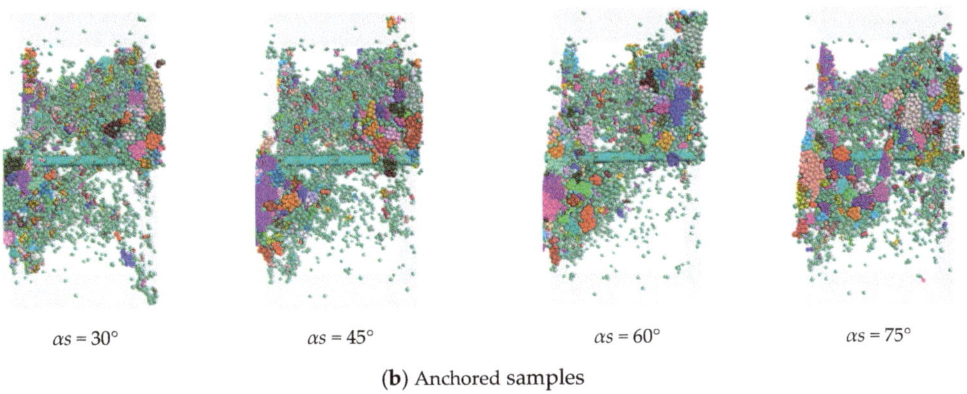

Figure 9. Variation in fragment field for numerical samples with different α_s.

4. Analysis of Force Chain Network Characteristics in Anchored Granite with Different Fissure Angles

Figure 10 demonstrates the internal force chain level cloud diagrams of unanchored and anchored samples under uniaxial compression for different inclination cracks. It can be clearly observed that the overall level of internal force chains in the anchored samples is greater than that in the unanchored samples.

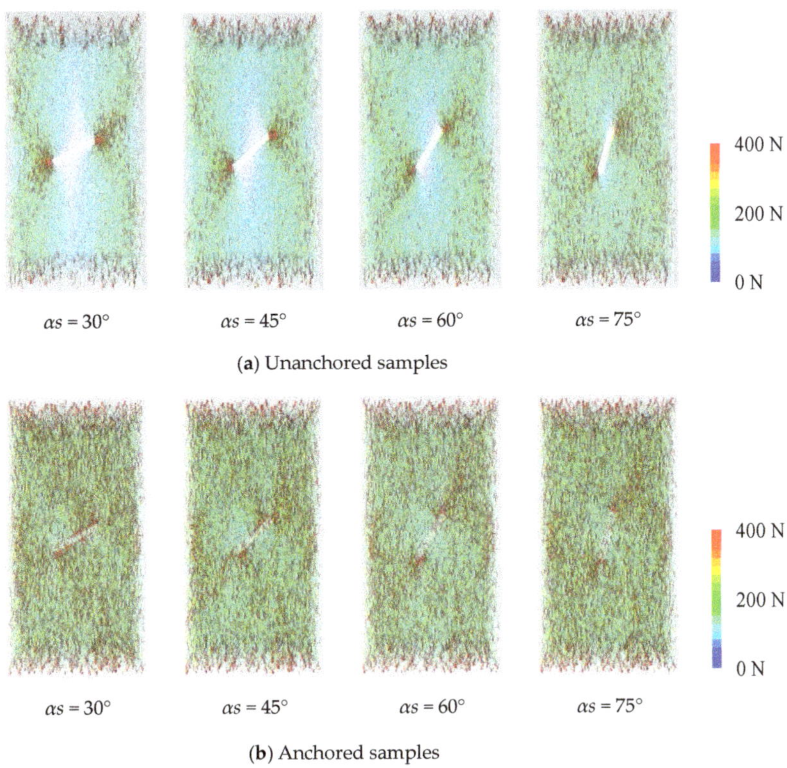

Figure 10. Evolution of force chain networks of the sample with different α_s under uniaxial compression.

The force chain values of the internal force chains in the fracture anchorage samples with different α_s were extracted, and all samples' internal force chains were arranged according to the magnitude of the force chain values from smallest to largest, as shown in Figure 11. It can be observed that the force chain values of the internal force chains for the fracture-anchored samples with different α_s are mostly at a very low level, with only a few having higher values, as indicated in the red-marked section of Figure 11. Upon zooming in on the red area, we can see that the horizontal axis of the curve, which represents the number of force chains, increases and then decreases with the increase in α_s. Meanwhile, the vertical axis, indicative of the force chain values, shows a trend of first decreasing and then increasing with the increase in α_s. This suggests that the anchoring system has varying degrees of anchoring effects on the anchored samples with fractures at different inclination angles, but further research is needed to explain this phenomenon.

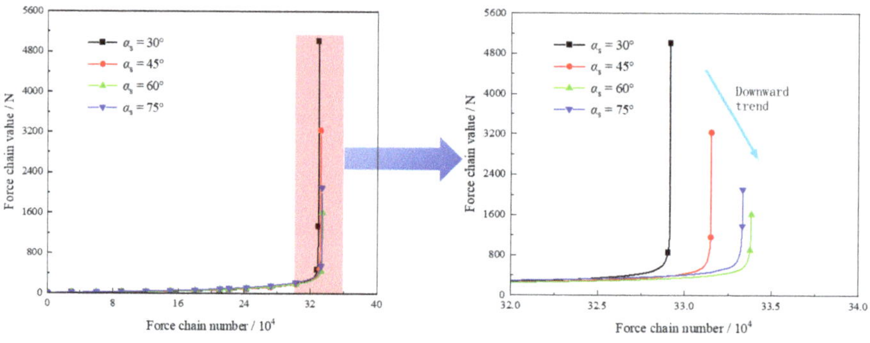

(**a**) Overall distribution (**b**) Enlarge the regional section

Figure 11. Value distribution of force chains in samples with different α_s under uniaxial compression.

Figure 12 shows a comparative plot of the variation in the eigenvalues of the internal force chains of the anchored samples of the fissures containing different α_s. It can be clearly seen that the mean and sum of the internal force chains of the unanchored and anchored samples have a similar trend. Specifically, as the α_s increase, there is a tendency for the increase in the mean value of the force chains and the sum of the force chains of the anchored samples compared to the unanchored samples to decrease gradually.

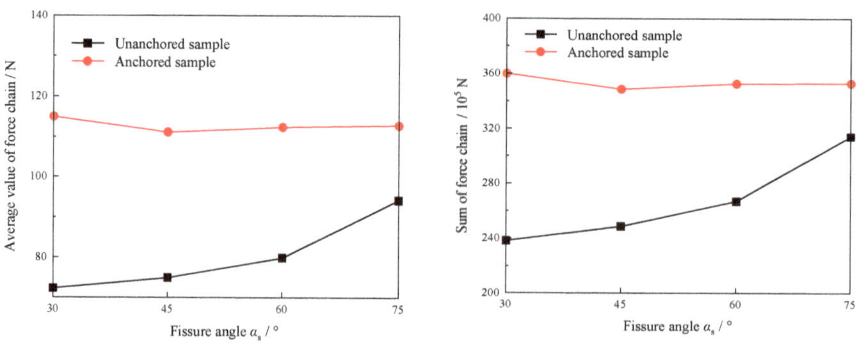

(**a**) Average value of force chain (**b**) Sum of force chain

Figure 12. Variations in characteristic values of force chains in samples with different α_s under uniaxial compression.

Specifically, in Figure 12a, as the α_s increase from 30° to 75°, the mean force chain values of the unanchored samples are 72.38 N, 75.01 N, 79.95 N, and 94.21 N. The mean force chain values of the anchored samples were 114.93 N, 111.14 N, 112.38 N, and 112.83 N, which increased by 58.79% compared to the unanchored samples, respectively, 48.17%, 40.56%, and 19.76%. In Figure 12b, as the α_s increase from 30° to 75°, the sum of force chains of the unanchored samples are 238.26 × 10^5 N, 248.65 × 10^5 N, 266.90 × 10^5 N, and 314.01 × 10^5 N. The mean values of force chains of the anchored samples were 359.80 × 10^5 N, 348.60 × 10^5 N, 352.50 × 10^5 N, and 352.80×10^5 N, with increases of 51.01%, 40.20%, 32.07% and 12.35% compared with the unanchored samples.

Next, all force chains inside the fissure unanchored and anchored samples containing different α_s were screened by setting thresholds, which were also selected as 0 N, 50 N, 100 N, and 150 N. Figure 13 illustrates the rose diagrams of the force chains inside the fissure unanchored and anchored samples containing different α_s for different screening thresholds.

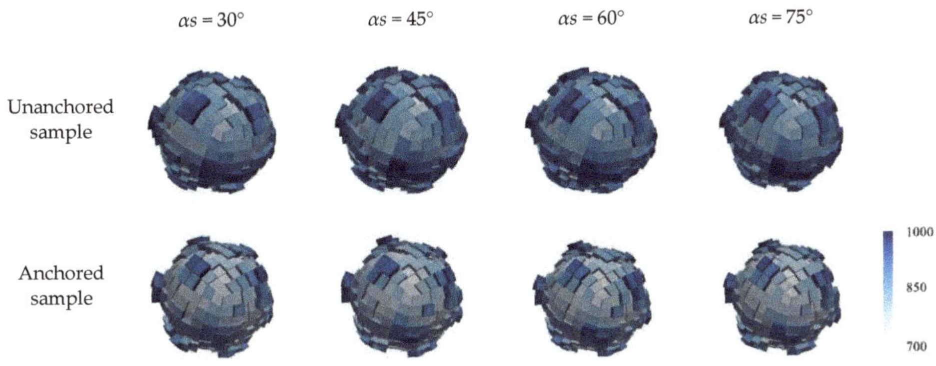

(**a**) 0 N

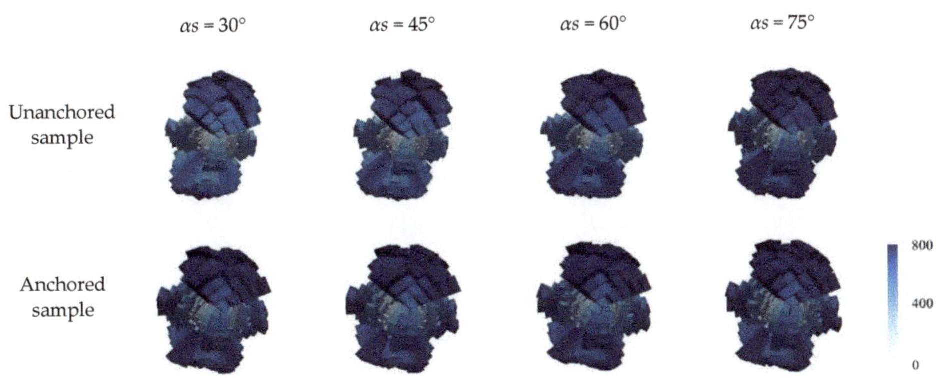

(**b**) 50 N

Figure 13. *Cont.*

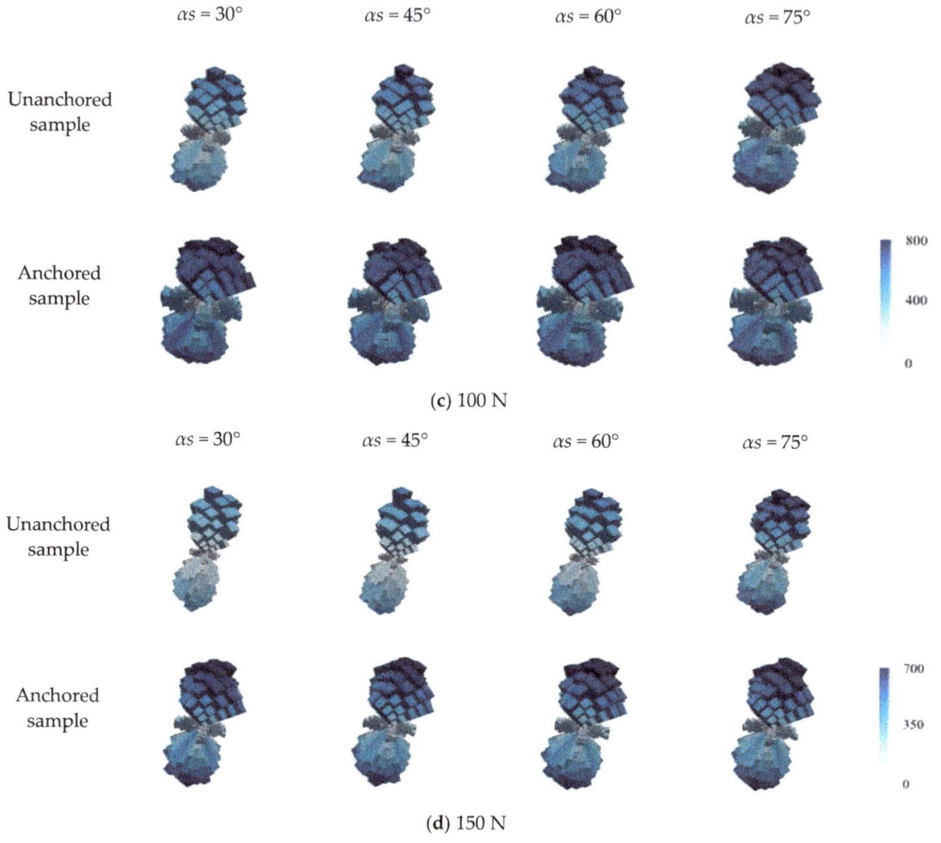

Figure 13. Variations in orientation distribution of force chains in samples with different α_s under different filtering thresholds.

Figure 13a shows the force chain rose diagrams for all force chains that were not threshold screened for force chains. Overall, the force chain rose diagrams for both the anchored and unanchored samples closely resemble a spherical shape, and the volume of the force chain rose diagrams for samples with inclined fractures, whether anchored or unanchored, is similar, with no significant differences in coloration. However, the volume of the rose diagrams of the anchored samples is reduced and the color is lightened to some extent from the unanchored samples compared to the unanchored samples. This implies that the number of force chains inside the unanchored sample is more.

As the sieving threshold increases, the force chain rose diagrams for both the unanchored and anchored samples no longer exhibit a spherical shape. Among samples with the same α_s, when the sieving threshold is increased from 50 N to 150 N, the volume of the force chain rose diagrams for both types of samples gradually decrease. This indicates that only a small number of force chains within the samples are influenced by the external load. Additionally, Figure 13b–d show that after threshold sieving, the volume of the force chain rose diagrams for the anchored samples is larger than that for the unanchored samples. Moreover, as the sieving threshold increases, this phenomenon becomes increasingly evident. This suggests that compared to the unanchored samples, the anchored samples carry a higher level of load within the sample under external loading and are able to retain more force chains with higher force chain values, thereby possessing greater load-bearing capacity.

Figure 14 illustrates a comparative plot of the number of force chains in the unanchored and anchored samples of the fissure containing different α_s screened with different thresholds. As can be observed from Figure 14a, the number of force chains inside the unanchored sample is also consistently larger than that of the anchored sample without threshold screening.

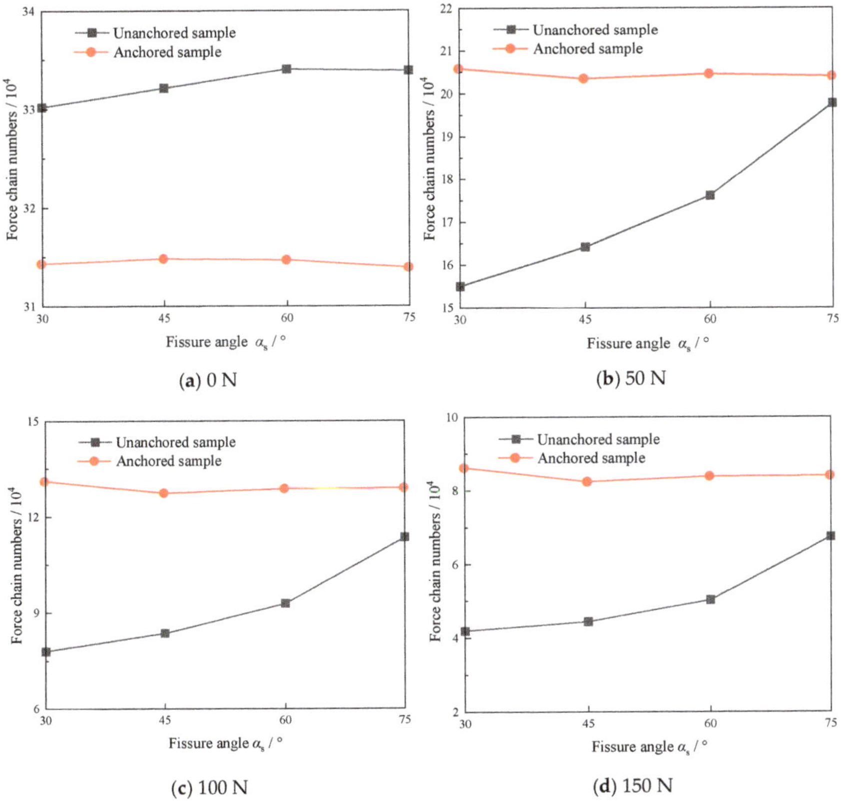

Figure 14. Variations in force chain numbers in samples with different α_s under different filtering thresholds.

In Figure 14b–d, it can be clearly observed that the number of force chains inside the unanchored sample is always lower than that of the anchored sample at different screening thresholds. The increase in the number of force chains of the anchored sample compared to the unanchored sample tends to diminish with the increase in α_s.

Among them, when the force chain screening threshold is set to 50 N, as α_s increases from 30° to 75°, the number of force chains within the anchored samples increases by 32.81%, 23.98%, 16.18%, and 3.23% compared to the unanchored samples, respectively. When the force chain screening threshold is set to 100 N, the number of force chains within the anchored samples increase by 68.30%, 52.53%, 38.75%, and 13.66%, respectively; when the force chain screening threshold is set to 150 N, as α_s increases from 30° to 75°, the number of force chains within the anchored samples increases by 105.67%, 85.75%, 66.99%, and 24.59% compared to the unanchored samples, respectively. From the above results, it can be inferred that the increase in the number of force chains in the anchored samples relative to the unanchored samples gradually decreases, which may be related to the influence of the fracture dip angle on the anchoring effect. The mechanism by which the fracture dip angle affects the anchoring effect warrants further research.

5. Discussion

5.1. Effect of Fissure Angles on Crack Distribution

To further explore the anchoring and crack-stopping effect of rock bolts on fracture samples with different α_s, three load levels before peak load were selected for each of the four dip angle fracture samples. A comparison and analysis were conducted on the changes in crack density cloud diagrams, crack quantity, crack rose diagrams, and the force chain values required to generate individual cracks at different load levels.

Figure 15 displays the internal crack change cloud diagrams of unanchored and anchored samples containing fractures with different α_s under various load levels. It can be observed that as the load level increases for each, the crack density in the unanchored samples with different α_s gradually increases in the fracture tip area. However, at the same load level, the internal crack density of the anchored samples is relatively less compared to the unanchored samples. Additionally, the cloud diagrams clearly show that the anchored samples with a fracture of $\alpha_s = 30°$ have the greatest reduction in crack density compared to the unanchored samples, followed by the samples with a fracture of $\alpha_s = 45°$, while the change in the anchored samples with a fracture of $\alpha_s = 75°$ is the smallest. This phenomenon further suggests that the anchoring and crack-stopping effect of the rock bolts is most pronounced in the samples with a fracture of $\alpha_s = 30°$.

Figure 16 presents the crack number curves and crack rose diagrams of unanchored and anchored samples with fractures of different α_s at their respective different load levels. It can be observed that the internal crack number in the unanchored samples is always higher than that in the anchored samples; hence, the volume of the crack rose diagrams for the unanchored samples is larger, and the color is correspondingly deeper towards blue. However, the increase in the volume of the crack rose diagrams for the anchored samples with different α_s compared to the unanchored samples varies; visually, the increase in volume and color deepening of the crack rose diagrams for the anchored samples with a fracture of $\alpha_s = 30°$ is the most significant, while the increase for the samples with a fracture of $\alpha_s = 75°$ is the smallest.

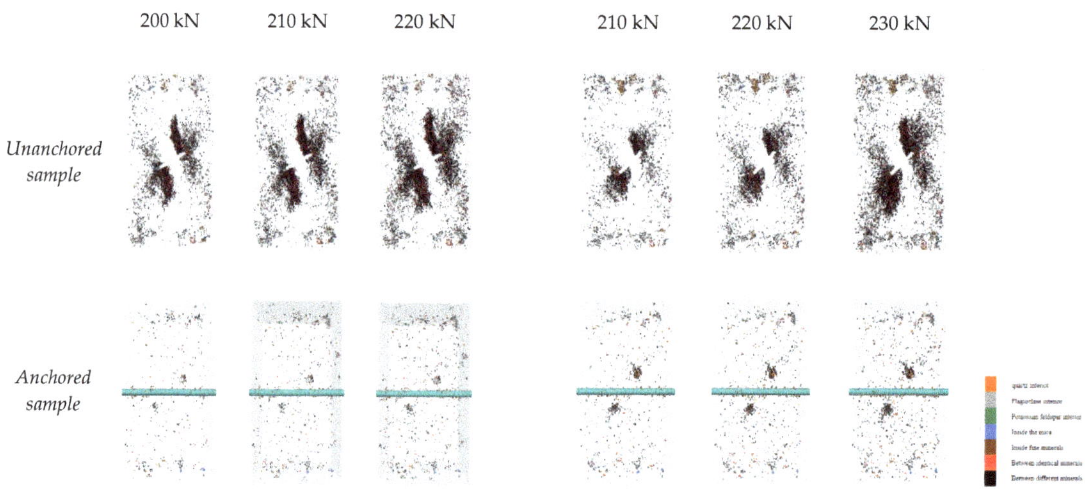

(**a**) $\alpha s = 30°$ (**b**) $\alpha s = 45°$

Figure 15. *Cont.*

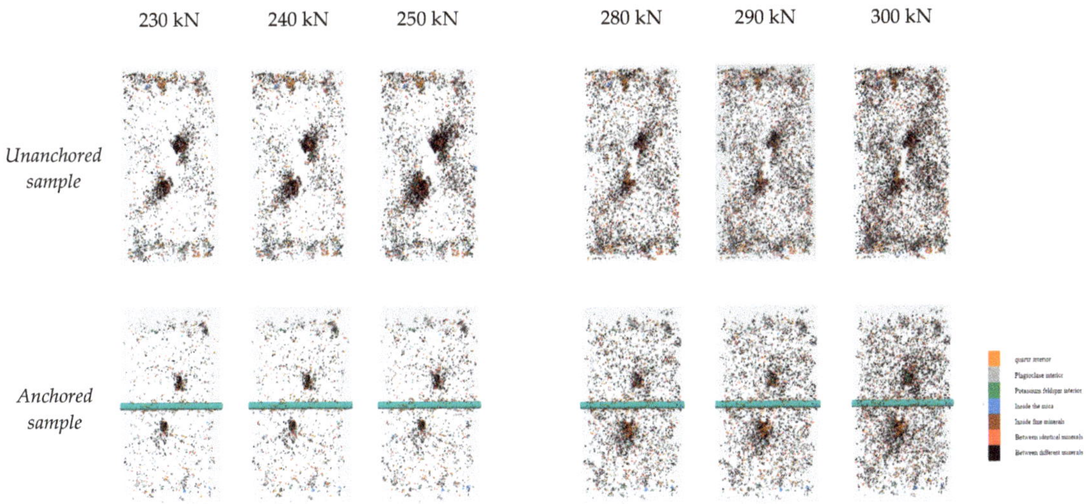

(**c**) αs = 60° (**d**) αs = 75°

Figure 15. Evolution of crack distribution in samples with different α_s before peak load.

The curve diagrams also show that at their highest load levels, the reductions in crack number for the anchored samples with fractures of α_s = 30°, 45°, 60°, and 75° compared to the unanchored samples are 88.25%, 81.99%, 57.85%, and 25.17%, respectively. This further confirms that the anchoring system can significantly reduce the number of cracks and enhance the stability of the rock mass, and also indicates that the fracture dip angle has a certain impact on the anchoring effect.

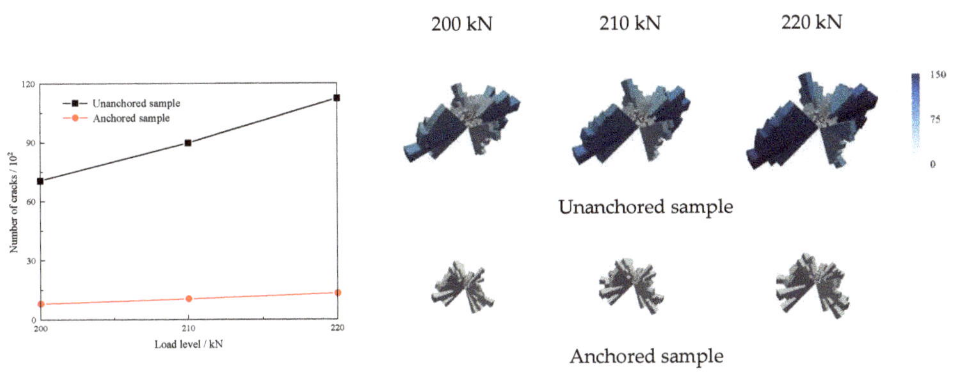

(**a**) αs = 30°

Figure 16. *Cont.*

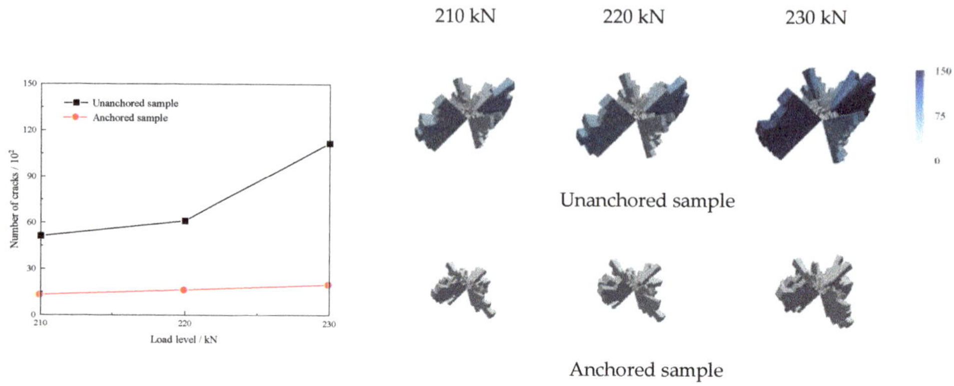

(**b**) $\alpha_s = 45°$

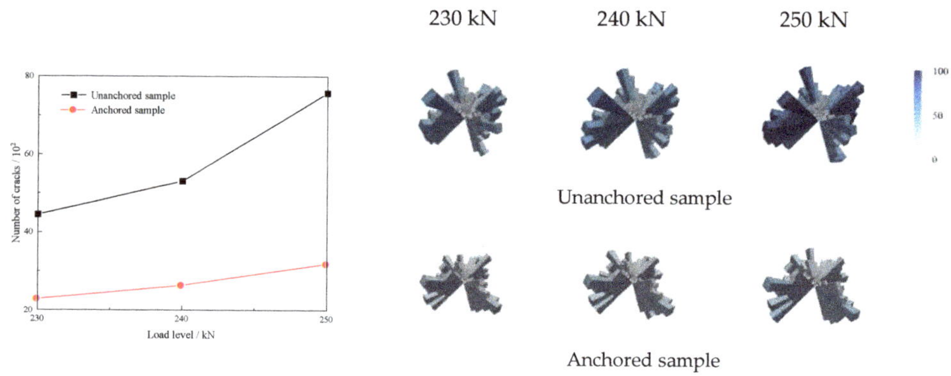

(**c**) $\alpha_s = 60°$

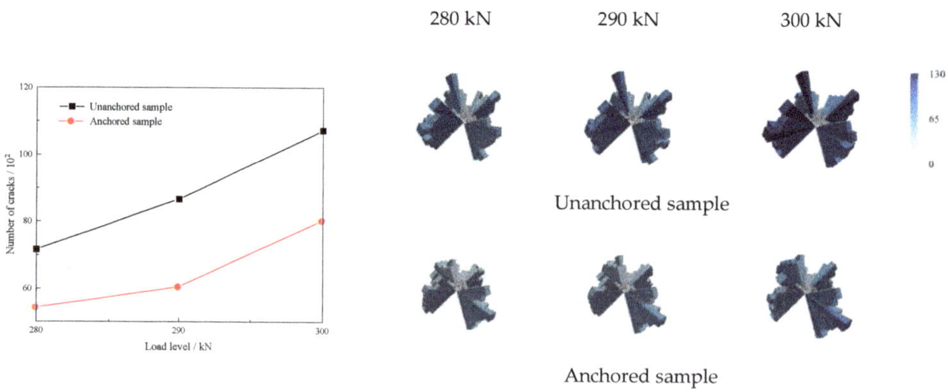

(**d**) $\alpha_s = 75°$

Figure 16. Variations in orientation distribution and numbers of force chains in samples with different α_S under different load levels.

5.2. *Effect of Fissure Angles on the Force Chain Values*

Figure 17 illustrates the change curves of the force chain values required to generate individual cracks within unanchored and anchored samples with different α_S under various load levels. It can be directly observed that the increase in the force chain values needed to produce a single crack within the anchored samples with a fracture of $\alpha_s = 30°$ is the greatest compared to the unanchored samples, as indicated by the area of the red annotation, while the samples with a fracture of $\alpha_s = 75°$ show the smallest increase. These above analysis results show that the crack inclination angle significantly affects the anchoring performance of the specimen. As the crack inclination angle decreases, the anchoring and crack-stopping effects become more pronounced.

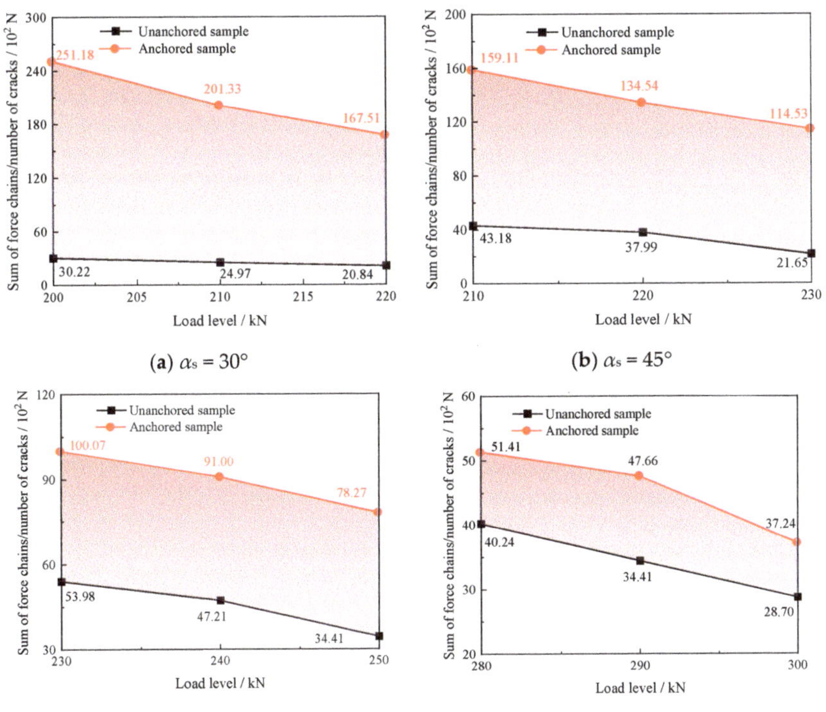

(a) $\alpha_s = 30°$ (b) $\alpha_s = 45°$ (c) $\alpha_s = 60°$ (d) $\alpha_s = 75°$

Figure 17. Variation in force chain values required for single crack initiation in samples with different α_s.

6. Conclusions

In this paper, granite anchorage test samples were prepared and a numerical model of anchored fractured granite samples was constructed. A series of indoor experiments and numerical simulation studies were undertaken, exploring from the perspectives of macroscopic mechanical properties, deformation, and fracture patterns, and the evolution of the force chain network information. The ultimate goal was to reveal the differences in the mechanical behavior under uniaxial compression between unanchored and anchored fractured granite samples with varying α_s, as well as to elucidate the influence of rock bolts on the anchoring and crack-stopping effects in these samples with different α_s. The main conclusions of this paper are as follows:

(1) It was found that there is a strong consistency in the trend of load–displacement results between experimental and simulation results. The macroscopic fracture patterns of unanchored samples all show obvious shear failure, with debris fields concentrated at the fracture tips and distributed along the diagonal of the samples, while the distribution of debris in anchored samples is more uniform. This, in turn, validates the effectiveness of the numerical model parameter calibration.

(2) The number of force chains, the average force chain value, and the total force chain count within the anchored samples all exceed those of the unanchored samples. This suggests that there are fewer cementation fractures within the anchored samples, which implies fewer cracks. Furthermore, as the α_s increase, the increase in force chain characteristics of the anchored samples compared to the unanchored samples decreases.
(3) The results show that the volume of the three-dimensional rose diagrams of the anchored samples with fractures at various dips is greater than that of the unanchored samples. As the α_s increase, the above phenomenon becomes less apparent, which means that the anchored samples have a higher load-bearing capacity. In addition, as α_s increase, the anchoring reinforcement effect of the rock bolts becomes less obvious.
(4) A quantitative exploration was conducted on the anchoring and crack-stopping effect of rock bolts on granite samples with different dip angles of fractures. The results show that at $\alpha_s = 30°$, the rock bolts have the best crack-stopping effect on the samples. As α_s increases, the crack-stopping effect gradually weakens and becomes less apparent. This suggests that in practical engineering, when dealing with fractured rock masses under different working conditions, targeted anchoring and support measures can be applied. This approach can help reduce costs to a certain extent.

Author Contributions: Conceptualization, J.G.; methodology, D.F.; validation, L.Y., T.Z. and B.H.; formal analysis, H.S.; investigation, M.S.; writing—original draft preparation, M.S.; writing—review and editing, B.H., D.F. and M.S. All authors have read and agreed to the published version of the manuscript.

Funding: This research was funded by the National Natural Science Foundation of China (52274100, 42307238, 52409148), the Excellent Postdoctoral Program of Jiangsu Province (Grant No. 2023ZB602) and the China Postdoctoral Science Foundation (Grant No. 2023M733773).

Institutional Review Board Statement: Not applicable.

Informed Consent Statement: Not applicable.

Data Availability Statement: The original contributions presented in the study are included in the article, further inquiries can be directed to the corresponding author.

Conflicts of Interest: Author Jiangfeng Guo was employed by the company Shanxi Huaning Coking Coal Co., Ltd., China Coal Group Huajin Co., Ltd. The remaining authors declare that the research was conducted in the absence of any commercial or financial relationships that could be construed as a potential conflict of interest.

References

1. Kang, H.P.; Gao, F.Q. Evolution of Mining-Induced Stress and Strata Control in Underground Coal Mines. *Chin. J. Rock Mech. Eng.* **2024**, *43*, 1–40.
2. Jiao, K.J.; Fang, Q.B. Optimization Design Study of Bolt Support for Tunnel with High Ground Stress and Steep Bedding Surrounding Rock. *Chin. J. Undergr. Space Eng.* **2019**, *19* (Suppl. S2), 813–818.
3. Zhao, L.L.; Yang, W.B.; Pan, W.T.; Wu, F.Y.; Wang, Z.L.; Yang, Z.C. Study on model test and numerical simulation of layered soft rock tunnel construction with different large deformation grades. *Chin. J. Rock Mech. Eng.* **2024**, *43*, 454–467.
4. Long, J.K.; Zu, Z.Y.; Wang, Y.S.; Luo, W.D. Experimental Study on Linkage anchorage Technique of the prestressed anchors. *J. Min. Saf. Eng.* **2012**, *29*, 662–667.
5. Chen, W.Z.; Wang, L.Y.; Tan, X.J.; Yang, D.S.; Yuan, J.Q.; Yang, J.P. State-of-the-art and development tendency of the underground engineering stability of fractured rock mass. *Chin. J. Rock Mech. Eng.* **2021**, *40*, 1945–1961.
6. Bjurstrom, S. Shear Strength of Hard Rock Joints Reinforced by Grouted Un-Tensioned Bolts. In Proceedings of the 3rd International Congress on Rock Mechanics, Denver, CO, USA, 1–7 September 1974; Volume 2, pp. 1194–1199.
7. Littlejohn, G.S.; Bruce, D.A. *Rock Anchors: State-of-the-Art Part I, Design, Ground Engineering*; Foundation Publications Ltd.: London, UK, 1975; pp. 163–175.
8. Egger, P.; Spang, K. Stability Investigations for Ground Improvement by Rock Bolt at a Large Dam. In Proceedings of the VI International Conference ISRM, Montreal, QC, Canada, 30 August–3 September 1987; pp. 349–354.
9. Spang, K.; Egger, P. Action of fully-grouted bolt in jointed rock and factors of influence. *Rock Mech. Rock Eng.* **1990**, *23*, 201–229. [CrossRef]

10. Bezuijen, A. Compensation Grouting in Sand: Experiments, Field Experiences and Mechanisms. Ph.D. Thesis, Delft University of Technology, Delft, The Netherlands, 2010.
11. Chen, Y. Experimental study and stress analysis of rock bolt anchorage performance. *J. Rock Mech. Geotech. Eng.* **2014**, *6*, 428–437. [CrossRef]
12. Zong, Y.J.; Han, L.J.; Qu, T.; Yang, S.Q. Mechanical properties and failure characteristics of fractured sandstone with grouting and anchorage. *Int. J. Min. Sci. Technol.* **2014**, *24*, 165–170. [CrossRef]
13. Ren, M.Y.; Zhang, Q.Y.; Chen, S.Y.; Zhang, L.Y.; Xiang, W. Experimental study on mechanical properties of anchored rock-like material with weak interlayer under uniaxial compression. *Geotech. Geol. Eng.* **2020**, *38*, 4545–4556. [CrossRef]
14. Ding, S.; Gao, Y.; Jing, H.; Shi, X.; Qi, Y.; Guo, J. Influence of weak interlayer on the mechanical performance of the bolted rock mass with a single free surface in deep mining. *Minerals* **2021**, *11*, 496. [CrossRef]
15. Li, Y.; Zhou, H.; Zhang, L.; Zhu, W.S.; Li, S.C.; Liu, J. Experimental and numerical investigations on mechanical property and reinforcement effect of bolted jointed rock mass. *Constr. Build. Mater.* **2016**, *126*, 843–856. [CrossRef]
16. Yang, W.D.; Luo, G.Y.; Bo, C.J.; Wang, L.; Lü, X.X.; Wang, Y.N.; Wang, X.P. Mechanical properties and reinforcement effect of jointed rock mass with pre-stressed bolt. *J. Cent. South Univ.* **2020**, *27*, 3513–3530. [CrossRef]
17. Yang, S.Q.; Chen, M.; Huang, Y.H.; Jing, H.W.; Ranjith, P.G. An experimental study on fracture evolution mechanism of a non-persistent jointed rock mass with various anchorage effects by DSCM, AE and X-ray CT observations. *Int. J. Rock Mech. Min. Sci.* **2020**, *134*, 104469. [CrossRef]
18. Li, Y.Z.; Tannant, D.D.; Pang, J.Y.; Su, G.S. Experimental and analytical investigation of the shear resistance of a rock joint held by a fully-grouted bolt and subject to large deformations. *Transp. Geotech.* **2021**, *31*, 100671. [CrossRef]
19. Zhang, T. *Investigation of Microscopic Mechanism of Fracture Process of Granite Based on Three-Dimensional GBM and Multi-Level Force Chain Network*; China University of Mining and Technology: Xuzhou, China, 2023.
20. Li, W.; Yu, L.Y.; Zhang, T.; Su, H.J.; Mi, X.Z.; Fan, D.D.; Jin, B. Quantitative Analysis of Grain Size Effect on Tensile Mechanical Behavior of Granite Based on Multi-Level Force Chain Networks. *Comput. Part. Mech.* 2024, *in press*. [CrossRef]
21. Han, G.S.; Jing, H.W.; Zhu, T.T.; Su, H.J. Study on numerical simulation of effect of rock mass fracture on load transfer mechanism of rockbolt. *Coal Technol.* **2015**, *34*, 45–48.
22. Kang, Z.Q.; Zhang, X.Y.; Zhao, J.M.; Zhang, S.Q. Effects of "S" type fractured rock anchoring and the distribution rule of stress field. *Met. Mines.* **2014**, *1*, 46–49.
23. Saadat, M.; Taheri, A. A numerical approach to investigate the effects of rock texture on the damage and crack propagation of a pre-cracked granite. *Comput. Geotech.* **2019**, *111*, 89–111. [CrossRef]
24. Wu, D.Y.; Yu, L.Y.; Su, H.J.; Wu, J.Y.; Liu, R.C.; Zhou, J. Experimental study and PFC3D simulation on crack propagation of fractured rock-like specimens with bolts under uniaxial compression. *Rock Soil Mech.* **2021**, *42*, 1681–1692.
25. Chen, M. *Study on the Deformation Failure behavior and Reinforcement Effect of Bolts of Non-Persistent Jointed Rock*; China University of Mining and Technology: Xuzhou, China, 2019.
26. Xu, Y.; Dai, F.; Xu, N.W.; Zhao, T. Numerical investigation of dynamic rock fracture toughness determination using a semi-circular bend sample in split Hopkinson pressure bar testing. *Rock Mech. Rock Eng.* **2016**, *49*, 731–745. [CrossRef]
27. Huang, Y.H.; Yang, S.Q.; Ranjith, P.G.; Zhao, J. Strength failure behavior and crack evolution mechanism of granite containing pre-existing non-coplanar holes: Experimental study and particle flow modeling. *Comput. Geotech.* **2017**, *88*, 182–198. [CrossRef]
28. Zhang, X.P.; Wang, G.; Jiang, Y.J.; Wu, X.Z.; Wang, Z.C.; Huang, N. Simulation research on granite compression test based on particle discrete element model. *Rock Soil Mech.* **2014**, *35*, 99–105.
29. Du, H.B.; Dai, F.; Xu, Y.; Liu, Y.; Xu, H.N. Numerical investigation on the dynamic strength and failure behavior of rocks under hydrostatic confinement in SHPB testing. *Int. J. Rock Mech. Min. Sci.* **2018**, *108*, 43–57. [CrossRef]
30. Zhang, Y.; Ding, Z.Q.; Fang, C.; Cong, B.T. Three-dimensional Discrete Element Simulation on Macro-Micro Mechanical Properties of Granite under Triaxial Cyclic Loading and Unloading. *Min. Res. Dev.* **2019**, *39*, 149–156.
31. Cowie, S.; Walton, G. The effect of mineralogical parameters on the mechanical properties of granitic rocks. *Eng. Geol.* **2018**, *240*, 204–225. [CrossRef]
32. Luo, X.Y.; Cao, P.; Liu, T.Y.; Zhao, Q.X.; Meng, G.; Fan, Z.; Xie, W.P. Mechanical behaviour of anchored rock containing weak interlayer under uniaxial compression: Laboratory test and coupled DEM-FEM Simulation. *Minerals* **2022**, *12*, 492. [CrossRef]
33. Fan, D.D.; Zhang, T.; Yu, L.Y.; Hu, L.H.; Su, H.J.; Wei, J.B. A numerical study based on GBM3D-DEM model: Multi-level force chain analysis in dynamic flexural tensile strength test. *J. Cent. South Univ.* **2023**, *30*, 3821–3839. [CrossRef]
34. Peng, Y.X.; Zhang, T.; Yu, L.Y.; Li, J.; Gao, Y.A.; Tian, W.L. Numerical investigation on the effect of intergranular contact bonding strength on the mechanical properties of granite using PFC3D-GBM. *Int. J. Numer. Anal. Methods Geomech.* **2023**, *47*, 694–716. [CrossRef]
35. Potyondy, D.O.; Cundall, P.A. A bonded-particle model for rock. *Int. J. Rock Mech. Min. Sci.* **2004**, *41*, 1329–1364. [CrossRef]
36. Potyondy, D.O. The bonded-particle model as a tool for rock mechanics research and application: Current trends and future directions. *Geosyst. Eng.* **2015**, *18*, 1–28. [CrossRef]
37. Peng, J.; Wong, L.N.Y.; Teh, C.I.; Li, Z.H. Modeling micro-cracking behavior of Bukit Timah granite using grain-based model. *Rock Mech. Rock Eng.* **2018**, *51*, 135–154. [CrossRef]
38. Li, X.F.; Zhang, Q.B.; Li, H.B.; Zhao, J. Grain-based discrete element method (GB-DEM) modelling of multi scale fracturing in rocks under dynamic loading. *Rock Mech. Rock Eng.* **2018**, *51*, 3785–3817. [CrossRef]

39. Tian, W.L.; Yang, S.Q.; Huang, Y.H. Particle flow analysis of mechanical behavior and meso-mechanism of sandstone under unloading conditions. *Rock Soil Mech.* **2016**, *37*, 775–782.
40. Zhang, L. Numerical simulations of the direct shear test. *Chem. Eng. Technol.* **2003**, *26*, 109–228.
41. Leśniewska, D.; Nitka, M.; Tejchman, J.; Pietrzak, M. Contact force network evolution in active earth pressure state of granular materials: Photo-elastic tests and DEM. *Granul. Matter* **2020**, *22*, 71. [CrossRef]
42. Wang, S.D.; Miao, Y.H.; Wang, L.B. Effect of grain size composition on mechanical performance requirement for particles in aggregate blend based on photoelastic method. *Constr. Build. Mater.* **2023**, *363*, 129808. [CrossRef]
43. Zhang, T.; Yu, L.Y.; Peng, Y.X.; Jing, H.W.; Su, H.J.; Wei, J.B. Effect of the mineral spatial distribution heterogeneity on the tensile strength of granite: Insights from PFC3D-GBM numerical analysis. *J. Rock Mech. Geotech. Eng.* **2023**, *15*, 1144–1160. [CrossRef]
44. Zhang, T.; Yu, L.Y.; Wu, B.B.; Tan, Y.Z.; Su, H.J.; Zhou, L.J. Influence of grain-to-particle size ratio on the tensile mechanical response of granite based on a novel three-dimensional grain-based model. *Eng. Fract. Mech.* **2022**, *29*, 108161. [CrossRef]
45. Wang, G.; Yuan, K.; Jiang, Y.J.; Wu, X. Macro-micro mechanical on joint -grout-bolt interaction in rockmass subjected to shear loading. *J. Cent. South Univ. Sci. Technol.* **2015**, *46*, 2207–2215.
46. Huang, Y.H.; Yang, S.Q.; Tian, W.L.; Wu, S.Y. Experimental and DEM study on failure behavior and stress distribution of flawed sandstone samples under uniaxial compression. *Theor. Appl. Fract. Mech.* **2022**, *118*, 103266. [CrossRef]

applied sciences

MDPI

Article

Numerical Simulation and Engineering Application of Synergistic Support Effect of Bolt–Mesh–Cable Support in Gob-Side Entry of Deep Soft Coal Seam

Haifeng Ma, Shuo Zhang *, Huaiyi Zhai, Zenghui Liu and Chuang Jie

School of Mining Engineering, Anhui University of Science and Technology, Huainan 232001, China; hfma@aust.edu.cn (H.M.); 2022200301@aust.edu.cn (H.Z.); 2023201967@aust.edu.cn (Z.L.); 2022200335@aust.edu.cn (C.J.)

* Correspondence: 2023201997@aust.edu.cn; Tel.: +86-18155430963

Abstract: Aiming at solving the problem of support failure caused by a large deformation of roadway surrounding rock in a deep soft coal seam, and taking the surrounding rock control of the roadway in the 11-2 coal seam in Zhujidong Coal Mine as the research background, numerical simulation and field industrial test and inspection methods were used to study the support effect of a supporting system of gob-side entry in deep soft coal seam. The deformation characteristics of various supporting systems of metal mesh, diamond mesh, metal mesh with anchor rod, steel ladder beam, M-shaped steel belt, 14#b channel steel, and 11# I-steel in the goaf supporting body of deep soft coal seam were studied under vertical load. The supporting effect of effective compressive stress zone generated by bolt and cable under different row spacings and lengths was analyzed, and the law of variation in the compressive stress field generated by supporting members with supporting parameters was explored. The length and interrow distance of bolt and cable were compared, respectively, and reasonable supporting parameters were selected. Based on the abovementioned research results and the geological conditions of the 1331 (1) track roadway, the support scheme of the 1331 (1) track roadway was designed, and the industrial test was carried out. The results show that the surrounding rock of the roadway is within the effective anchorage range of the supporting body, the active support function of the supporting components has been fully brought into play, and the overall control effect of the surrounding rock of the roadway is good, which can ensure the safety and stability of the goaf roadway. The maximum displacement of the roof and floor of the roadway is 86 mm, the maximum displacement of the solid coal side is 50 mm, the maximum displacement of the coal pillar side is 70 mm, and the maximum separation of layers is 22 mm. There is no failure phenomenon in relation to the anchor bolt and cable, and the overall deformation of the roadway surrounding the rock is good, which can provide some references for roadway-surrounding-rock control under similar conditions in deep coal seams.

Keywords: gob-side entry; combined components of anchor bolt and anchor cable support; stability of roadway; deep soft coal seam; numerical simulation

1. Introduction

Coal plays an important role in China's economic development. For a long time in the future, China's primary energy consumption will be dominated by coal. With the increase in mining depth, the geo-stress will also increase. Due to the combined effect of various factors, the deformation of both sides of the roadway is large, the bottom bulge is serious also, and it will be difficult to meet the requirements of transportation and ventilation, even causing accidents such as spalling, caving, and collapse. The application of large-section gob-side roadways in deep coal seams will cause excessive deformation of the surrounding rock of the roadway due to the excessive ground stress and roadway section, causing a series of support problems. The support technology of gob-side roadway in deep soft coal seams has always been one of the main research focuses in the field of coal mining.

Scholars at home and abroad have put forward many classical theories on the control of the surrounding rock of roadway and achieved a lot of research results. Gao and Wang [1,2] conducted a statistical analysis on the diversity of dangerous sources of deep-coal-mining accidents, established a comprehensive safety index system for deep-well roadway rock accidents, and evaluated the stability of deep-roadway surrounding rock. Plastic deformation and failure first appear on the side of the roadway and then expand to other parts to explore the stability of the surrounding rock during deep mining. A control scheme for the surrounding rock of deep roadways is proposed and verified through on-site industrial tests [3]. The stress concentration area of the surrounding rock of deep roadways is far away from the surrounding area and has a larger size. The maximum concentration coefficient is slightly smaller than that of shallow roadways. There is a large range of broken zones in the surrounding rock of deep soft rock roadways [4]. Aiming to solve the support problem of rock burst in the deep roadway of the 401,111 fully mechanized caving working face, a control scheme for the surrounding rock of the return air roadway is proposed with large-diameter pressure relief and deep hole blasting as the main pressure-relief means [5].

A comprehensive surrounding-rock management method of anchor–grit–cable-grouting coupled support to address the difficulties in supporting deep-well high-stress broken soft rock roadways is proposed, and the interaction mechanism between coupled support and surrounding rock is analyzed [6]. The composite support mode of 'shotcrete + grouting bolt + anchor bolt + grouting cable + anchor cable' is put forward, and the rationality of the co-reinforcement technology of layer and double arch is verified [7]. Based on the instability mechanism of roadway, the control technology of high preload bolt + deep and shallow-hole crack filling is put forward, and this technology ensures the stability of the roadway surrounding rock by inhibiting the expansion of deep and shallow cracks and reinforcing the surrounding rock [8]. The composite failure mechanism of the surrounding rock of coal roadway based on the stress difference between deep and shallow anchor cables is revealed, and an anchoring and unloading coupled control technology is put forward [9]. Based on the aging characteristics of plastic zone expansion and the control effect of bolt support on plastic zone of surrounding rock, the control technology of surrounding rock stability is put forward [10].

A mechanical model of the lateral roof structure of the gob-side roadway is established, the mechanical state of the roadway support body at different mining stages is analyzed, and the dynamic instability mechanism of the support body caused by the fracture and sinking of the lateral basic roof is revealed [11]. The impact-damage mechanism of roadways in extra-thick coal seams is studied, the main influencing factors of top coal damage are revealed, and the anti-impact support technology of roadways in extra-thick coal seams is determined [12]. Taking the common problems of surrounding rock strength deterioration, stress environment deterioration, and structural instability and large deformation of surrounding rock in deep roadways as the starting point, the response mechanism of the crack evolution scale in the macroscopic surrounding rock damage process is discussed [13,14]. The key concerns of experts and scholars on the butterfly failure theory are carried out in research, and the applicability of butterfly failure theory under different roadway cross-sectional shapes and layered surrounding rock conditions is analyzed [15].

The deformation and failure characteristics of the surrounding rock of rock-burst roadways and the main influencing factors are analyzed, and the selection principle of the support form of rock-burst roadways is proposed [16]. A mechanical model of full-length anchor bolt is established, and the stability of the mechanical bearing structure of the surrounding rock of high-stress soft rock roadways after full-length anchor bolts were anchored is studied [17]. The differential spatiotemporal evolution law of the number of cracks and development inflection points at different depths in the surrounding rocks of the two roadways is clarified, and the differential deformation mechanism of the surrounding rocks of the two roadways caused by the differences in stress state and crack development is revealed [18]. Uniaxial compression and true triaxial tests to address the problem of surrounding rock stability are conducted, the method of energy characterization and significance of rocks under uniaxial compression conditions is studied, and the temperature effect of granite cavern rock burst under true triaxial test conditions is explored [19,20].

A new type of large deformation resistance-enhancing anchor cable is developed, and the developed new anchor cable to the actual roadway is applied successfully [21]. The maximum convergence deformation of the two sides is about 300 mm, and a good support effect is achieved. The local instability problem caused by anchor fracture failure in deep roadways under impact dynamic loads is revealed, and the reinforcement support technology to reduce anchor fracture and improve the stability of roadway support body is explored [22]. Combining anchor support with grouting technology, an effective solution for deep soft rock and broken roadway surrounding rock control is provided [23,24]. An intelligent real-time detection and positioning method for coal mine roadway support steel belt anchor hole based on deep learning model and depth camera is proposed, and the feasibility of this method in the unstructured environment of coal mine underground is verified [25].

Based on structural mechanics theory, the constraint stress in the plastic extrusion deformation zone of the anchor is regarded as a rectangular distribution mode, and the rationality of the model is verified by comparing the model calculation results with experimental data [26]. Taking the roadway of 10,607 gas extraction as the research object, the large deformation and damage phenomenon of roadways under dynamic pressure is studied, active and passive support technologies are proposed, the engineering application is carried out, and a good control effect of roadway surrounding rock is obtained [27]. The results of the tensile and shear tests of anchor cables are discussed. By establishing a mechanical model of the interaction between anchor cables and jointed rock mass, the failure mechanism of anchored jointed rock mass under tension and shear was revealed, providing a practical and convenient method for improving the shear strength of the anchor beam anchor cable system and enhancing the stability of the roadway [28]. The changing trends of the physical and mechanical properties of coal rock mass in the strong wind oxidation zone are revealed, the deformation and failure mechanism are analyzed, and a coupling support scheme of 'pre-grouting + anchor mesh shotcrete + inverted arch structure + U-shaped steel + high and low pressure, deep and shallow hole reinforcement grouting' based on the deformation and failure characteristics of the wind-oxidized surrounding rock roadway is proposed [29]. The failure mode of a steel tube–concrete structure is studied, and the elastic center method to obtain the internal force of the steel tube–concrete structure with cable is used. The results show that the cable can ensure that it has a high bearing capacity [30].

In view of the harsh environment of deep mechanized mining face and the seriously unbalanced distribution of driving, supporting, and anchoring time, a new type of highly adaptable advanced support of the mechanized mining roadway with strong adaptability, a large support strength, and a large working space is designed [31]. The problem of stability control of surrounding rock in deep soft rock roadways is explored, the mineral composition and microstructure of the surrounding rock is analyzed, the microscopic mechanism of roadway damage and the deformation and damage mechanism of soft rock cross-cutting roadways are obtained, and a high prestressed compensating support countermeasure based on constant resistance large deformation anchor bolts is proposed [32]. The problems of poor support effect and serious deformation and damage of surrounding rock in mining roadways under deep-mining stress are analyzed. Taking the upper roadway 2424 of Suncun Coal Mine as the background, the stress, deformation, and damage laws of surrounding rock in mining roadways under single support and multi-level support were compared and analyzed. It was found that multi-level support technology has obvious advantages. The key parameters of multi-level support were determined, and field industrial tests were carried out. The results show that the overall deformation of surrounding rock is significantly reduced after multi-level support [33]. The difficulty of stability control in deep roadways is explored, and the coupled support technology of a high prestressed constant resistance and large deformation anchor cable is put forward [34]. It is discussed how the joint distributed in the rock stratum is often affected by complex stress and interlayer dislocation in the roadway with a high ground stress or large buried depth. The anchor bolts and anchor cables anchored in the rock strata are subject to tension and shear forces [35].

The stability and control of the surrounding rock of deep broken soft rock roadways, which are located in the underground mine of Jinfeng Gold Mine in Guizhou Province, China, are studied. A combined support system of high-strength 'cable anchor fiber shotcrete + steel mesh split sleeve resin anchor rod cement grouting' was proposed, which significantly improved the integrity and bearing capacity of the effective bearing structure [36]. The problems of complex advance support technology for roadways in super-kilometer deep mines, the impact on rapid advancement of the working face, and the damage of roof anchors (cables) by advance single hydraulic props are explored, and the deformation characteristics of surrounding rock are analyzed. An active advanced support technology scheme of replacing existing single hydraulic props with grouting anchors within the advance influence range of the working face in super-kilometer deep mines was proposed, which solved the problem of safe and efficient production faced by super-kilometer-deep mines [37].

The stability of mining roadways in deep mines is studied, and the coordinated control effect of long anchor support for deep roadways under strong mining disturbance of adjacent working faces in Wenjiapo Coal Mine in West China is analyzed. The results show that extending the anchor length can reduce the consumption of roof anchors while ensuring the support effect, and it can reduce the bottom heave of the roadway by 5.5% [38]. Aiming at addressing the influence of mining stress on the stability of the surrounding rock of deep inclined-roof mining roadways, full-length anchor bolt support technology based on the mining roadway of the 17,102 (3) working face of Pansan Coal Mine was studied, and it achieved good control effect [39]. Three different support technologies according to the surrounding rock deformation and engineering technology characteristics are adopted [40]. The monitoring results showed that the fully enclosed U-shaped steel ring support technology had the best support effect, the roadway deformation met the production requirements, the roadway renovation time was longer, and it could meet the surrounding rock requirements of roadways with a burial depth of 1000 m.

The stress evolution, displacement field, local deformation, overall distribution, and failure characteristics of the surrounding rock anchoring structure with different anchor spacing were studied [41]. The influence of anchor preload and spacing on the support strength of the surrounding rock anchoring structure was analyzed. A support scheme for the excavated roadway was designed, and the effectiveness of the support scheme was verified by roadway displacement measurement [42]. A design method for roof anchoring support using the anchor potential design method based on the analysis of roadway roof bending failure is proposed. The support problem of deep soft rock roadways in Chaohu Coal Mine is analyzed, and the main reasons for the instability of the original U-shaped steel support roadways in Chaohu Coal Mine are analyzed [43]. The instability of the original support is the result of the passive support of deep high-stress soft rock roadways being unable to adapt to the deformation of the surrounding rock. The failure of the original support is mainly caused by unreasonable support parameters. In response to the problem of surrounding rock deformation in deep soft rock roadways, a high-strength and high-stability secondary anchor net support technology was proposed. Due to the tensile force generated by the deformation and movement of the roof surrounding rock, the axial force of the anchor cable first increases at a decreasing rate as the calculation time increases, and then it reaches an almost-constant value [44].

Many scholars and experts have carried out a lot of research on the surrounding rock control of roadways in coal mine, and many research results were achieved, which promoted the research progress of stability control of the roadway effectively. However, there are certain differences in the geological environment, stress environment, and other factors where the roadway is located in different coal mines, and even great differences. To ensure the stability of the roadway, it is necessary to determine it based on the specific geological conditions of the roadway, the mining disturbance of the adjacent working face, etc. Numerical simulations, industrial tests, and other methods are used to study the stability of gob-side roadways in deep soft coal in order to provide reference for the reasonable support design of deep mine roadways.

With the background of surrounding rock control of goaf roadway of 11-2 coal seam in Zhujidong Coal Mine, the deformation characteristics of various supporting systems, such as metal mesh, diamond mesh, metal mesh with anchor rod, and M-shaped steel belt in goaf roadway supporting body of deep soft coal seam under vertical load are studied by using numerical simulation and on-site industrial test and inspection methods. The supporting effect of effective compressive stress zone of bolt and cable under different spacings and lengths is analyzed, and reasonable supporting parameters are selected. The research results can provide reference values for the stability control of goaf roadway surrounding rock in deep soft coal seams.

2. Engineering Background

2.1. Relationship between the Roadway and the Working Face

Zhujidong Mine is located in Panji District, Huainan City, China. The 1321 (1) working face is located in Dongsanpan District in Zhujidong coal mine, which is the first mining face of Dongsanpan District. It reaches the boundary of the mine field in the east and the main root roadway in Dongsanpan District in the west. The mining strike length of the 1321 (1) working face is 1734 m, the inclination length is 200 m, the elevation of the floor of the roadway ranges from −932 to −983.6 m, and the average buried depth is about 950 m. The 1331 (1) working face extends to the boundary of the mine field in the east, to the west to the root roadway of Dongsanpan District, and to the north to the 1321 (1) working face; there is no mining activity in the corresponding overlying 13-1 coal seam.

The mining strike length of the 1331 (1) working face is 1707.9 m, the inclination length is 220 m, the elevation of the floor of the roadway ranges from −981.8 to −894.1 m, and the average buried depth is about 950 m. After the completion of mining of 1321 (1) working face, the 1331 (1) track roadway is driven along the gob, and the width of coal pillar is 6 m between the 1331 (1) track roadway and the 1321 (1) haulage roadway.

The position relationship between the 1331 (1) track roadway and the 1321 (1) working face is shown in Figure 1. The design length of the 1331 (1) track roadway is 1776 m, rectangular section, using anchor–beam–mesh–cable support; the dimensions of the section are 5.4 m wide and 3.6 m high, and the width of the small coal pillar along the adjacent goaf is 6 m. The driving mode of the 1331 (1) track roadway is to use a comprehensive excavator for full section, and the roadway is driving along the 11-2 coal seam during excavation.

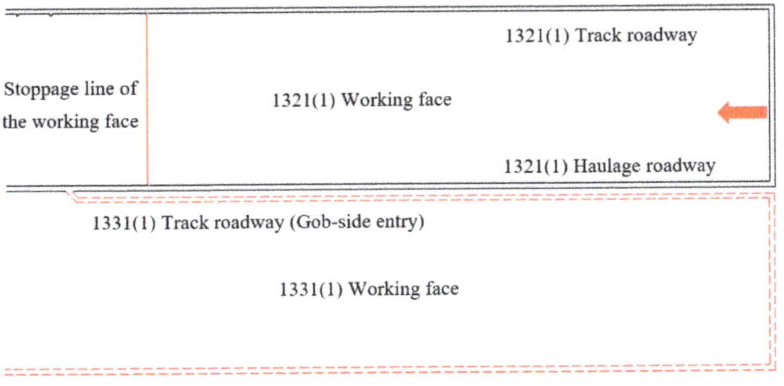

Figure 1. Layout of working face and roadway.

2.2. Occurrence Condition of Coal Seam and Roof and Floor

The overall structural shape of this working face is monoclinic structure, which is high in the west and low in the east. The dip angle of most of the coal seam is from 0° to 5°, with an average of 3°. The color of 11-2 coal seam is black, mainly massive, a small amount of granular, powdered, containing dark coal and vitritic coal strips, belongs to semi-dark~semi-bright briquette, mainly bright coal.

(1) State of coal seam

The coal seam is generally monoclinal structure, the general trend is high in the west and low in the east, the occurrence of the normal coal (rock) layer is from 80° to 160°, the dip angle is from 0° to 5°, the average is 2°, and the Platts coefficient is from about 0.6 to 0.65. The coal seam is mainly black and massive, with a small amount of granular and powdery, containing dark coal and vitritic coal strips, belonging to semi-dark~semi-bright briquette, mainly bright coal.

(2) Condition of roof and floor

The upper roof is fine sandstone, gray, fine texture structure, dense, massive, the main components are feldspar, quartz, argillaceous cement.

The direct roof is mudstone, gray, muddy structure, dense, massive, containing silty and fossil fragments, developed 1-to-2 layers of unstable coal line.

The direct bottom is mudstone, gray, muddy structure, locally intercalated with thin layer of fine sand.

The bottom is fine sandstone, light gray to gray, fine-grained structure, with a few coarse grains at the bottom, dense, massive, mainly composed of feldspar, quartz.

The lithological characteristics of the top and bottom plates of the coal seam are shown in Table 1.

Table 1. Lithologic characteristics of coal seam roof and floor.

Name of Roof and Floor	Lithology	Thickness (m)	Lithologic Character
Upper roof	Fine sandstone	3.4~11.9 6.0	Gray, fine-grained structure, dense, massive, the main components of feldspar, quartz, argillaceous cement.
Direct roof	Mudstone and coal line interlayer	1.6~4.5 4.2	Gray, muddy structure, dense, massive, containing silty and fossil fragments, developed 1 to 2 layers of unstable coal line.
Direct bottom	Mudstone	0.4~3.6 1.7	Gray, muddy structure, locally intercalated with thin layer of fine sand.
Bottom	Fine sandstone	15.3~48.7 27.2	Light gray to gray, fine-grained structure, with a few coarse grains at the bottom, dense, massive, mainly composed of feldspar, quartz.

2.3. Geology of the Working Face

According to the analysis of the actual field data of the working face, the roof sandstone is relatively rich in water, which belongs to the static storage type. There may be a small amount of dripping water in the structural or fracture development block section, but it has no effect on the mining of the working face. The original gas content of coal seam face is from 3.48 to 5.40 5.40 m^3/t, the maximum original gas pressure is 0.91 MPa, and the absolute gas emission in mining process is 55 m^3/min. Coal dust has an explosion risk. The spontaneous combustion tendency of coal seam is from not easy to spontaneous combustion, which belongs to class II spontaneous combustion. The ground temperature is greater than 37 °C, which is in the state of secondary heat damage. The working face is located in deep depth and has a large ground stress.

The geological structure of 1321 (1) working face is as follows:

(1) The overall structure of the working face is monoclinal, high in the west and low in the east, and the rock dip angle is from 1° to 6°, with an average of 2°.
(2) According to the surrounding roadway disclosure data and three-dimensional seismic interpretation analysis, there are 19 faults developed in this working face, with the middle fall, there are two faults whose fall is greater than or equal to 5 m, there are 5 faults with a drop of 2-to-5 m, and 12 faults with a drop of less than 2 m. According to the design of the roadway, it is expected that the 1321 (1) haulage roadway will pass through 16 faults, and the 1321 (1) track roadway will pass through 7 faults.

The specific parameters are shown in Table 2.

Table 2. Fault parameters of 1321 (1) working face.

Construction Name	Trend (°)	Tendency (°)	Dip Angle (°)	Falling Head (°)
DF326-4	43	313	44~64	1.2
DF326	13~37	283~307	30~60	2.2
DF326-1	11~25	101~115	57~65	2~3.5
DF326-2	20	110	42	0.7
DF327-1	19~42	109~132	37~65	1.1~6.0
DF327	24~47	114~137	45	5.0
DF327-4	41~72	131~162	43~63	1.0~3.0
DF319-1	40	310	65	1.3
DF319-2	46	136	59~70	1.3
DF319-5	11	101	57	1.0
DF319	1~25	271~295	45	2~2.5
DF322-1	15	105	45~80	1.7
DF326-3	80	170	75	1.2
DF327-3	15	105	80	1.5
DF327-5	42	132	47~59	2.0
DF322-1-1	162	252	41	1.6
DF322-2	19	109	40	3.0
DF322-2-1	85	175	80	2.0
DF322-3	18	108	75	1.0

3. Analysis of Deformation and Failure Characteristics of Deep Gob-Side Entry

The reasons for the deformation of deep roadway are diverse, involving geological conditions, local structures, the design and construction of the tunnel itself, etc. The on-site investigation of the deformation of the roadway along goaf in deep soft coal seam in Zhujidong Coal Mine and the roadway in Dongtan Mine under similar conditions found that, due to the large depth of the working face and the low strength of the raw coal, the roof and side of the roadway have deformed to different degrees, and the phenomena of steel strip collapse and anchor bolt failure have appeared in some sections of the roadway. As shown in Figure 2.

According to the on-site investigation of the roadway along the goaf in Zhujidong Coal Mine, the following reasons are the main causes.

(1) The roadway section is large, the roof strength is low, and it is difficult to set up top coal support.

The excavation cross-sectional area of the roadway along the goaf is close to 20 square meters, the roadway span is large, the roof is prone to flexural deformation, and the probability of roof rock separation, sinking, and collapse increases. The overall strength of the direct roof is low, mainly due to the complex rock properties of the direct roof and the development of cracks, thus greatly reducing the stability of the roadway surrounding rock, and part of the roof has netted phenomenon, which is difficult to support.

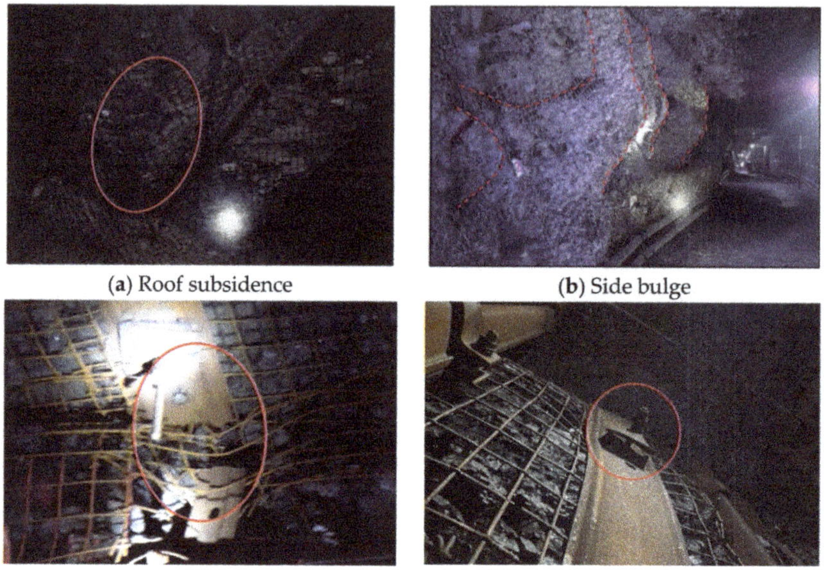

(**a**) Roof subsidence (**b**) Side bulge

(**c**) Steel strip breakage (**d**) Bolt failure

Figure 2. In situ deformation of roadway [45].

(2) The size of the coal pillar is small, the coal quality is relatively soft, and it contains interlayers.

The size of the coal pillar left is small, with a width of 7 m, and the coal quality is weak, soft and broken. The asymmetric deformation of the two sides is obvious, especially the side squeezing of the small coal pillar is serious. The coal seam contains interlayers, and the roadway excavation causes the interlayers to be exposed, and the weak surface is easily disturbed.

(3) The depth of the working face is large, the geo-stress is high, the existing support control effect is poor, and the deformation of the roadway is large.

The average buried depth of the roadway is 950 m. The long-term rheology of the roadway under high stress will increase the deformation rate of the roadway and affect the stability of the roadway surrounding rock. Under the existing support scheme, the support components have collapsed and failed, losing the surface protection ability, and the investment and loss of support materials are serious.

Deep rock formations usually have high ground stress, and the rock strength and integrity are relatively poor, which is easy to deform. For example, weak rock formations such as mudstone and shale have weak bearing capacity and are prone to large deformation under the influence of mining.

In summary, this type of roadway deformation is serious, and roof control is difficult. The existing roadway support means do not match the support needs, which seriously affects the daily safety production of the mine. Therefore, the impact of mining on the coal pillar should be analyzed, and a reasonable support scheme should be designed to achieve the stability of the roadway surrounding rock, thereby achieving safe and efficient production.

4. Simulation of Combined Components of Anchor Bolt and Anchor Cable Support

4.1. Establishment of Combined Components of Anchor Bolt and Anchor Cable Support Model

The displacement and deformation of steel mesh support system is a complex three-dimensional problem, and the displacement deformation of each part is difficult to be continuously tested by laboratory monitoring. In this paper, the anti-deformation ability of six kinds of support systems is numerically simulated and analyzed by finite element analysis software Ansys Workbench 2020. By extracting the displacement deformation of different nodes and fitting the overall displacement deformation curve of the steel mesh support system, the displacement constraint effect of the bolt support composite components on the steel mesh is further analyzed.

The simulation model is established by using SolidWorks modeling software 2020, in which the row distance between anchors is set to 800 mm, the mesh size of metal mesh is 100 mm × 100 mm, and the diameter of wire is 6 mm. The size of diamond mesh is 40 mm × 40 mm, and the diameter of wire is 3.25 mm. The thickness of M5 section steel strip is 5 mm and the width is 173 mm. The diameter of the reinforced ladder beam is 12 mm, 14 mm, and 16 mm. The material parameters of 14#b channel steel and 11# I-beam are shown in Table 3. The numerical model is imported into Ansys workbench for calculation, metal mesh and diamond mesh, the material is set to Q235 structural steel, the elastic modulus is set to 200 GPa, the yield strength is set to 235 MPa, the ultimate tensile strength is 400 MPa, and the Poisson's ratio is 0.3. The channel and strip materials are set to Q345 structural steel, the elastic modulus of the material is set to 200 GPa, the yield strength is set to 380 MPa, the ultimate tensile strength is 550 MPa, and the Poisson's ratio is 0.3. The I-beam is 20 MnK structural steel, the elastic modulus of the material is set to 200 GPa, the yield strength is set to 400 MPa, the ultimate tensile strength is 550 MPa, and the Poisson's ratio is 0.3. Among them, the numerical model is meshed by solid elements, and the contacts between the models are treated as 'binding'; that is, it is considered that the mesh wire located in the combined component of bolt support and the lower part of the pallet will not produce horizontal slip phenomenon. The boundary fixing method is the cross section of metal mesh wire and the fixed constraint at the anchor hole of the tray, which simulates the boundary binding wire fixation and anchor constraint of the underground steel mesh, respectively. The constitutive model is selected as the equal strength hardening multilinear elastic–plastic model, and the analysis method is the nonlinear large deformation static analysis method. The displacement and deformation of 6 supporting systems of metal mesh, diamond mesh, steel ladder beam, M-shaped steel belt, 14#b channel steel and 11# I-beam under 15,000 newtons (15 KN) vertical load is simulated.

Table 3. Material parameters of 14#b channel steel and 11# mine I-beam.

Parameter Number		14#b Channel Steel	T3 Steel Strip	11# MiningI-Beam
Material		16 Mn (Q345)	Q235 (Q345)	20 MnK
Sectional area (cm^2)		21.3	18.3	33.2
Theoretical weight (kg/m)		16.7	14	26.1
Mechanical property	Section coefficient, W_y (cm^3)	14.1	8.1	28.4
	Moment of inertia, I_y (cm^4)	61.1	13.8	127.7
	Yield strength, σ_s (MPa)	≥325	≥345	≥355
	Tensile strength, σ_b (MPa)	≥490	≥490	≥510
	Allowable bending moment M (kN · m)	4.23	2.63	10.1
	Carrying capacity, P (kN)	16.9	10.52	40.4

The material parameters of 14#b channel steel and 11# I-beam used in the numerical simulation were provided by the Technical Department of Zhujidong Coal Mine. The material parameters used in the numerical simulation are shown in Table 3.

In ANSYS model, the translational and rotational degrees of freedom are constrained by the nodes at the holes fixed by the anchor bolt.

The convergence criterion of ANSYS simulation is used to control the convergence, and the 2-norm of residual force is used to control the convergence. The convergence accuracy is 0.1% by default, but it can be relaxed to 5% in general to improve the convergence speed.

The supporting component model is shown in Figure 3.

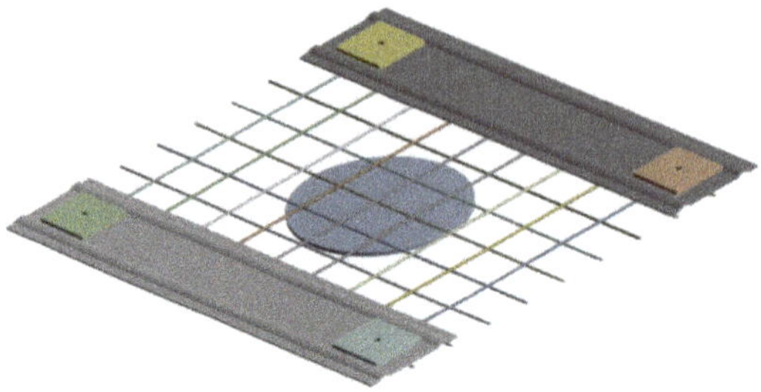

(**a**) Model of M5 steel strip

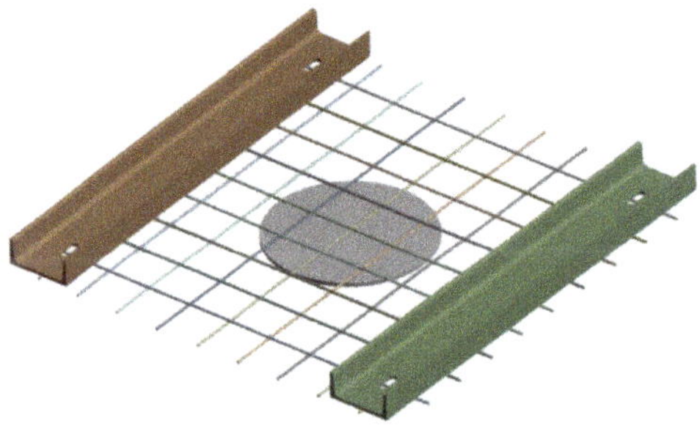

(**b**) Model of 14#b channel steel

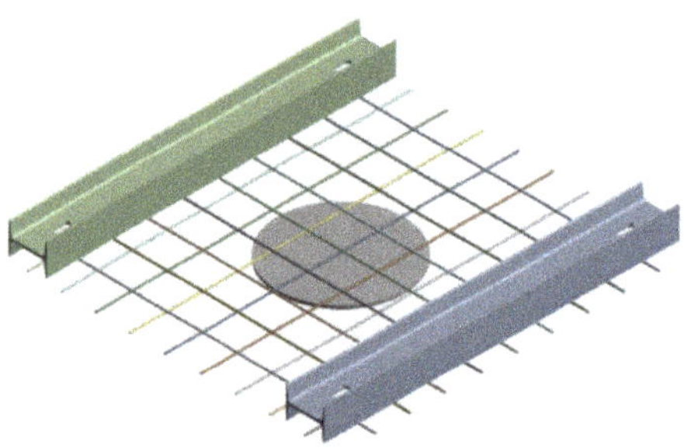

(**c**) Model of 11# I-beam

Figure 3. Model of supporting component.

4.2. Simulation Results and Analysis

(1) Deformation analysis of M5 steel strip structure under load

Figure 4 shows the deformation diagram of M5 steel strip under load. The thickness of M5 steel strip is 5 mm, the width is 173 mm, the material of the steel strip is set as Q345 structural steel, the elastic modulus of the material is set as 200 GPa, the yield strength is set as 350 MPa, the ultimate tensile strength is 550 MPa, and the Poisson's ratio is 0.3.

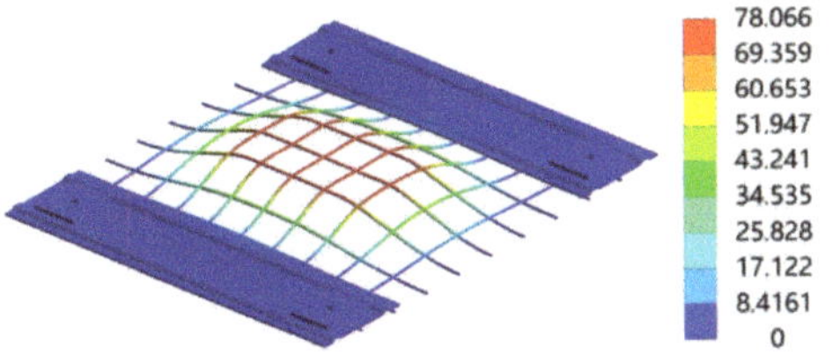

Figure 4. Deformation diagram of M5 steel strip under load (unit: mm).

In M-shaped steel strip support system, there is a great difference between longitudinal wire deformation and transverse wire deformation. Compared with the longitudinal wire, the arch width of the deformation curve of the transverse wire is narrower and the arch curve is more convex. As a result, it can be concluded that with the addition of bolt support composite components, the deformation of steel mesh shows asymmetry in longitudinal and transverse direction, and the restraint effect on the deformation of transverse mesh wire is obviously higher than that of longitudinal mesh wire. Compared with the reinforcement joist, the asymmetric characteristic of M-shaped steel belt is more obvious. The displacement cloud diagram of M-shaped steel strip under vertical load shows that there is a large deformation in the middle of the span, which gradually decreases during the diffusion process from the middle of the span to both sides, and there is basically no displacement deformation at the long hole due to the constraint of the anchor, and with the gradual increase in the thickness of the steel strip, the deformation area decreases gradually. Under the action of fixed load, the stress value of M-shaped steel strip is higher in the installation position of anchor tray and the loading position in the middle of span, in which the stress concentration in the position of pallet makes it easier for the steel strip to tear and break down there. The stress value of the loading part in the middle of the span is higher, indicating that the vertical displacement of the steel strip is mainly caused by the plastic deformation in the middle of the span.

(2) Deformation analysis of reinforced ladder beam under load

Figure 5 shows the deformation diagram of the reinforced ladder beam under load. The diameters of the reinforced ladder beam are, respectively, 12 mm and 14 mm.

In the steel ladder beam combination system, the arch width of the longitudinal wire deformation curve is narrower than that of the transverse wire deformation curve. In the reinforced ladder beam support system, the longitudinal mesh of measuring point No. 2 has a displacement increase of 18 mm compared with measuring point No. 1, the displacement of measuring point No. 3 increased by 18 mm compared with measuring point No. 2, while the transverse mesh wire corresponding to the measurement point displacement increase is 15 mm and 14 mm, the displacement increase compared to the longitudinal mesh wire is reduced by 3 mm and 4 mm, respectively.

When the tension of the steel ladder beam on the fixed position of the anchor is greater than the maximum static friction, the steel ladder beam gradually produces the phenomenon of horizontal slip in the lower part of the pallet, the rising speed of the load curve is slow, and the initial support stiffness is low. When the lateral transverse bar of the steel ladder beam is in direct contact with the anchor, and there is no slip space, the model of the steel ladder beam under load becomes a three-point bending model with both sides fixed and loaded in the middle of the span. In this stage, the rising speed of the load–displacement curve is constant, the double longitudinal bars of the reinforced ladder beam are stretched at the same time under the vertical load, and the components produce a lot of elastic–plastic deformation. With the thickening of the diameter of round steel, the ultimate load carrying capacity of the reinforced ladder beam is gradually increased, in which the two adjacent models with 14 mm diameter are 32.15 mm more resistant to deformation than the 12 mm diameter reinforced ladder beam. The bearing capacity of reinforced ladder beam shows a nonlinear increasing trend, and the increasing range decreases gradually, in which the ultimate bearing capacity of steel ladder beam with round steel diameter of 12 mm and 14 mm has a small difference. With the increase in round steel diameter, the supporting stiffness of reinforced ladder beam shows a nonlinear increasing trend, and the increasing range decreases gradually.

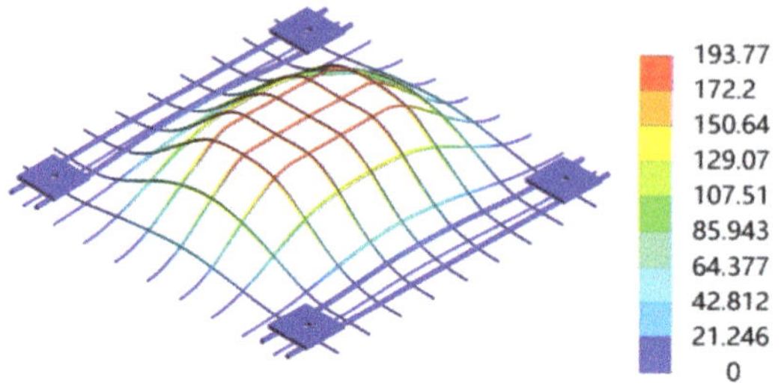

(**a**) The diameter of the reinforced ladder beam is 12 mm

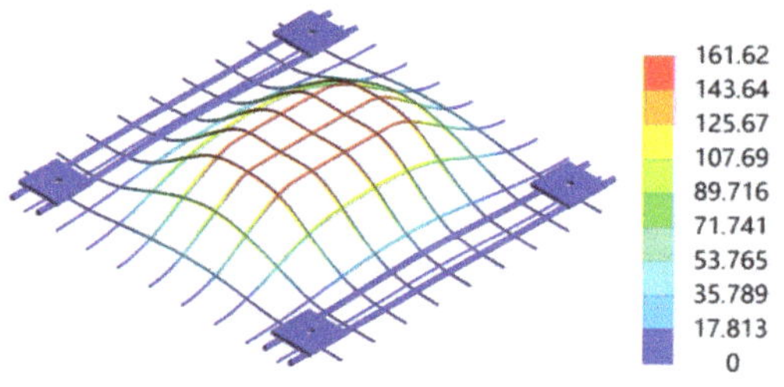

(**b**) The diameter of the reinforced ladder beam is 14 mm

Figure 5. Deformation diagram of reinforced ladder beam under load (unit: mm).

(3) Deformation analysis of 14#b channel steel structure under load

Figure 6 shows the deformation diagram of 14#b channel steel structure under load. The material of channel 14#b is set as Q345 structural steel, the elastic modulus of the material is set as 200 GPa, the yield strength is set as 380 MPa, the ultimate tensile strength is 550 MPa, and the Poisson's ratio is 0.3.

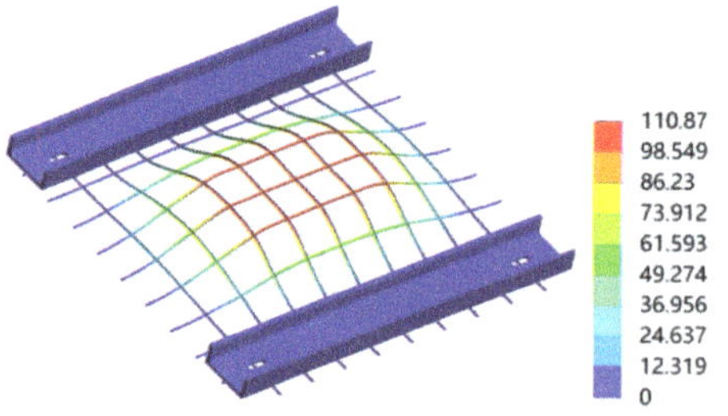

Figure 6. Deformation diagram of 14#b channel steel structure under load (unit: mm).

In the 14#b channel steel support system, the deformation of longitudinal mesh wire is different from that of transverse mesh wire. Compared with the longitudinal wire, the arch width of the deformation curve of the transverse wire is narrower and the arch curve is more convex. As a result, the addition of support composite components of bolt makes the deformation of steel mesh show an asymmetry in vertical and transverse direction, and the restraint effect on the deformation of transverse mesh wire is obviously higher than that of longitudinal mesh wire.

According to the displacement cloud diagram of channel 14#b steel under vertical load, it can be seen that the 14#b steel has a large amount of deformation in the mid-span position, and the deformation gradually decreases in the process of diffusion from the mid-span position to both sides. There is basically no displacement deformation at the long hole due to the constraint of the anchor rod, and the 14#b channel steel has its own displacement distribution rule under the action of vertical load. An 'elliptical' displacement distribution area is formed in contact with the pallet, and a higher stress value is produced at the inner edge of the anchor pallet, but the displacement deformation is small, and the displacement increases gradually in the process of expanding to the center. The mid-span displacement of 14#b channel steel is the largest and has a wide distribution range, which is the main plastic deformation area. In the 14#b channel steel, the bottoms of the two pressure slots are subjected to greater load in the installation position of the bolt tray and the middle part of the channel steel span, and the displacement and deformation are also the largest.

(4) Deformation analysis of 11# I-steel structure under load

Figure 7 shows the deformation diagram of 11# I-beam structure under load. 11# I-beam is 20 MnK structural steel. The elastic modulus of the material is set at 200 GPa, the yield strength is set at 400 MPa, the ultimate tensile strength is 550 MPa, and the Poisson's ratio is 0.3.

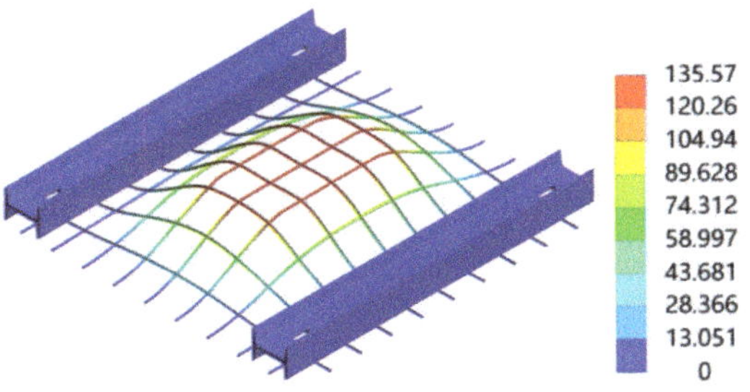

Figure 7. Deformation diagram of 11# I-beam structure under load (Unit: mm).

In the 11# I-beam support system, the deformation curve of longitudinal mesh wire is different from that of transverse mesh wire, which is similar to that of 14#b channel steel. Compared with the longitudinal mesh wire, the arch width of the transverse mesh deformation curve is narrower, and the arch curve is more convex. With the addition of support composite components of bolt, the deformation of steel mesh shows an asymmetry in longitudinal and transverse direction, and the restraint effect on the deformation of transverse mesh is obviously higher than that of longitudinal mesh. The deformation of the 11# I-steel belt directly in contact with the bolt tray is in the surrounding area, and in the process of spreading to the surrounding area, the deformation gradually increases, showing an 'elliptical' displacement distribution state. The displacement of 11# I-steel appears at the sharp corner of the bolt tray, and the displacement gradually increases during the diffusion process along the edge of the tray to the center. During the loading process of 11# I-steel, the stress value of the mid-span part is large, and the distribution range is wide, and the metal material in this area enters the plastic yield stage, and the material causes a lot of plastic deformation.

To sum up, under the action of vertical fixed load, the 11# I-steel is subjected to higher stress values and the largest displacement at the installation position and mid-span loading position of the anchor bolt tray, respectively. The stress concentration phenomenon at the position of the tray makes the 11# I-steel more likely to tear and crush damage there, resulting in the largest displacement. The results show that the vertical displacement of 11# I-steel is mainly caused by the plastic deformation in the middle span.

5. Simulation of Synergistic Support Effect of Bolt–Cable Support

5.1. Numerical Simulation Scheme of Bolt–Cable Support Parameters Model

By using the FLAC three-dimensional finite element difference method, the support effects of different support parameters are numerically simulated and compared to find the optimal support solution. During the simulation, due to the large, buried depth of the roadway, the surrounding rock stress is much greater than the additional stress field generated by the anchor bolt (anchor cable) support, resulting in the formation of additional stress field diffusion is not obvious, so in the support simulation process, the original rock stress will be cancelled to study the additional stress field formed by the support construction. In order to obtain the best support solution, the support simulation parameters are shown in Table 4.

Table 4. Support parameters of numerical simulation.

Anchor Bolt on Roof					Anchor Cable	
Roof Bolt		Side Bolt		Array Pitch/mm	Roof Cable/Root	Side Cable/Root
Length/mm	Spacing/mm	Length/mm	Spacing /mm		2	0
2700		2400			3	1
2800	700~1000	2500	700~1000	800~1000	4	2
2900		2600			5	3

The convergence standard of the numerical simulation model is that the maximum unbalance force reaches 1×10^{-5}, and the default convergence value of this model is also 1×10^{-5}. The rule of grid division is that the boundary grid of the model is relatively sparse, and the closer the coal pillar and the two sides of the working face, the denser the grid. The grid size of the sparse boundary part is 3 m, and the grid size is gradually reduced, and the grid near the coal pillar and on the two sides of the working face is compressed to 0.5 m. The purpose of this division is to better observe the stress, strain, and plastic zone. The left and right borders and the bottom are fixed, and the top vertical direction is applied with self-weight stress.

5.2. Supporting Effect Simulation Results and Analysis

(1) Optimization of bolt spacing and length in the roof of the roadway

When the length of bolt in the roof of the roadway is 2700 mm, the distribution of the vertical stress field of the surrounding rock with different spacings is shown in Figure 8.

When the length of bolt in the roof of the roadway is 2800 mm, the distribution of the vertical stress field of the surrounding rock with different spacings is shown in Figure 9.

When the length of bolt in the roof of the roadway is 2900 mm, the distribution of the vertical stress field of the surrounding rock with different spacings is shown in Figure 10.

As can be seen from the above figures, when the roof of the roadway is supported only, the maximum vertical stress appears at the tail of the anchor, and the stress is transmitted to the surrounding rock of the roadway. The area supported by the anchor bolt in the roof is affected by the stress superposition between the anchor bolts, and the mutual influence during diffusion forms an effective rectangular stress field. The compressive stress gradually decreases during the diffusion process, forming an ear-shaped compressive stress area in the horizontal direction, and the stress is reduced to zero when approaching the top of the anchor in the vertical direction, thus converting into tensile stress. A circular tensile stress area appears at the top of each anchor bolt.

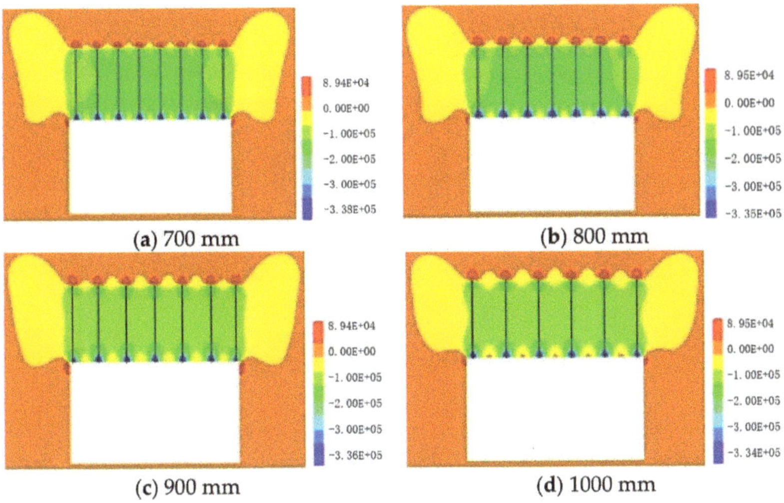

Figure 8. Stress-field diagram of roof bolt at 2700 mm.

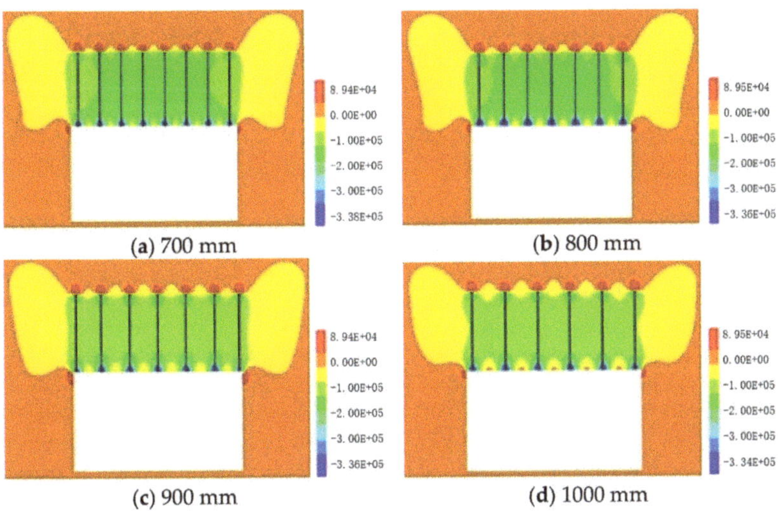

Figure 9. Stress-field diagram of roof bolt at 2800 mm.

When the length of the anchor bolt is the same and the spacing between the anchor bolts is small, the compressive stress area and continuity effect formed by the stress superposition of the anchor bolts are better. As the spacing between the anchors gradually increases, the range of action and diffusion trend of the additional stress field formed by the support components through stress diffusion basically do not change, but the stress-field area formed by the anchor bolts tends to be independent, and the area of stress superposition gradually decreases.

When the spacings between the anchor bolts in the roof are 700 mm and 800 mm, an effective compressive stress range of no less than 0.1 MPa is basically formed in the action area of the anchor bolts at the bottom of the top plate, and the stress is fully exerted in the diffusion and superposition process. When the spacings between the anchor bolts in the roof are 900 mm and 1000 mm, the effective compressive stress range of no less than 0.1 MPa in the action area of the anchor bolts in the roof gradually decreases. The compressive stress between two adjacent anchor bolts cannot be superimposed into a continuous piece due to the large spacing. The compressive stress between the two anchor bolts is too small, which will cause a significant decrease in the active support effect of the anchor bolts, and the compressive stress area and the bearing structure range will also decrease. Therefore, the most effective support effect of the anchor bolts can be exerted by selecting a reasonable spacing of anchor bolts. According to the simulation results of the spacing of anchor bolts in the roof, it can be concluded that the reasonable spacing of anchor bolts should be 800 mm.

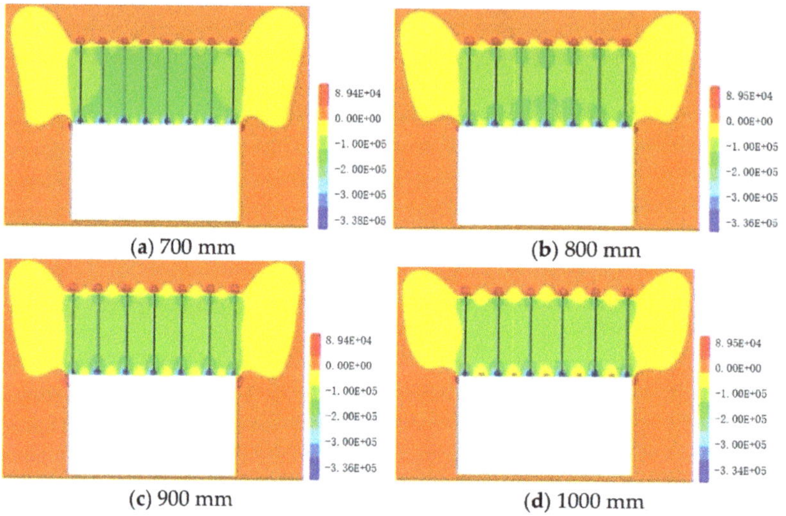

Figure 10. Stress-field diagram of roof bolt at 2900 mm.

Under the condition of the same spacing of anchor bolts, within a certain range, as the anchor length increases, the area of compressive stress will increase, and the range of the bearing structure will also increase accordingly. When the length of the anchor bolt increases further outside this range, the compressive stress value between the anchor bolts will gradually decrease due to the increase in the prestress diffusion distance, resulting in the stress field-segmentation phenomenon of the two parts of the stress of a single anchor bolt. When the length of the anchor bolt is 2700 mm and 2800 mm, the effective compressive stress range of the anchor bolts continues to expand, increasing the effective bearing area of the roof, but when the length increases to 2900 mm, the compressive stress formed by the two parts of the single anchor bolt gradually separates, which will lead to a decrease in the stability of the bearing structure, uneven stress on the surrounding rock of the roof of the working face, and thus the stability cannot be effectively guaranteed. Therefore, when selecting the length of anchor bolt, a compressive stress between the two ends of the anchor bolt should be selected that does not cause separation and can increase the effective bearing area of the anchor bolt. It can be concluded that the length of anchor bolt should be selected as 2800 mm.

In summary, through numerical simulation analysis, it can be concluded that the optimal spacing of anchor bolts in the roof of 1331 (1) track roadway in Zhujidong Mine should be 800 mm and the length should be 2800 mm.

(2) Optimization of bolt spacing and length on both sides of the roadway

When the length of bolt on both sides of the roadway is 2400 mm, the distribution of the horizontal stress field of the surrounding rock with different spacings is shown in Figure 11.

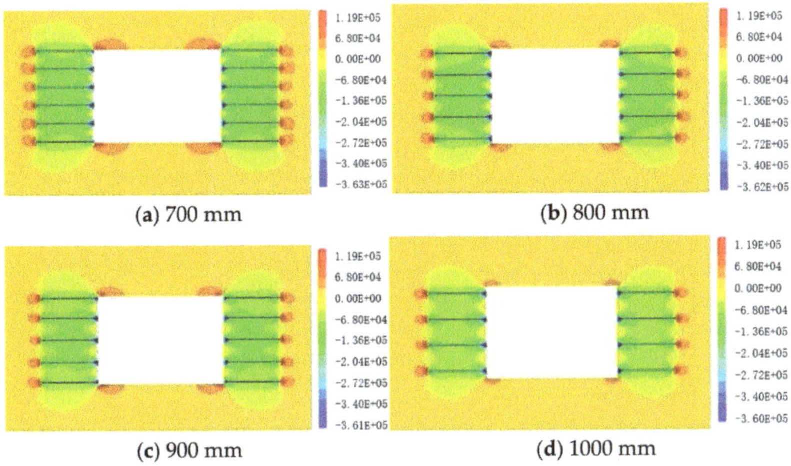

Figure 11. Stress-field diagram of side bolt at 2400 mm.

When the length of bolt on both sides of the roadway is 2500 mm, the distribution of the horizontal stress field of the surrounding rock with different spacings is shown in Figure 12.

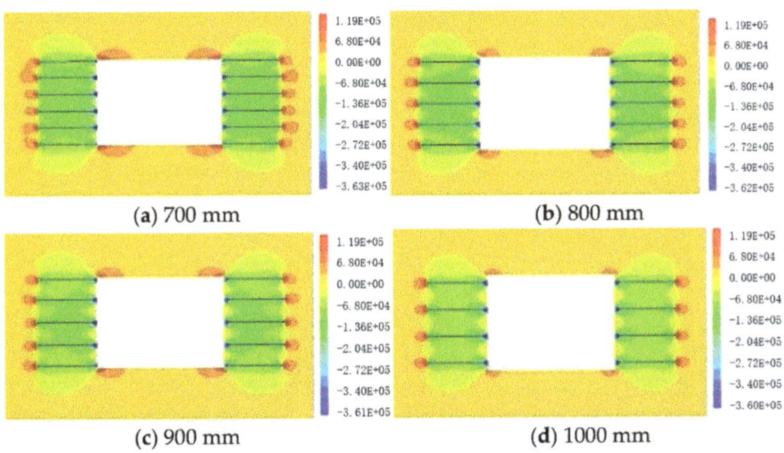

Figure 12. Stress-field diagram of side bolt at 2500 mm.

When the length of bolt on both sides of the roadway is 2600 mm, the distribution of the horizontal stress field of the surrounding rock with different spacings is shown in Figure 13.

As can be seen from the above figures, when both sides of the roadway are supported only, without the influence of ground stress, the distribution characteristics of stress field formed by the anchor bolt on both sides of the roadway are similar to those of anchor bolt in the roof. The prestress diffuses from both sides of the roadway to the surrounding rock, forming a crescent-shaped stress zone. As the stress on both sides of the crescent gradually decreases until it approaches zero, a stress superposition zone is formed at the side of the anchor bolt, providing effective support for the surrounding rock of the roadway.

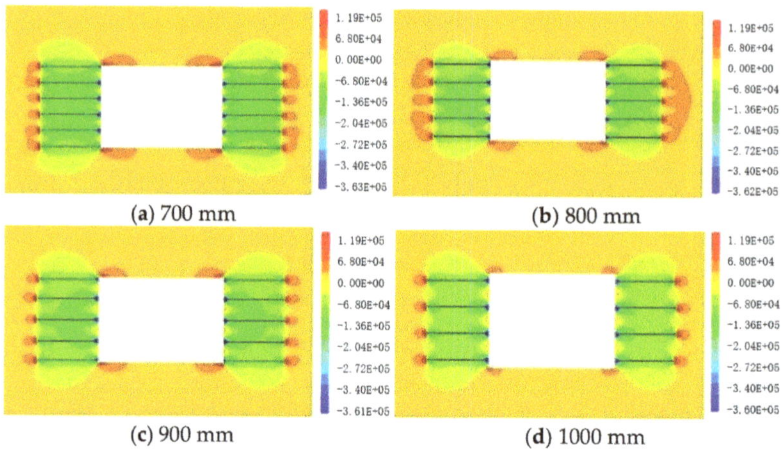

Figure 13. Stress-field diagram of side bolt at 2600 mm.

Under the same length of anchor bolts, the effective compressive stress area formed by the superposition of anchor bolts decreases as the spacing of the anchor bolts increases. The reduced position of the superimposed stress area starts from the heads of adjacent anchor rods and diffuses to both sides, resulting in the anchor bolts being unable to form an effective support compressive stress field when the spacing of the anchor bolts is too large. When the spacing is expanded to 900 mm, the compressive stress between two adjacent anchor bolts is small, and an effective stress bearing area cannot be formed, resulting in a significant reduction in the active support effect of the anchor bolts. It can be concluded that the spacing of the anchor bolts on both sides of the roadway should be less than 900 mm. When the spacing is 800 mm, the number of anchor bolts is less than that of 700 mm, but the difference in effective compressive stress formed by the spacing of 700 mm is not large. Based on the above analysis, the reasonable spacing of anchor bolts on both sides of the roadway is 800 mm.

Under the same spacing of anchor bolts, as the length of the anchor bolts increases, the area of the bearing compressive stress area formed by the anchor bolts will also increase, but the degree of stress diffusion will decrease with the increase in the diffusion distance. When the length of the anchor bolt is too large, the compressive stress at the head of the anchor bolts is too small, and the stress distribution is discontinuous, which affects the support effect. The stress distribution characteristics of the roof and both sides of the roadway are similar. In order to ensure that the stress can be fully diffused and that the compressive stress distribution will not separate at the head of the anchor rod, the length of anchor bolts is 2500 mm. Effective stress superposition can form a greater bearing effect, so that a larger area of surrounding rock in the sides can be actively supported. Therefore, the reasonable length of the anchor bolts on both sides of the roadway is 2500 mm.

In summary, as the length of the anchor bolts increases, the additional stress-field area will also increase, and the stress diffusion effect will decrease with the increase in the diffusion distance. Therefore, it can be determined that the optimal length of anchor bolts in the roof is 2800 mm, and the optimal length of anchor bolts in both sides of the roadway is 2500 mm. As the spacing of anchor bolts increases, the compressive stress area range and continuity formed by stress superposition will gradually decrease, but considering the integrity of the rock formation and economy, the optimal spacing of the anchor bolts in the roof is determined to be 800 mm, and the optimal spacing of the anchor bolts on both sides of the roadway is 800 mm.

(3) Optimization of row spacing of anchor bolts

On the basis of the above numerical simulation of reasonable spacing of anchor bolts, the distribution law of effective stress formed when the row spacing of anchor bolts is 800 mm, 900 mm, and 1000 mm is studied, aiming to obtain reasonable row spacing of anchor bolts. The compressive stress effect formed by numerical simulation of roof bolt support is shown in Figure 14.

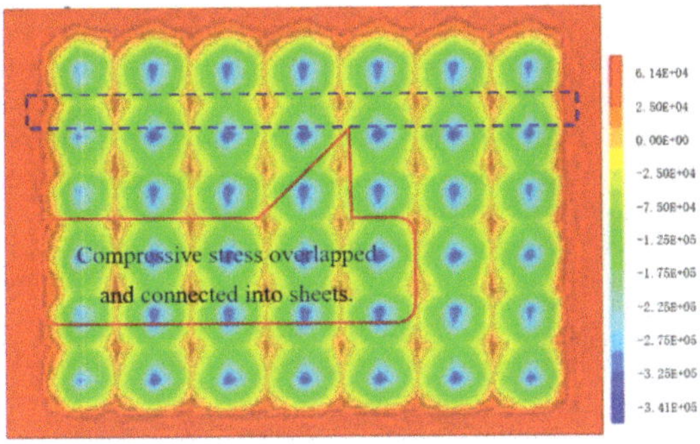

(**a**) Row spacing is 800 mm

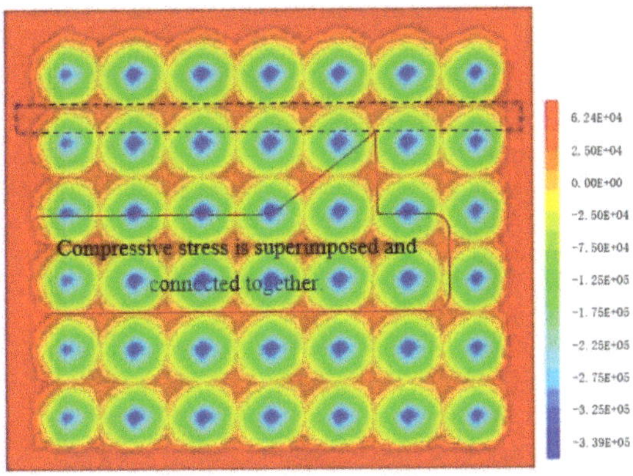

(**b**) Row spacing is 900 mm

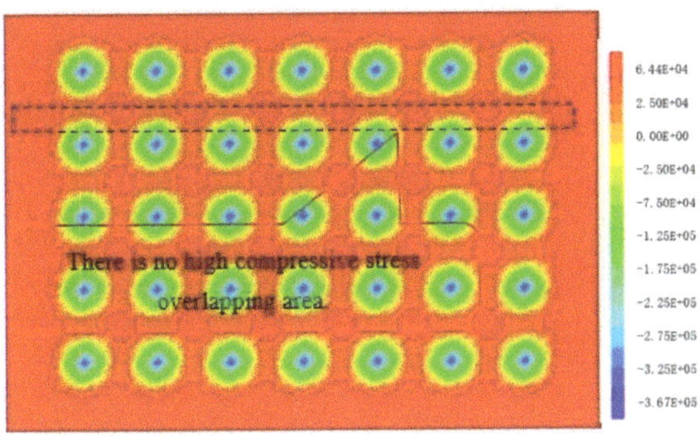

(**c**) Row spacing is 1000 mm

Figure 14. Cloud diagram of vertical stress under different row spacings of anchor bolts.

As can be seen from the figures above, the compressive stress field formed by the anchor bolts diffuses from the center of the anchor bolt to the surrounding area and gradually decreases, and the stress diffusion rate is the same. When the row spacing is 800 mm and 900 mm, the stress field formed by the anchor bolts is superimposed and connected together to form a relatively continuous bearing structure. The maximum compressive stress is at the center of a single anchor bolt, and the maximum compressive stress is 0.037 MPa. When the row spacing is 800 mm, the minimum compressive stress in the middle between two adjacent anchor bolts is 0.075 MPa. When the row spacing is 900 mm, the minimum compressive stress in the middle between two adjacent anchor bolts reaches 0.05 MPa, which is not much different from the row spacing of 800 mm. As the row spacing of the anchor bolts increases to 1000 mm, the maximum compressive stress at the middle of a single anchor bolt undergoes no significant alteration, but the area of compressive stress close to 0 appeared in the middle of the two anchor bolts. In this case, the anchor bolts cannot provide effective active support to this area, affecting the continuity of the bearing structure. Therefore, in order to ensure effective support of the surrounding rock of the roadway and reduce support costs, the reasonable row spacing of anchor bolts is 900 mm.

6. Stability Control and Assessment of Gob-Side Entry

6.1. Stability Control of Gob-Side Entry

(1) Roof support

The combined support of grade IV rebar prestressed anchor bolt (MG400) and M5 steel strip and 10# rhombic metal mesh was applied in the roof. The anchor bolts in the form of a '7-7' arrangement were applied to the roof, the length of the M5 steel strip was 5.1 m, the diameter of the bolt was 22 mm, the length was 2800 mm, and the spacing between the bolts was 800 mm and the row spacing was 900 mm. The full length of the bolt was anchored with one roll k2860 and two roll z2860 resin roll, the angle between the anchor bolts on both sides of the roof and the vertical direction was 15°, the torque of the bolt nut should not be less than 200 Newton-meters, and the anchoring force should not be less than 150,000 newtons. The lap connection of the metal mesh should be no less than 200 mm, the connection between the networks should not be greater than 200 mm, the stubble should not be pressed under the steel strip, and the double strands and double rows of diamond iron wires of not less than 1 inch should be used for binding.

The anchor cable and W5 steel belt (4 m in length) in the '4-4' configuration was adopted, with two rows of 11# I-beams (2.4 m in length) and anchor cables along the strike. The diameter of the anchor cable, which matching the butterfly-shaped large tray (280 mm in length, 280 mm in width, and 16 mm in thickness), on the W5 steel belt was 21.8 mm, the length was 7300 mm, and the spacing between the anchor cables was 1200 mm and the row spacing was 900 mm. The diameter of the anchor cable on the 11# I-steel beam was 21.8 mm, the length was 6300 mm, and the spacing was 900 mm. Each hole was anchored with one roll of k2860 and two rolls of z2860 resin drug rolls. The preload of anchor cable should be no less than 180,000 newtons and the anchoring force should be no less than 350,000 newtons. The support design of the anchor bolt and anchor cable on the roof is shown in Figure 15.

A group of anchor cable clusters was constructed in the middle of the roadway roof every 15 m, and the preload of the anchor cable was no less than 180,000 newtons, and the anchoring force was no less than 350,000 newtons.

(2) Support of both side of the roadway

① A combined support consisting of five full-threaded anchor bolts (MG335); two M3 steel strips, each 1.8 m in length with vertical lap arrangement; 10# rhomboid metal mesh; and one row of 11# I-steel (with a length of 2.4 m) in strike arrangement-reinforced anchor cables was applied to the solid coal side. The diameter of the anchor bolt was 22 mm, the length was 2500 mm, the spacing between the bolts was 800 mm, and the row spacing was

900 mm. The diameter of the anchor cable was 21.8 mm, the length was 4300 mm, and the row spacing was 900 mm.

The angle between the upper anchor bolt and the horizontal plane was 15°, and the angle between the lower anchor bolt and the horizontal plane was negative 15°, and the other anchor bolts were arranged vertically. The drill bit with a diameter of 28 mm was used in the drilling of the hole of the anchor bolt, and two rolls of Z2350 resin cartridge were applied to each anchor bolt. The torque of anchor bolt nut was not less than 180 Newton-meters, and the pre-tightening force of anchor bolt was not less than 125,000 newtons. The lap connection of the metal mesh should be no less than 200 mm, and the connection between the networks should not be greater than 200 mm. The distance between the anchor cable and the floor was 1.6 m, and the cable was arranged along the roadway. Two rolls of Z2860 resin cartridge were used for each hole, and the pre-tightening force of the anchor cable was not less than 150,000 newtons.

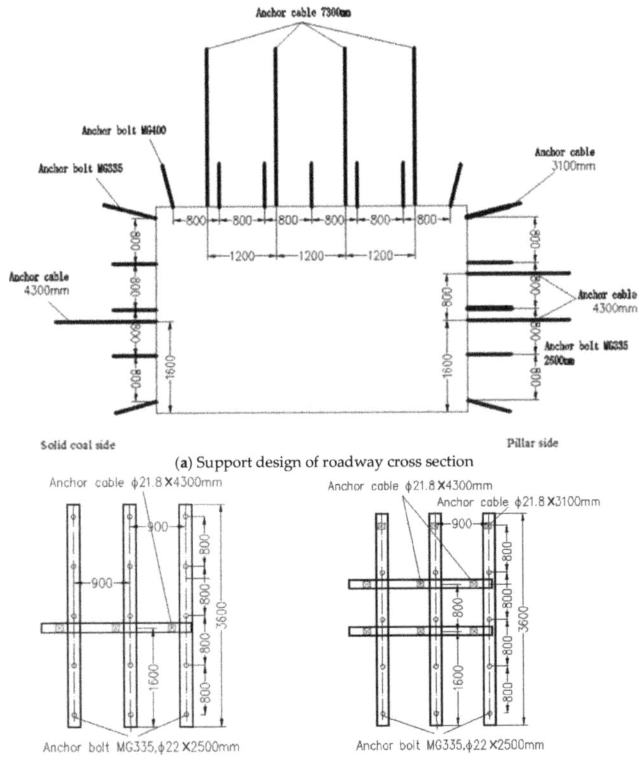

(a) Support design of roadway cross section

(b) Solid coal side

(c) Pillar side

Figure 15. *Cont.*

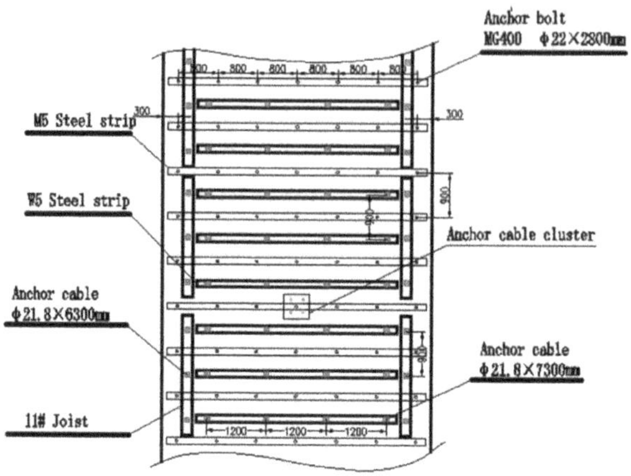

(**d**) Top view of roof support

Figure 15. Support design drawing of roadway.

② A combined support consisting of four full-threaded anchor bolts (MG335), one replacement anchor bolt at the shoulder joint of the roadway, two M3 steel strips each 1.8 m in length with vertical lap arrangement, 10# rhomboid metal mesh, and two rows of 11# I-steel (with a length of 2.4 m) in strike arrangement reinforced anchor cables was applied to the coal pillar side. The diameter of the bolt was 22 mm, and the length was 2500 mm; the diameter of the anchor cable was 21.8 mm, and the length was 4300 mm; and the diameter of the alternative anchor cable was 21.8 mm, and the length was 3100 mm. The spacing between the bolts (anchor cables) was 800 mm and the row spacing was 900 mm.

The drill bit with the diameter of 28 mm was used in the drilling of the hole of the anchor bolt on the coal pillar side, and two rolls of Z2350 resin cartridge were applied to each anchor bolt. The diameter of the anchor cable at the upper corner of the roadway was 21.8 mm, and the length was 3100 mm. The angle between the anchor cable at the upper corner and the horizontal plane was 15°, and the ball lock matched the anchor cable at the upper corner. The diameter of the other anchor cables was 21.8 mm, and the length was 4300 mm, the distance between the two rows of anchor cables arranged along the roadway was 1.6 m and 2.4 m, respectively, and the pre-tightening force of the anchor cable should not be less than 150,000 newtons. The angle between the anchor bolt at the lower corner of the roadway and the horizontal plane was 15°, and the other anchor bolts were arranged vertically. The torque of the bolt nut was not less than 180 Newton-meters, and the anchoring force of the side anchor bolts were not less than 125,000 newtons.

The supporting design diagram of the roadway is shown in Figure 15.

(3) Support of special sections

During the period when the roadway passes through the fault or in the section where the roadway deformation is large, the roof support should be strengthened. In order to strengthen the roof support strength, three rows of 11# I-beams (2.4 m in length) with anchor cables were arranged along the strike; the diameter of the anchor cables on the 11# I-beam on both sides of the roadway roof was 21.8 mm, and the length is 6300 mm. The diameter of the anchor cables on the 11# I-beam near the middle of the roadway roof was 21.8 mm, and the length was 8300 mm. The anchor cable preload force is not less than 180,000 newtons, and the anchoring force is not less than 350,000 newtons.

A group of anchor cable clusters were constructed every 10 m on the top plate of the roadway near the middle of the roadway.

In the case of broken top plate of the roadway or serious deformation of the roadway, 36 U type shed was used for support, and five groups of 11# I-beams and 8.3 m anchor cables were reinforced and excavated row by row. When shed support was used, the spacing of the sheds was 700 mm.

6.2. Mine-Pressure Observation Scheme of Gob-Side Entry

In order to obtain the mine-pressure condition of the 1331 (1) track roadway during the trial excavation, the observation scheme of mine pressure was set up. The arrangement of the measuring stations is shown in the Figure 16. Observation station 1 is arranged 10 m inside the stoppage line of the 1321 (1) working face, and observation station 2 and observation station 3 are arranged 30 m and 60 m back, respectively, as shown in Figure 16. The main contents of the mine pressure monitoring of the track roadway are the surface displacement of the roadway, the deep displacement of the surrounding rock, the roof separation, and the working load of the anchor and cable.

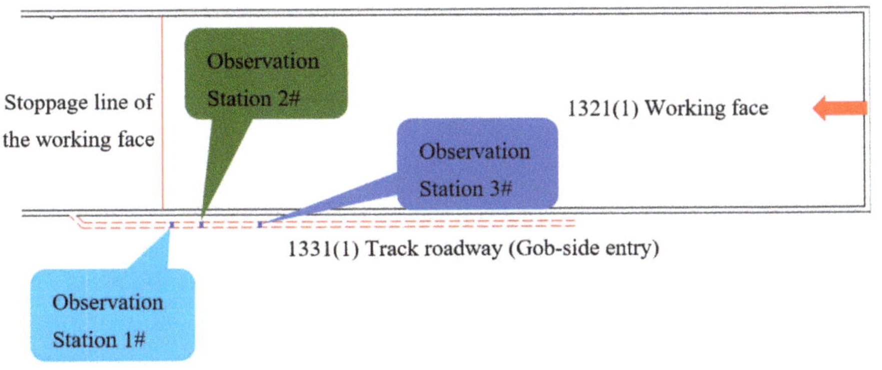

Figure 16. Layout map of stations.

6.3. Stability Assessment of Gob-Side Entry

1. Displacement change in surrounding rock on roadway surface

The displacement data obtained from the observation station arranged on site are collated, and the displacement changes in the surrounding rock on roadway surface are shown in Figure 17. After the monitoring, the site survey was carried out on 1331 (1) track roadway in Zhujidong Mine, and the on-site condition of surrounding rock on the roadway surface was shown in Figure 18.

As can be seen from the figure, the deformation rate of the surrounding rock on the surface of the track roadway is the largest within about 10 days of the formation of the roadway. At this stage, it is mainly due to the fact that the mining stress generated by the formation of the roadway has not reached equilibrium. Within 10 to 15 days of the formation of the roadway, the displacement growth rate of the surrounding rock on the surface of the roadway gradually decreases until it tends to be stable, and the disturbance effect of roadway excavation on the observation point is becoming less and less; About 20 days after the formation of the roadway, the displacement of the surrounding rock on the surface of the roadway tends to be stable and basically does not change. During the monitoring period, the maximum displacement of the top and bottom plates is 86 mm, the maximum displacement of the solid coal side of the roadway is 50 mm, and the maximum displacement of the coal pillar side is 70 mm. The displacement of the coal pillar side of the roadway is greater than the displacement of the solid coal side. It can be seen from the on-site conditions of the track roadway in the later stage that the overall control effect of the roadway surrounding rock is good.

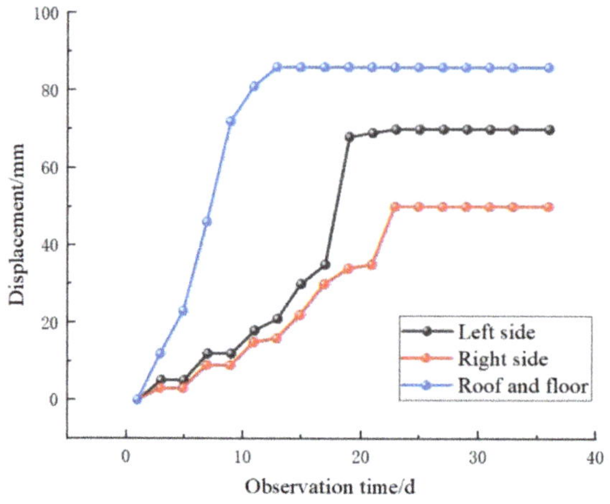

(**a**) Observation Station 1#

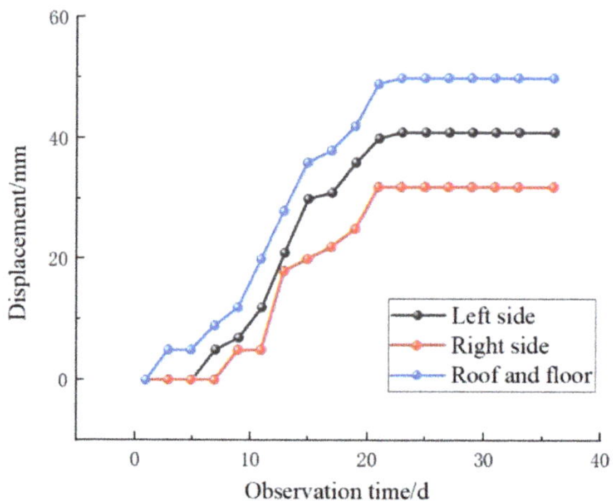

(**b**) Observation Station 2#

Figure 17. *Cont.*

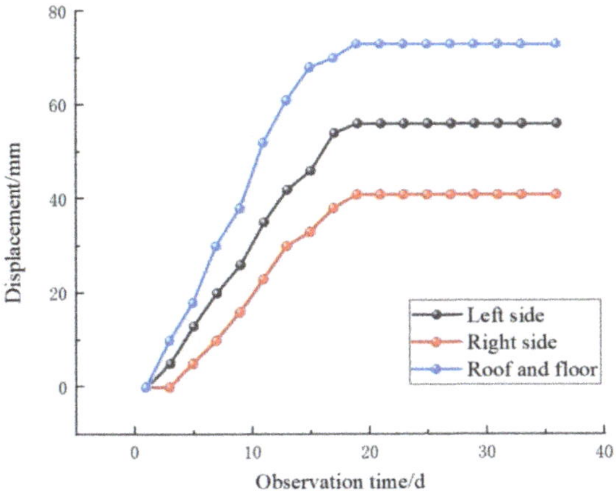

(c) Observation Station 3#

Figure 17. Change in displacement of surrounding rock on the roadway surface.

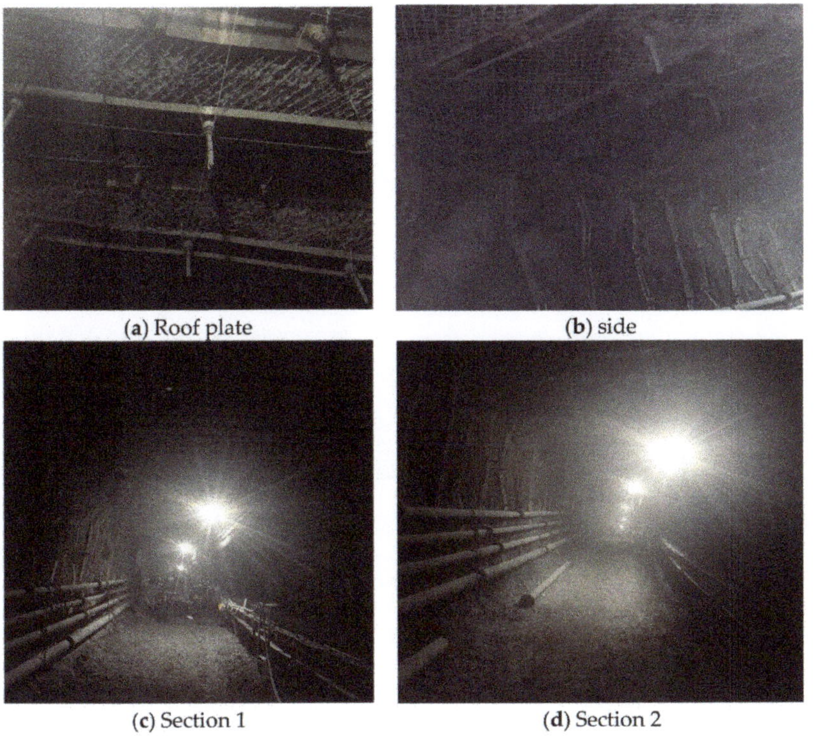

(a) Roof plate (b) side

(c) Section 1 (d) Section 2

Figure 18. The on-site condition of the roadway.

2. Separation of the roof of the roadway

The roof separation of 1331 (1) track roadway was monitored by a 4-point separation meter, and the depths of observation points 1#~4# were 2.7 m, 4.0 m, 7.3 m, and 8.0 m, respectively. The separation of the roof of the track roadway was shown in Figure 19.

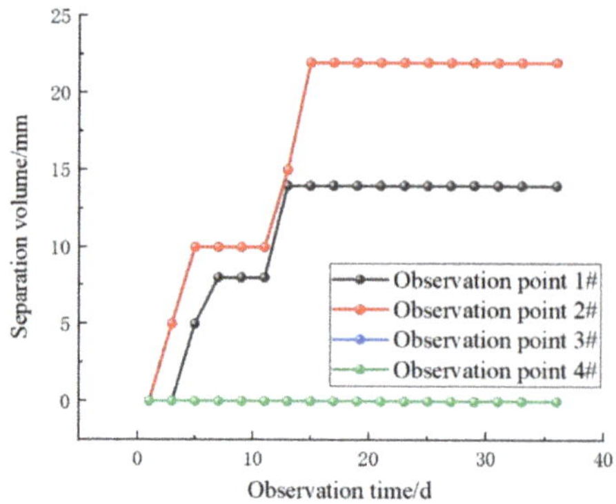

(**a**) Observation Station 1#

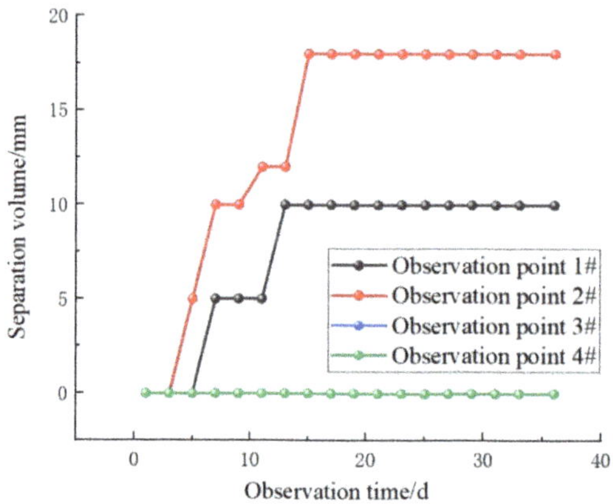

(**b**) Observation Station 2#

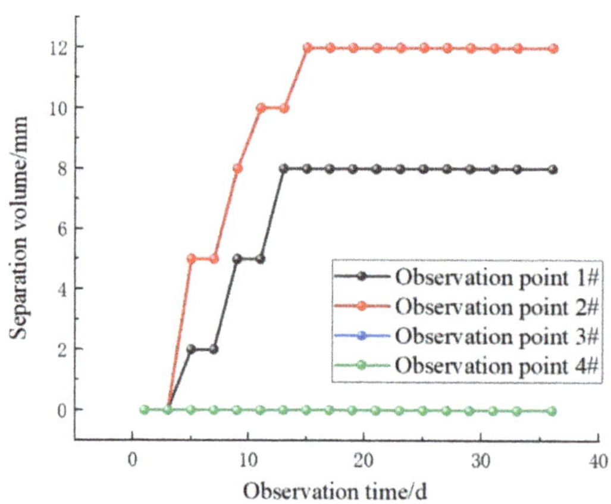

(**c**) Observation Station 3#

Figure 19. Roof separation curve.

According to the data of the separation meter at the on-site measuring station, the data of the roof separation meter at each observation station are 0 mm at the deep measuring point. A small separation occurred at the shallow measuring point within about 12 days after the track roadway was completed. After that, the separation amount tended to be stable and basically did not change. The maximum separation amounts of observation point 1# and 2# at the observation station 1# were 14 mm and 22 mm, respectively, and the roof separation amount at the observation station 1# was small. The maximum separation amounts of observation point 1# and 2# at the observation station 2# were 10 mm and 18 mm, respectively, and the roof separation amount at the observation station 2# was small. The maximum separation amounts of observation point 1# and 2# at the observation station 3# were 8 mm and 12 mm, respectively, and the roof separation amount at the observation station 3# was small.

The monitoring data of the roof of 1331 (1) track roadway show that the roof separation amount is small, and the surrounding rock of the roof is relatively stable.

3. Analysis of load on anchor bolt and cable

During the excavation of 1331 (1) track roadway, the load conditions of some anchor bolts and cables on the coal pillar side, solid coal side and the roof of the roadway is monitored, and the data are summarized and sorted to obtain the stress changes in the anchor bolts and cables, as shown in Figure 20.

As can be seen from the above figures, in the early stage after driving the track roadway, due to the influence of driving disturbance and residual stress of the adjacent gob area on the surrounding rock of the roadway, the load on anchor bolts and cables gradually increased within about 12 days, and then the load tended to be stable and basically did not change. The initial load on the anchor bolt on the left side of the observation station 1# was 43,000 newtons (43 KN), and then the load on the anchor bolt gradually increased. After about 13 days, the load on the anchor bolt tended to be stable and stabilized at 55,000 newtons (55 KN). The initial load on the anchor bolt on the right side was 29,000 newtons (29 KN), and the initial load on the anchor bolt on the roof was 25,000 newtons (25 KN). During the observation period, the load on the anchor bolt on the right side and the roof basically did not change. The initial load on the anchor cable on the roof was 123,000 newtons (123 KN), and then the load on the anchor cable gradually increased. After about 13 days, the load on the anchor cable tended to be stable and stabilized at 129,000 newtons (129 KN).

The initial load on the anchor bolt on the left side of observation station 2# was 31,000 newtons (31 KN), and then the load on the anchor bolt gradually increased. After about 10 days, the load on the anchor bolt tended to be stable and stabilized at 39,000 newtons (39 KN). The initial load on the anchor bolt on the right side was 22,000 newtons (22 KN), and then the load on the anchor bolt gradually increased. After about 10 days, the load on the anchor bolt tended to be stable and stabilized at 30,000 newtons (30 KN). The initial load on the anchor bolt on the roof was 37,000 newtons (37 KN), and then the load on the anchor bolt gradually increased; after about 9 days, the load on the anchor bolt tended to be stable and stabilized at 43,000 newtons (43 KN). The initial load on the anchor cable on the roof was 132,000 newtons (132 KN), and then the load on the anchor cable gradually increased; after about 15 days, the load on the anchor cable tended to be stable and stabilized at 155,000 newtons (155 KN).

The initial load on the anchor bolt on the left side of observation station 3# was 62,000 newtons (62 KN), and then the load on the anchor bolt gradually increased. After about 9 days, the load on the anchor bolt tended to be stable and stabilized at 86,000 newtons (86 KN). The initial load on the anchor bolt on the right side was 51,000 newtons (51 KN), and then the load on the anchor bolt gradually increased. After about 9 days, the load on the anchor bolt tended to be stable and stabilized at 70,000 newtons (70 KN). The initial load on the anchor bolt on the roof was 38,000 newtons (38 KN), and then the load on the anchor bolt gradually increased; after about 8 days, the load on the anchor bolt tended to be stable and stabilized at 45,000 newtons (45 KN). The initial load on the anchor cable on the roof was 139,000 newtons (139 KN), and then the load on the anchor cable gradually increased; after about 15 days, the load on the anchor cable tended to be stable and stabilized at 168,000 newtons (168 KN).

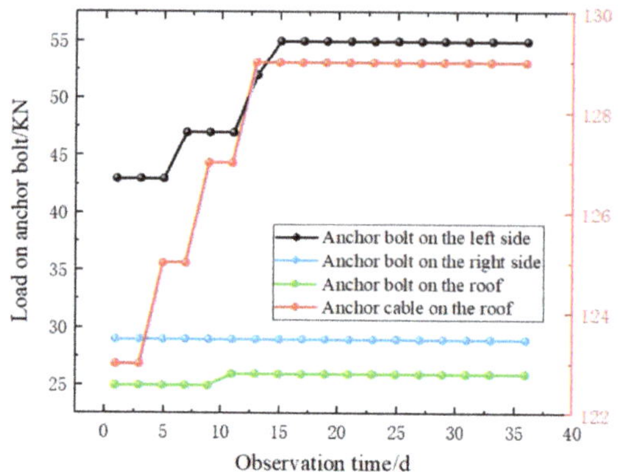

(**a**) Observation Station 1#

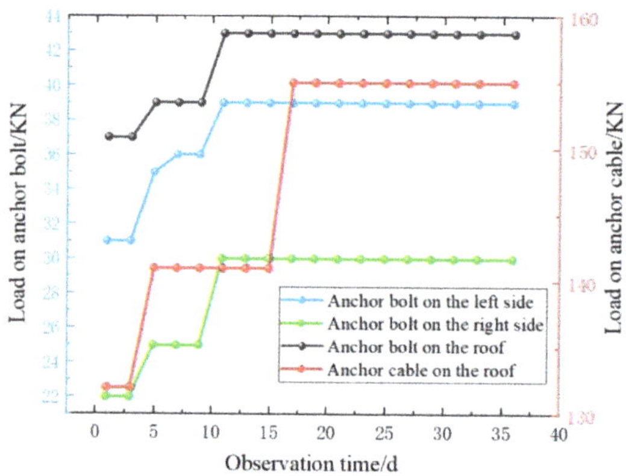

(**b**) Observation Station 2#

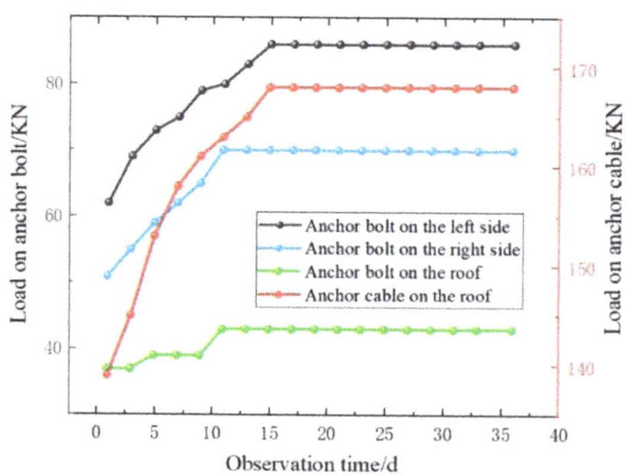

(**c**) Observation Station 3#

Figure 20. Monitoring curves of the load on the anchor bolt and cable.

During the monitoring period after the excavation of the track roadway, there was no failure phenomenon of anchor bolt and cable. The load of the anchor bolt and anchor cable was less than the critical warning value stipulated by the coal industry company. The support resistance of the roadway support components increased and stabilized at a stress value higher than the initial value, indicating that the roadway's surrounding rock was in a stable state, which was beneficial to the stability control of the surrounding rock of the roadway and played an active role in controlling the deformation of the surrounding rock of the roadway.

6.4. Deformation Characteristics of Gob-Side Entry

In order to obtain the deformation characteristics of the surrounding rock of the track roadway during excavation along the goaf, a simulation calculation model was established by using the simulation software Fast Lagrangian Analysis of Continua (FLAC 3 D 6.0) according to the geological conditions of 1321 (1) working face and 1331 (1) track roadway. The size of the model is 343 m long, 300 m wide, and 172 m high, with a total of 825,346 units. The bottom boundary of the model is fixed, the left and right boundary limits the horizontal displacement, and the top boundary imposes self-weight stress. The simulated coal-seam mining height is 2 m, and the simulated roadway size is 5.4 m wide and 3.6 m high. Coal and rock mass can be regarded as elastic–plastic material, which is simplified to elastic–plastic model in numerical calculation, and Molar–Coulomb failure criterion is selected. The working face was simulated by advancing 10 m each time, and the roadway was simulated by excavating 10 m each time. The numerical calculation model is shown in Figure 21, and the physical and mechanical parameters of coal and rock are shown in Table 5.

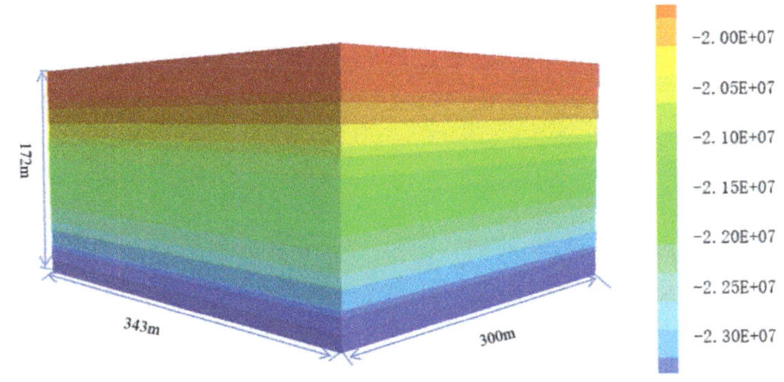

Figure 21. Stress-balance cloud map of primary rock.

Table 5. Physical and mechanical parameters of coal and rock.

Rock Stratum	Density d/kg·m^{-3}	Bulk Modulus K/GPa	Shear Elasticity G/GPa	Frictional Angle φ/°	Adhesion Stress C/MPa	Strength of Extension t/MPa
Fine sandstone	2650	9.5	5.6	38	5.2	5.5
Mudstone	2430	4.6	2.5	33	3.5	1.5
Coal	1350	3.5	1.2	28	1.2	0.5
Mudstone	2400	4.5	2.5	33	3.5	1.4
Siltstone	2750	9.8	5.6	36	5.5	5.6

Simulation process: initial ground stress balance → 1321 (1) waking face →1331 (1) track roadway → output result.

The equilibrium condition is reached when the geo-stress is calculated to 8430 steps, and the maximum unbalance force is less than 1×10^{-5}. The average geo-stress of coal seam and roadway is 21.3 MPa, and the geo-stress is generally distributed horizontally.

After the mining of 1321 (1) waking face, 1331 (1) track roadway was excavated. The displacement changes in roadway surrounding rock of track roadway under different excavation distance conditions were shown in Figure 22.

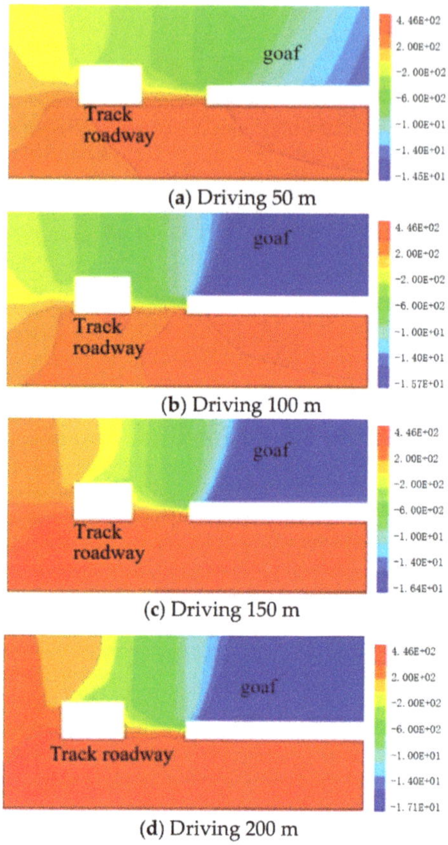

Figure 22. Distribution characteristics of vertical displacement under different excavation-distance conditions.

According to the analysis of the above figure, under the influence of residual mining stress of adjacent working face and disturbance during roadway excavation, the displacement of surrounding rock is mainly concentrated on both sides and the roof and floor of the roadway during roadway excavation. With the increase in excavation distance, the displacement of overlying rock gradually increases and tends to be stable, while the floor-heave phenomenon occurs in the underlying rock in a small range.

When the roadway is driven to 50 m, the maximum displacement variation in the roof and floor of the roadway is about 81 mm. When the roadway is driven to 100 m, the maximum displacement change in the roof and floor is 88 mm. When the roadway is driven to 150 m, the maximum displacement change in the roof and floor is 95 mm, and when the roadway is driven to 200 m, the maximum displacement change in the roof and floor is 97 mm.

In the direction of inclination, with the excavation of the roadway, the displacement changes in the stratification phenomenon appear on both sides of the roadway. However, since one side of the roadway is close to the goaf of the upper section, and the other side is a solid wall, the displacement changes on both sides are obviously asymmetric.

With the continuous excavation of the roadway, the displacement asymmetry deformation difference of surrounding rock on both sides of the roadway gradually decreases. With the gob roadway driving, the maximum displacement change in the strike direction occurs closer and closer to the roadway roof, and the displacement change becomes larger and larger and becomes stable. In the direction of the dip, the displacement range of the top strata gradually increases and becomes stable.

The numerical simulation results of the deformation of the roof and floor of the roadway show that the deformation of the roof and floor of the roadway is consistent with the field monitoring results.

7. Conclusions

Aiming at solving the problem of support failure caused by large deformation of roadway surrounding rock in deep soft coal seam, the support effect of the supporting system of gob-side entry in a deep soft coal seam was studied by using a numerical simulation and field industrial test and inspection. The deformation characteristics of various supporting systems of metal mesh, diamond mesh, metal mesh with anchor rod, steel ladder beam, and M-shaped steel belt in the goaf supporting body of deep soft coal seam were studied under vertical load. The supporting effect of the effective compressive stress zone generated by bolt and cable under different row spacings and lengths was analyzed, the law of variation in compressive stress field generated by supporting members with supporting parameters was explored, and the reasonable supporting parameters were optimized. The main conclusions are as follows:

(1) Based on the effective compressive stress zone generated by the anchor bolt and cable support components, the supporting effect of the support components was analyzed, and the law of the compressive stress field generated by the support components that varied with the supporting parameters was obtained. The supporting effect of bolt and cable under different length, spacing, and preload was simulated and analyzed, respectively. The turning point of compressive stress-zone change under different supporting parameters was determined, and the critical supporting parameters of effective compressive stress zone continuous and good supporting effect were selected.

(2) When different support systems were under the same fixed load, the maximum deformation of the metal mesh was 239.61 mm, the maximum deformation of the diamond mesh was 578.86 mm, the maximum deformation of the 12# steel ladder beam was 193.77 mm, the maximum deformation of the 14# steel ladder beam was 161.62 mm, the maximum deformation of the M-shaped steel strip was 78.066 mm, and the maximum deformation of 14#b channel steel was 110.87 mm, while that of 11# I-steel was 135.57 mm. The strongest resistance to deformation was observed in the M-shaped steel strip, and the worst resistance to deformation was observed in the diamond mesh.

(3) The support scheme of the 1331 (1) track roadway was designed, and the industrial test was carried out. The investigation results of the stability of the surrounding rock of the 1331 (1) track roadway showed that the maximum displacement of the roof and floor of the roadway was 86 mm, the maximum displacement of the solid coal side was 50 mm, the maximum displacement of the coal pillar side was 70 mm, and the maximum separation of strata was 22 mm. The load of the anchor bolt and anchor cable was less than the critical warning value stipulated by the coal industry company, there was no failure phenomenon of anchor bolt and cable, and the overall deformation of the roadway surrounding rock was small.

The results of the stability investigation of the surrounding rock of the 1331 (1) track roadway show that the overall deformation of the surrounding rock of the roadway was small, which indicated that the roadway support scheme in this study was reasonable, and the research results can provide a reference for other roadway-surrounding-rock control under similar conditions in the 11-2 coal seam of Zhujidong Coal Mine.

Author Contributions: Methodology, H.M. and Z.L.; software, C.J. and H.Z.; validation, S.Z.; investigation, S.Z., H.Z., C.J. and H.M.; data curation, H.Z. and C.J.; writing—original draft preparation, H.M. and C.J.; writing—review and editing, Z.L. and S.Z.; visualization, S.Z. and Z.L.; supervision, Z.L. All authors have read and agreed to the published version of the manuscript.

Funding: This research was funded by the National Natural Science Foundation of China Youth Program (52304074, 51974007), the Anhui University of Science and Technology 2022 Graduate Core (first-class) Course Construction Project (2022HX002), and the 2023 New Era Education Quality Project (2023zyxwjxalk070).

Institutional Review Board Statement: Not applicable.

Informed Consent Statement: Not applicable.

Data Availability Statement: The raw data supporting the conclusions of this article will be made available by the authors upon request.

Conflicts of Interest: The authors declare no conflicts of interest.

References

1. Gao, C.J.; Huang, D.M.; Chang, X.K.; Xi, H. Risk Analysis and Extension Assessment for The Stability of Surrounding Rock In Deep Coal Roadway. *Int. J. Environ. Res. Public Health* **2019**, *16*, 4752. [CrossRef] [PubMed]
2. Wang, X.F.; Zhang, Y.Y.; Zhang, Q.; Wei, Y.Y.; Liu, W.G.; Jiang, T. Space-Time Evolution Characteristics of Deformation and Failure of Surrounding Rock in Deep Soft Rock Roadway. *Sustainability* **2022**, *14*, 12587. [CrossRef]
3. Wu, C.; Qin, T.; Wang, L.; Liu, Z. Research on Surrounding Rock Control Technology of Dongbaowei Deep Mining Roadway. *Adv. Civ. Eng.* **2021**, *1*, 6660989. [CrossRef]
4. Li, S.Q.; Sun, X.F.; Ye, Z.Y.; Tan, Y.L. Analysis of Bearing Structures around Deep Soft Rock Roadway. *Disaster Adv.* **2014**, *1*, 250–255.
5. Liu, W.J.; Qian, D.Y.; Yang, X.G.; Wang, S.J.; Deng, J.P.; Cui, Q.; Li, Z.X. Stress Relief and Support for Stability of Deep Mining Roadway with Thick Top Coal in Hujiahe Coal Mine with the Risk of Rock Burst. *Shock. Vib.* **2021**, *9*, 3822336. [CrossRef]
6. Yuan, W.H.; Hong, K.; Liu, R.; Ji, L.J.; Meng, L. Numerical Simulation of Coupling Support for High-Stress Fractured Soft Rock Roadway in Deep Mine. *Adv. Civ. Eng.* **2022**, *6*, 7221168. [CrossRef]
7. Zhao, C.X.; Li, Y.M.; Liu, G.; Meng, X.R. Mechanism Analysis and Control Technology of Surrounding Rock Failure in Deep Soft Rock Roadway. *Eng. Fail. Anal.* **2020**, *115*, 104611. [CrossRef]
8. Wu, B.W.; Chang, J.C.; Li, C.M.; Wang, T.; Shi, W.B.; Wang, X.Y. Mechanism of Time-Dependent Instability of Deep Soft-Rock Roadway and Crack-Filling Reinforcement Technology. *Appl. Sci.* **2023**, *13*, 4641. [CrossRef]
9. Wang, E.; Yin, S.F.; Kang, Q.T.; Zhao, X.B.; Lan, Q.K.; Sheng, H.Y.; Liang, H.Y. Coupling control technology of anchoring and unloading in deep intense-mining and large-deformation roadway: A case study. *Sci. Rep.* **2024**, *14*, 12075. [CrossRef]
10. Wang, W.J.; Fan, L.; Zhao, Z.Q. Research progress of support theory and technology of the roadway surrounding rock based on the plastic zone control. *J. China Coal Soc.* **2024**, *49*, 320–336.
11. Liu, X.S.; Li, X.B.; Tan, Y.L. Instability Mechanism and Control Method of Energy-Reducing for Roadside Support in Deep Gob-Side Roadway. *J. China Coal Soc.* **2023**, *48*, 485–500.
12. Gao, M.S.; Xu, D.; Wang, H.C. Shock Failure Mechanism of Roadway in Extra Thick Coal Seam and Full Anchor Cable Support Technology. *J. China Coal Soc.* **2023**, *48*, 1943–1956.
13. Hou, C.J.; Wang, X.Y.; BAI, J.B. Basic theory and technology study of stability control for surrounding rock in deep roadway. *J. China Univ. Min. Technol.* **2021**, *50*, 1–12.
14. Wang, F.T.; Shang, J.J.; Zhao, B. Strengthened Anchor Cable Support Mechanism and Its Parameter Optimization Design for Roadway's Dynamic Pressure Section. *J. China Univ. Min. Technol.* **2022**, *51*, 56–66.
15. Guo, X.F.; Guo, L.F.; Ma, N.J. Applicability Analysis of The Roadway Butterfly Failure Theory. *J. China Univ. Min. Technol.* **2020**, *49*, 646–653.
16. Kang, H.P.; Wu, Y.Z.; He, J. Rock Bolting Performance and Field Practice in Deep Roadway with Rock Burst. *J. China Coal Soc.* **2015**, *40*, 2225–2233.
17. Wang, X.J.; Li, Y.M.; Zhao, G.M. Mechanical Response Analysis of Surrounding Rock Bearing Structure of Soft Rock Roadway Based on Full-Length Anchorage. *Coal Sci. Technol.* **2023**, *51*, 24–34.
18. Cao, C.; Xie, Z.; Zhang, N.; Han, C.; Yan, G.; Mu, F.; Zhang, W. Differential roof cutting for roadway support in dual gob-side entry retention on a single working face-Multilevel continuous anchor-grouting control technology: A case study. *Eng. Fail. Anal.* **2024**, *163*, 108475. [CrossRef]
19. Luo, Y.; Huang, J.C.; Si, X.F.; Lin, F.; Wu, W.X. An energy-based method for uniaxially compressed rocks and its implication. *J. Rock Mech. Geotech. Eng.* **2024**, *12*, 353–365. [CrossRef]
20. Si, X.F.; Luo, Y.; Gong, F.Q.; Huang, J.C.; Han, K.F. Temperature effect of rockburst in granite caverns: Insights from reduced-scale model true-triaxial test. *Geomech. Geophys. Geo-Energ. Geo-Resour.* **2024**, *10*, 26. [CrossRef]

21. Meng, B.; Yin, Q.; Jing, H.W.; Liu, S.D.; Jia, B.G. Stability Control of Deep-Buried Roadways Using Large Deformation and Increasing Resistance Anchor Cables. *J. Cent. South Univ.* **2024**, *31*, 558–575. [CrossRef]
22. Wu, Y.; Liu, X.; Tan, Y.; Wang, W.; Li, X.; Wang, X. Mechanism of Bolt Breakage in Deep Mining Roadway Under Dynamic Load and Advanced Strengthening Support Technology. *Eng. Fail. Anal.* **2024**, *161*, 108255. [CrossRef]
23. Kang, H.P.; Yang, J.W.; Jiang, P.F.; Gao, F.Q.; Li, W.Z.; Li, J.F.; Chen, H.Y. Theory, Technology and Application of Grouted Bolting in Soft Rock Roadways of Deep Coal Mines. *Int. J. Miner. Metall. Mater.* **2024**, *31*, 1463–1479. [CrossRef]
24. Wang, J.; Liu, P.; He, M.C.; Tian, H.Z.; Gong, W.L. Mechanical Behaviour of a Deep Soft Rock Large Deformation Roadway Supported by NPR Bolts: A Case Study. *Rock Mech. Rock Eng.* **2023**, *56*, 8851–8867. [CrossRef]
25. Wang, H.W.; Zhang, F.J.; Wang, H.R.; Li, Z.L.; Wang, Y.H. Real-Time Detection and Location of Reserved Anchor Hole in Coal Mine Roadway Support Steel Belt. *J. Real-Time Image Process.* **2023**, *20*, 6563–6581. [CrossRef]
26. Wang, H.P.; Song, Y.; Zhou, J.H.; Mao, J.H.; Zhang, W.D. Research On the Mechanical Model of Anchorage Resistance in Deeply Filled Jointed Rock Masses. *Eng. Fail. Anal.* **2024**, *163*, 108565. [CrossRef]
27. Shan, R.; Wei, Y.; Wang, C.; Li, Z.; Li, Y.; Liu, D.; Zhao, X. Research on the Failure Mechanism of Surrounding Rock in a Dynamic Pressure Roadway and Active and Passive Coordinated Support Technology. *Appl. Sci.* **2024**, *14*, 2352–2368. [CrossRef]
28. Feng, C.; Liu, S.; Fu, M.; Yao, B.; Jia, H.; He, D. Failure Mechanism and Strengthening Countermeasures of End-Anchored Cable Free Section Under Combined Tension and Shear. *Eng. Fail. Anal.* **2024**, *157*, 107949. [CrossRef]
29. Tian, M.L.; Gao, X.X.; Zhang, A.F.; Han, L.J.; Xiao, H.T. Study on The Deformation Failure Mechanism and Coupling Support Technology of Soft Rock Roadways in Strong Wind Oxidation Zones. *Eng. Fail. Anal.* **2024**, *156*, 107840. [CrossRef]
30. Zuo, J.P.; Liu, H.Y.; Liu, D.J.; Zhu, F.; Zhao, W.P.; Zhang, Q.; Li, J.H. Theoretical Analysis and Numerical Simulation on The Coupled Support Technology of Concrete-Filled Steel Tube and Bolt-Cable in Deep Roadway. *J. Cent. South Univ.* **2023**, *30*, 257–275. [CrossRef]
31. Ding, K.; Wang, L.G.; Tian, J.S.; Ren, B.; Jiang, C.Y.; Wang, S. Design of a Highly Adaptable Advance Support for a Deep, Fully Mechanized Roadway and Analysis of Its Support Performance. *Appl. Sci.* **2022**, *12*, 12728. [CrossRef]
32. Sun, X.M.; Zhao, W.C.; Shen, F.X.; Zhang, Y.; Jiang, M. Study on Failure Mechanism of Deep Soft Rock Roadway and High Prestress Compensation Support Countermeasure. *Eng. Fail. Anal.* **2023**, *143*, 865–878. [CrossRef]
33. Chu, H.B.; Li, G.Q.; Liu, Z.J.; Liu, X.S.; Wu, Y.H.; Yang, S.L. Multi-Level Support Technology and Application of Deep Roadway Surrounding Rock in the Suncun Coal Mine, China. *Materials* **2022**, *15*, 8665. [CrossRef] [PubMed]
34. Sun, X.M.; Zhu, M.Q.; Zhang, Y.; Zhao, C.W.; Miao, C.Y.; Zhang, S.K. Highly Prestressed Npr Cable Coupling Support Technology and Its Application in The Deep Roadway. *Eng. Fail. Anal.* **2022**, *142*, 457–463. [CrossRef]
35. Shan, R.L.; Zhang, S.P.; Xiao, S.C.; Liang, J.Q. Research on Anchor Cable and C-Shaped Tube Support Method in Deep Layers Roadway, Experimental Study and Numerical Simulation. *Shock. Vib.* **2021**, *2021*, 7537979. [CrossRef]
36. Zheng, L.J.; Zuo, Y.J.; Hu, Y.F.; Wu, W. Deformation Mechanism and Support Technology of Deep and High-Stress Soft Rock Roadway. *Adv. Civ. Eng.* **2021**, *4*, 1–14. [CrossRef]
37. Su, W.P.; Zheng, B.Y.; Jiang, P.F. Study on Anchor Cable instead of Single Hydraulic Prop Support in Advance Support of Deep Roadway. *Adv. Civ. Eng.* **2021**, *2021*, 6644832. [CrossRef]
38. Wang, S.; Xiang, Z.; Deng, J.; Yang, H.; Yang, X.; Jin, S. Analysis of Cooperative Control Effect of Pressure Relief and Long Bolt Support for Deep Roadway Under Strong Mining Disturbance of Adjacent Working Face. *Geotech. Geol. Eng.* **2020**, *4*, 2259–2268. [CrossRef]
39. Chang, J.C.; He, K.; Yin, Z.Q.; Li, W.F.; Li, S.H.; Pang, D.D. Study on the Instability Characteristics and Bolt Support in Deep Mining Roadways Based on the Surrounding Rock Stability Index: Example of Pansan Coal Mine. *Adv. Civ. Eng.* **2020**, *5*, 1–16. [CrossRef]
40. Shi, J.; Feng, J.; Peng, R. Study on Support Technology of Deep Mine Roadway in Xingdong Coal Mine. *Geotech. Geol. Eng.* **2020**, *39*, 1683–1688. [CrossRef]
41. Jing, H.; Wu, J.; Yin, Q.; Wang, K. Deformation and Failure Characteristics of Anchorage Structure of Surrounding Rock in Deep Roadway. *Int. J. Min. Sci. Technol.* **2020**, *30*, 593–604. [CrossRef]
42. Li, H.; Zhao, W.; Zhou, K.; Liu, Y.; An, X.; Gao, G. Study on the Effect of Bolt Anchorage in Deep Roadway Roof Based on Anchorage Potential Design Method. *Geotech. Geol. Eng.* **2019**, *37*, 4043–4055. [CrossRef]
43. Fan, Z.X. The Secondary Anchor-Cable Network Support Technology in Deep Soft-rock Roadway in Chaohua Coal Mine. *IOP Conf. Ser. Earth Environ. Sci.* **2018**, *189*, 062055.
44. Li, J.B. Analysis of stability in pedestrian downhill roadway at Yindonggou 11th coal mine area: The impact of anchor cable support density. *Front. Energy Res.* **2024**, *12*, 326–342.
45. Yao, B. Master of Engineering. Research on Reasonable Width of Coal Pillar and Surrounding Rock Control Technology of Deep Large Section Gob-Side Entry. Master's Thesis, China University of Mining and Technology, Xuzhou, China, 2022.

applied sciences

MDPI

Article

Dynamic Response and Rock Damage of Different Shapes of Cavities under Blasting Loads

Xuejiao Cui [1,2], Mingsheng Zhao [2,3,*] and Qiyue Li [1]

1 School of Resources and Safety Engineering, Central South University, Changsha 410083, China
2 Poly Xinlian Blasting Engineering Group Co., Ltd., Guiyang 550002, China
3 School of Mining, Guizhou University, Guiyang 550025, China
* Correspondence: zjm377413454@126.com

Abstract: In order to investigate the dynamic response and rock mass damage characteristics of cavities with different shapes under blasting loads, this paper, through a combination of model tests and numerical simulations, studies the stress distribution, strain, failure modes, and blasting fragment size distribution of cavities with different shapes subjected to blasting loads. The results show that under the action of blasting loads, the presence of cavities with different shapes significantly affects the blasting effects and rock mass damage. Spherical cavities exhibit excellent blast resistance, whereas rectangular and triangular cavities are prone to stress concentration at their tips, which in turn promotes rock mass damage and failure. Subsequent analysis of the blasting fragment sizes reveals that rectangular and triangular cavities yield more favorable blasting results than spherical cavities. The research findings provide important theoretical foundations and practical guidance for the design and construction of underground engineering blasting, contributing to enhancing engineering safety and promoting the sustainable development of the underground engineering industry.

Keywords: cavity; cross-sectional shape; blasting load; failure mode; fragment size distribution

1. Introduction

With the rapid pace of urbanization in our country, there has been a significant increase in the construction and development of underground engineering projects. Against this background, the construction of underground facilities such as tunnels, water-diversion tunnels, and hydroelectric power projects often necessitates passage through complex and variable geological environments [1,2]. However, in actual engineering practice, traditional blasting techniques frequently encounter issues such as over-excavation and inaccurate excavation pit formation [3–5]. Under such complex geological conditions, the presence of rock mass with different-shaped caves, is of paramount importance for the stability and safety of underground engineering projects [6]. The action of impact loads may lead to excessive excavation damage, inaccurate cutting, or overdevelopment of fractures, which can lead to a sudden water inflow accident [7,8]. Therefore, the stability of rock mass containing different-shaped cavities is not only directly related to the safety and efficiency of construction projects but also plays an indispensable and vital role in ensuring the safe operations of workers on the construction site [9]. Therefore, conducting in-depth research on the dynamic responses and rock mass damage characteristics of cavities with different shapes under the action of impact loads is of great theoretical significance and practical application value for guiding the design and construction of underground engineering projects and enhancing their safety.

Scholars both at home and abroad have conducted in-depth research on the dynamic response and damage characteristics of rock masses under impact loads [10–13]. Impact loads refer to time-varying loads such as explosions, vibrations, stress shocks from adjacent rockbursts, and earthquakes [14]. The research primarily focuses on the macroscopic response characteristics of rock masses under impact loads, such as stress, strain, and displacement, as well as the influence of impact vibrations on the surrounding environment [15–18].

Tao et al. [19] investigated the behavior of rock with preset circular holes under static and dynamic loads using both theoretical and laboratory testing methods. The results showed that when an underground cavity is subjected to both static and dynamic loads, the combined effect of static and dynamic stress concentration can induce primary and ultimate macroscopic failure of the rock. Li et al. [20] studied the dynamic response surrounding a circular tunnel subjected to concentrated blast stress waves through theoretical analysis and numerical simulation, thus revealing the reasons for the dynamic stress concentration around pre-set holes.

In recent years, with the development of numerical simulation technology, scholars have started to use numerical simulation methods to study the damage and failure mechanisms of rock masses under blasting loads. Li et al. [21] investigated the failure mode of tunnels subjected to blasting loads when there are cracks nearby, using indoor tests and LS-DYNA numerical simulation. The study found that the direction of the blasting load has a significant impact on the final failure mode of the tunnel in the fractured rock mass. Lin et al. [22] explored the interaction mechanism between circular holes and cracks induced by explosions through experiments and numerical simulation. The research results showed that pre-set circular holes do not simply promote or inhibit the propagation of blast-generated cracks but are determined by the relative position between the circular holes and the blastholes. Ma et al. [23] analyzed the foundation of blast-fill dams using discrete element software, with the results indicating that appropriate construction progress and flexible waterproofing materials are key characteristics of the waterproofing structure. Liu et al. [8] conducted research on potential water inflow disasters caused by blasting vibrations during the excavation of submarine tunnels, combining experiments with numerical simulation and establishing a model for the evolution of seepage characteristics during the surrounding rock damage process. Despite numerous studies that have explored the dynamic response and damage issues of rock masses under blasting loads, there are still the following issues and controversies: (1) the dynamic response laws of cavities with different shapes under blasting loads are not clearly defined; (2) existing research is mostly focused on macroscopic response characteristics, lacking studies on the micro-damage mechanisms of cavities with different shapes.

In response to the aforementioned issues and controversies, this paper aims to analyze the dynamic response patterns of cavities with different shapes under blasting loads, providing a theoretical basis for underground engineering construction. The study seeks to investigate the damage mechanisms and blast fragment size distributions of cavities with different shapes under blasting loads, explore the influence of cavity geological bodies on the size of blast fragments and rock mass damage characteristics, and reveal the mechanisms of rock mass damage and failure. The goal is to provide guidance for underground engineering design and construction to enhance project safety. Addressing the cavity structural conditions in mining geology, this study combines field tests with numerical simulations. Firstly, concrete model specimens with pre-set spherical, triangular, and rectangular holes (representing cavities with different shapes) were cast. A series of field experiments were designed and conducted to study the dynamic response and damage of cavities with different shapes under blasting loads, deeply investigating the relationship between cavity geological structures and blasting effects and providing important information on blast fragment size distributions for engineering practice and geological research. Subsequently, using the LS-DYNA finite element software, numerical simulation models of cavities with different shapes were established to simulate their dynamic responses and damage processes under blasting loads. By simulating the blasting processes of cavities with different shapes, the results can validate the field tests and further reveal how these geological structures affect rock fracturing and fragmentation and subsequently influence the block size distribution of blast products. This is of guiding significance for blasting construction safety in engineering projects. In the experiments, it is possible to predict the blasting effects under different geological structures, including block size distribution and degree of fragmentation, thereby providing predictions and references for actual projects. The experimental results can provide important data support for rock blasting and excavation in engineering projects, helping engineers make reasonable decisions and reduce uncertainty and risk.

This study contributes to enriching the theory of rock mass dynamic response and damage under blasting loads, providing new perspectives and methods for subsequent research. (1) It offers a theoretical basis for underground engineering design and construction, enhancing project safety; (2) it aids in optimizing blasting parameters, reducing the impact of blasting vibrations on the surrounding environment; and (3) it provides technical support for China's underground engineering construction, fostering the sustainable development of the underground engineering industry. In summary, the research on the dynamic response and rock mass damage of caves with different shapes under blasting loads holds significant theoretical importance and practical application value.

2. Experimental Methodology

2.1. Specimen Making

Based on the experimental scheme for cavity structures, ten concrete casting molds were prepared. Before casting the samples, the molds were first checked to ensure their integrity. After the mold inspection, the molds were moistened with water to prevent the wooden boxes from absorbing moisture from the concrete, which could lead to an imbalance in the water-cement ratio and affect the strength and quality of the concrete, potentially causing cracking or detachment. Following the mold wetting, the concrete casting process commenced. The concrete was mixed using fine river sand, cement, and water in a ratio of 1:2:0.5 for cement, sand, and water, respectively. A small amount of clay was added to one batch, as shown in Figure 1. After thorough mixing, the concrete was placed into the molds. Blastholes were pre-made using PVC pipes with a diameter equal to the blasthole size. The strength of the PVC pipe is negligible compared to the strength of the concrete and the power of the explosive, and thus the pipes were left inside during the explosion experiment. For some models that required the placement of spheres, cubes, or other pre-made cavity shapes, foam objects were placed at the designated positions as the concrete was filled. After filling the mortar with concrete, a vibrating rod was used to expel air bubbles from the cement mortar, improving the compactness of the concrete. The vibration rod should be moved uniformly within the wooden box during compaction. Once the casting was completed, the samples were cured under standard conditions for 28 days. The finished concrete casting products are shown in Figure 1.

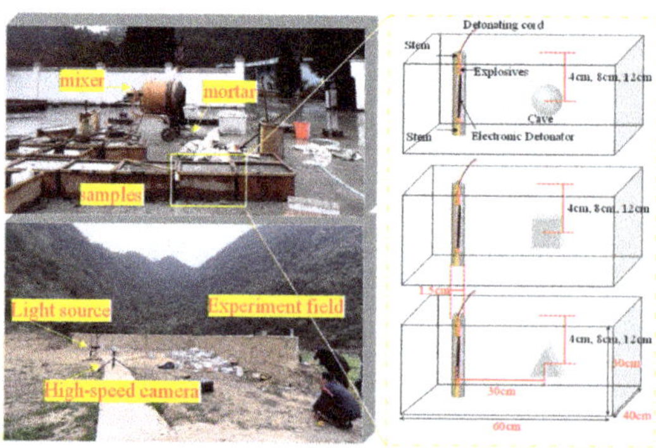

Figure 1. Specimen preparation and design scheme.

After the concrete samples were cured for 28 days, the process of demolding the samples began. Due to the adhesion between the concrete and the mold, tools such as pry bars and hammers were needed. For some molds with a nine-parallelepiped shape, the demolding was done in a sequence of first the edges and then the center. During the demolding process, iron tools such as hammers should not be used directly on the concrete model to avoid breaking the samples and ensure the accuracy of the experiment. After the samples were separated from the mold, they should be properly labeled to facilitate the identification of sample conditions and proportions during testing. As some concrete may enter the pre-made blastholes during the casting process, after demolding, it is necessary to check each blasthole to ensure that it is not obstructed. In cases of obstruction, a rod should be used to clear the hole. For more severe blockages, a drill should be used to clear the hole. When using a drill, the drilling direction must align with the direction of the blasthole, without deviation, to prevent damage to the sample.

2.2. Experimental Program

Due to the large volume of the samples and the relatively large amount of explosives used during the blast, the large samples were all placed in the blast pit, arranged in order as shown in Figure 2. During the explosion experiments, the blasts were conducted sequentially to ensure that each fragment came from the same sample. High-speed cameras and associated equipment were used in the experiment, with the high-speed cameras being used to record the destruction process of the samples and to measure key parameters during the blast, such as the propagation velocity of the blast waves and the fragmentation and separation process of the explosive material.

Figure 2. Experimental program and materials.

These parameters are crucial for evaluating the blasting effect and studying the blasting mechanism. The various experimental equipment was connected to set up the explosion test system, with the high-speed camera aimed at the sample to be blasted, adjusting the angle and focal length, and checking through the computer screen whether the adjustments were successful. Lighting was added to enhance the lighting effects for better photography, and it was necessary to turn off the lights promptly after each experiment to prevent damage to the light sources. The process of loading the explosive is shown in Figure 2. Considering the large volume of the samples, to achieve a better blasting effect, the single-hole loading quantity was kept at 20 g. An instantaneous electronic detonator was placed in each blasthole. The explosive was first loaded into the blasthole, followed by the digital electronic detonator, to ensure that the explosive would be initiated fully and effectively.

3. Experimental Results

3.1. Specimen Fracture and Break

Due to the frame rate of the high-speed camera being set at 20,000 Hz, the interval between each image is 50 μs. Key time-point images were selected for an overall analysis of the damage characteristics of the samples containing cavities under blasting loads. The damage and destruction process is shown in Figure 3. Generally speaking, the process of damage and failure of the rock mass under blasting loads is as follows: Firstly, the electronic detonator initiates the explosive charge, causing the PVC pipe in the blasthole to be propelled out by the blast wave, but no cracks were observed on the side of the sample. Subsequently, a vertical crack appeared at the position closest to the blasthole, and the crack extended further until it pierced through the entire sample, with a branching crack also forming at the middle of the side. As the damage progressed, the branching crack pierced through the sample, with the entire crack forming an "X" shape. At this point, the explosive gas began to be ejected, and the blocked sand and earth were continuously thrown out. New cracks continued to form at the top of the side and gradually expanded towards the left free surface. An additional branching crack also formed in the middle of the "X" crack, expanding towards the right. Finally, as the action of the explosive gas intensified further, the width of the cracks kept increasing, and no new cracks were produced.

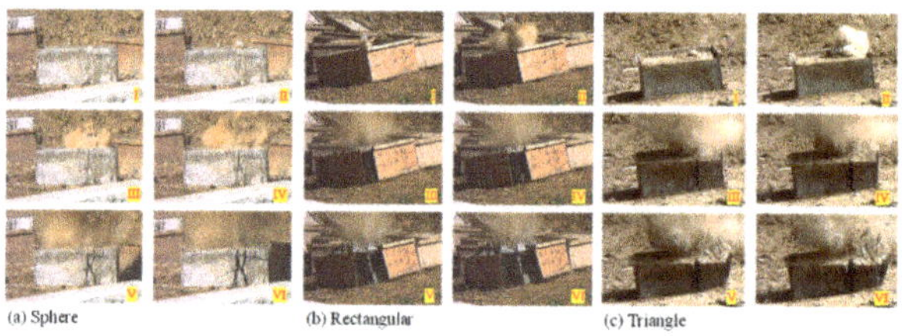

Figure 3. Fracture process of different shapes of cavities under blasting loads.

Figure 4 shows the crack propagation process of the control group (i.e., with no structural components) under blasting loads. From the figure, it can be observed that with the significant displacement of the PVC pipe, obvious deformation was also seen on the upper surface of the sample. Subsequently, a noticeable crack propagation was observed, and the propagation speed was relatively fast. After the basic formation of cracks, the width of the cracks continued to increase. Throughout the explosion process, the emission of explosive gases from the blast lasted for a considerable period. When no new cracks were produced, the process of width increase began to see the efflux of explosive gases and debris. This indicates that the action time of the explosive gas was prolonged, and the energy of the explosive was fully utilized, which is conducive to the fragmentation of the sample.

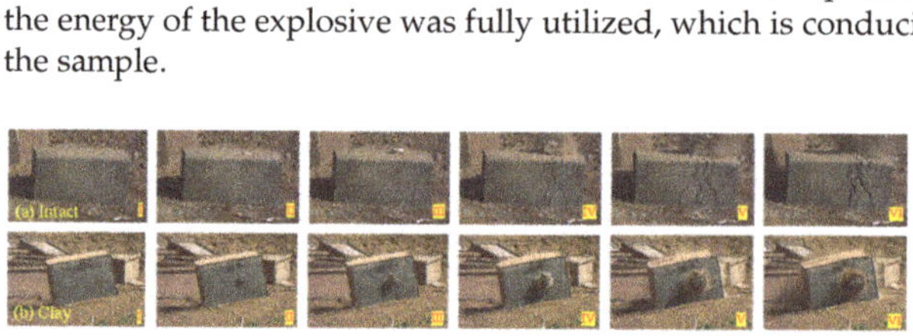

Figure 4. Rock fracture process of intact control specimen and clay-containing specimen under blast loading.

The results of the aforementioned research indicate that under the intense shock wave generated by the explosion of the explosive in the rock mass, the shock and compression of the surrounding rock mass cause local pulverization. After passing through the crushed zone, the compressive stress wave continues to propagate outward, but its intensity has decreased to the point where it can no longer directly cause rock fracturing. When the compressive stress wave reaches the free surface, it reflects from the free surface as a tensile stress wave. Since the tensile strength of rock is significantly lower than its compressive strength, if the strength of the reflected tensile waves exceeds the dynamic tensile strength of the rock, the rock will begin to produce layer-by-layer tensile fracture damage from the free surface [24]. This mode of fracture is also known as "falling in pieces". As the reflected wave propagates towards the explosion source, the falling in pieces continues until the rock within the explosion range is completely pulled apart. High-speed camera images clearly show that the free surface cracks are primarily generated in the area closest to the blasthole, as the stress waves arrive there first, producing a tensile effect and concentrating stress, which in turn leads to crack formation. Subsequently, under the action of the explosive gases, the cracks continue to propagate, and their widths keep increasing. At this point, it is clearly visible that explosion gases and debris are erupting from the blasthole.

3.2. Blasting Block Size and Effectiveness

The presence of cavities within concrete structures affects the propagation of stress waves, and the impact of these cavities varies depending on their location. For blasting experiments conducted on structures with cavities of different shapes, a corresponding size analysis is carried out. The analysis focuses on the blasting block sizes X_{50} and X_{100} at distances varying from the upper free surface. X_{50} reflects the average size of the fragments, while X_{100} represents the maximum size of the fragments, which effectively describes the proportion of large pieces generated by the explosion. Additionally, due to limitations in the experimental conditions, the spherical cavity model only yielded block size data at distances of 8 cm and 12 cm from the upper free surface. The block size distribution is shown in Figure 5.

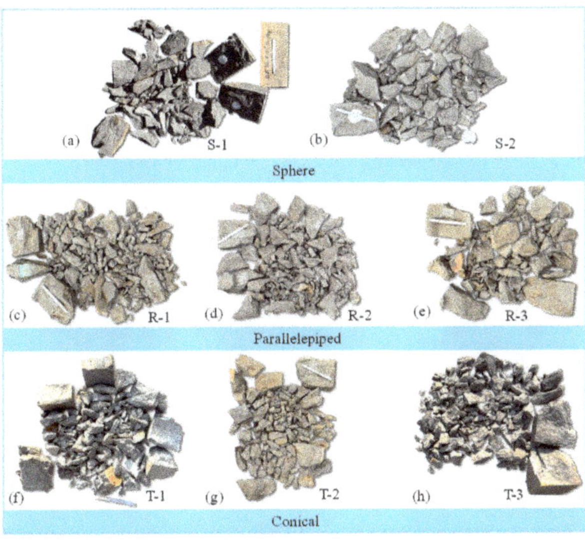

Figure 5. Fragments images for different distances and shapes of the rock mass: (**a**,**b**) the shape of the cavity is spherical, and the distances from the upper surface are 8 and 12 cm, respectively; (**c**–**e**) the shape of the cavity is parallelepiped, and the distances from the upper surface are 4, 8, and 12 cm, respectively; (**f**–**h**) the shape of the cavity is conical, and the distances from the upper surface are 4, 8, and 12 cm.

A large amount of measured data and literature indicate that the five-parameter Swebrec modified distribution function can well describe the distribution law of rock blasting fragments [25]. Therefore, this article, after obtaining the sizes of each rock block through ImagJ (https://imagej.net/ij/) image processing, uses the Swebrec modified distribution function to obtain the size distribution curves for each group of rock blasting tests. The basic expression of this function is as follows:

$$P(x) = 1 / \left\{ 1 + A[\ln(x_{\max}/x) / \ln(x_{\max}/x_{50})]B + (1-a)\left[\left(\frac{x_{\max}}{x} - 1\right) / \left(\frac{x_{\max}}{x_{50}} - 1\right)\right]^{C} \right\} \quad (1)$$

where $x_{\max}$ is the maximum size at $P(x_{\max})$ = 100%, which can be obtained by directly measuring the size of the largest fragment; x_{50} is the median size at $P(x_{50})$ = 50%; A is the grading coefficient, with a range of values from 0 to 1; and B, C are the fitting parameters, with a range of values around 2.

The distribution patterns of blasting fragment sizes under different shapes and distances from the upper free surface are illustrated in Figure 6a–c showing good fitting effects with R^2 values greater than 0.99. Figure 6a reveals that when the spherical cavity is 8 cm away from the upper free surface, the X_{50} value (average fragment size) is 10.64 cm, and the X_{100} value (maximum fragment size) is 35.97 cm. When the cavity is 12 cm away from the upper free surface, the X_{50} value is 6.84 cm, and the X_{100} value is 19.25 cm. The fragment size analysis indicates that a spherical cavity closer to the free surface yields a better blasting effect, resulting in smaller overall fragment sizes and a reduced proportion of large pieces.

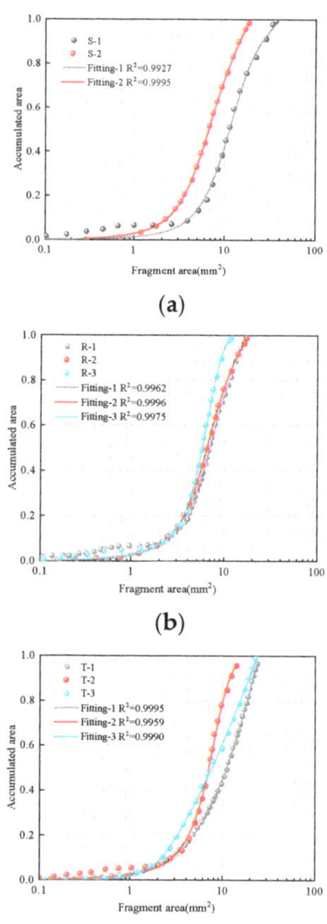

Figure 6. Fitted curves of the distribution pattern of the blasting block size for each group of specimens: (**a**) spherical cavities; (**b**) parallelepiped cavities; (**c**) conical cavities.

Similarly, for concrete structures with parallelepiped cavities at different positions, different blasting effects were observed. Figure 6b shows the blast fragments from different positions of a cube. In this case, the cavity is located in three different positions. The best blasting effect was achieved when the cavity was 12 cm away from the upper free surface, with X_{50} = 5.98 cm and X_{100} = 12.73 cm, resulting in no obvious large pieces, as shown in Figure 6b. When the cavity was 4 cm or 8 cm away from the upper free surface, the blasting effect was poorer, with values of X_{50} and X_{100} being relatively close to each other.

In summary, there is little difference in the average fragment size among the three samples. However, for the maximum fragment size, the sample with the spherical cavity 12 cm away from the upper free surface had significantly smaller values compared to the other two groups. Figure 6c displays the fitting curves of the blasting fragment sizes for concrete structures with conical cavities under explosive shock loads. For the three different positions of the conical structures, the best blasting effect was obtained when the cavity was 8 cm away from the upper free surface, with the smallest average fragment size, X_{50} = 6.99 cm and X_{100} = 14.59 cm. The worst effect was achieved when the cavity was 4 cm away from the upper free surface, with X_{50} = 10.84 cm and X_{100} = 25.54 cm. The fitting curve was relatively flat when the cavity was 12 cm away from the upper free surface, indicating a large discrepancy in blasting fragment sizes, with X_{50} = 7.93 cm and X_{100} = 22.62 cm. Overall, the conical cavities exhibited consistent blasting fragment size distribution patterns with other shapes, meaning that the best blasting effect is achieved when the conical cavity structure is located in the middle of the sample, while the worst effect is observed when the cavity is in the middle upper position. The shape of the cavity affects the propagation of explosion stress waves, and for spherical, parallelepiped, and conical cavities, the average fragment size after blasting for spherical cavities was significantly larger than for the other two shapes. The overall average sizes of the parallelepiped and conical cavities were relatively close, suggesting that these two shapes have similar effects on the blasting fragment sizes.

4. Numerical Simulation Study on the Response and Fracture Characteristics of Cavity Blasting

In this section, corresponding numerical simulation analyses are carried out for different blasting conditions in similar model tests, and the blasting model, intrinsic relationship and parameter selection of numerical simulation are calibrated by comparing numerical simulation and similar model test results to verify the accuracy of numerical simulation. On this basis, a two-dimensional mechanical response and three-dimensional dynamic damage numerical simulation are carried out to further obtain the influence of different cross-section shapes on the stress redistribution and damage process around the cavity, so as to provide a theoretical basis and practical guidance for the on-site engineering blasting and the stability of the cavity. The software selected for numerical simulation is the dynamic finite element program ANSYS/LS-DYNA R7.0.

4.1. Parameter Calibration and Cavity Stress Response

In order to obtain the material parameters, samples were taken on-site in the model test to make standard tests for mechanical property tests, and atriaxial testing machine was used to carry out uniaxial compression, Brazilian splitting and straight shear tests, and the mechanical property test data are shown in Table 1. In order to study the mechanical response characteristics of the cavity, the rock material is set as an elastic material, the keyword *MAT_ELASTIC is used, and the parameter settings are consistent with the mean values in Table 1. The foam material is used to simulate the cavity—ANSYS/LS-DYNA software simulation foam has a variety of materials available. In this paper, the MAT_SOIL_AND_FOAM material is used and the specific parameters determined by reference [26] are shown in Table 2.

Table 1. Experimental test data for similar materials.

Sample	Density ρ/(kg·m^{-3})	Static Compressive Strength σ_s/MPa	Dynamic Compressive Strength σ_d/MPa	Tensile Strength σ_t/MPa	Modulus of Elasticitye/GPa	Poisson's Ratio μ	Cohesion c/MPa	Angle of Internal Friction φ/°
1	2069.76	4.83	8.13	0.47	2.68	0.23	0.53	20.43
2	2103.71	4.66	8.48	0.45	2.59	0.24	0.51	20.29
3	1882.84	4.46	8.58	0.49	2.53	0.22	0.55	21.95
Average value	2018.77	4.65	8.4	0.47	2.60	0.23	0.53	20.89

Table 2. Parameter settings for simulated cavity materials.

Density ρ/(kg·m^{-3})	Tensile Strength σ_t/MPa	Volume Modulus/MPa	a_0	a_1	a_2
400	0.64	3	0.0034	703.3	0.3×10^9
vcr	ref	lcid	eps1	eps2	eps3
0	0	0	0	−0.104	−0.161
eps4	eps5	eps6	eps7	eps8	eps9
−0.192	−0.224	−0.246	−0.271	−0.283	−0.290
eps10	p1	p2	p3	p4	p5
−0.4	2×10^5	2×10^5	4×10^5	6×10^5	1.2×10^6
p6	p7	p8	p9	p10	
2×10^6	4×10^6	6×10^6	8×10^5	41×10^5	

The two-dimensional elastic numerical modeling is shown in Figure 7, which is consistent with the top-view cross-section size of the similar model test, and the model is a rectangular structure with the size of 60 cm × 30 cm, and the total number of model units is about 130,000 units. The explosive size is 2 cm × 2 cm, and the equivalent blasting curve is used to apply the load and generate the blasting stress wave propagating to the cavity. The shapes of the cavities are circular, parallelepiped, and conical, and the diameters and side lengths are 4 cm and 12 cm, respectively, so as to realize the numerical simulation of the process of blasting stress wave incident on cavities with different cross-sectional shapes. The needle-shaped wave is used to represent the blast stress wave, in which the waveform rises rapidly and falls slowly, with the energy concentrated primarily at the front, effectively approximating the stress wave generated by blasting. Additionally, the pressure is applied along the contour of the blasthole.

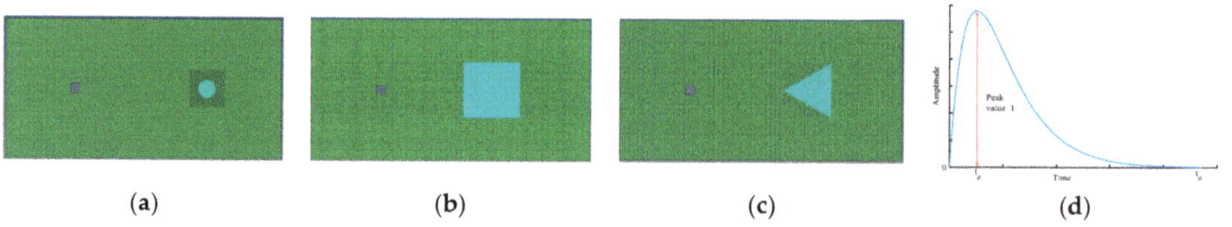

Figure 7. 2D model of the cavity with different cross-section shapes and the needle-shaped wave [27] (**a**) circular shape; (**b**) parallelepiped shape; (**c**) conical shape; (**d**) equivalence load curve.

The stress variation and effective strain cloud diagrams of different cross-sectional shapes of the cavities under blasting loads are shown in Figures 8–10. From Figure 9a, it can be seen that the blasting load produces a significant stress concentration at the bottom of the top of the cavity when it propagates to the circular cavity. As the stress wave propagates to the right free surface, a conical region of small effective stresses appears around the cavity due to the presence of the upper and lower free surfaces. This means that the circular cavity absorbed a large amount of energy during the incidence of the blast stress wave and played a good anti-detonation role. Subsequently, due to the free boundaries on the left and right sides, the reflected wave and the incident wave acted together on the cavity, resulting in the cavity being subjected to the stresses on both sides and an X-shaped high-stress distribution appeared around the circular cavity, which means that the cavity is prone to butterfly damage in the second half of the propagation of the blast stress wave, which is in line with the conclusions of the existing studies [28]. The effective strain cloud in Figure 9b is also consistent with the above analysis. The circular cavity first shows significant compressive strains on the wave-facing side, and with the propagation of the stress wave, compressive strains also appear on the back-blast side of the circular cavity. Finally, under the combined effect of the incident wave and the reflected wave, significant compressive strains appear in the whole cavity interior, but the strains on the periphery of the cavity are larger, which is due to the X-shape of the high-stress distribution.

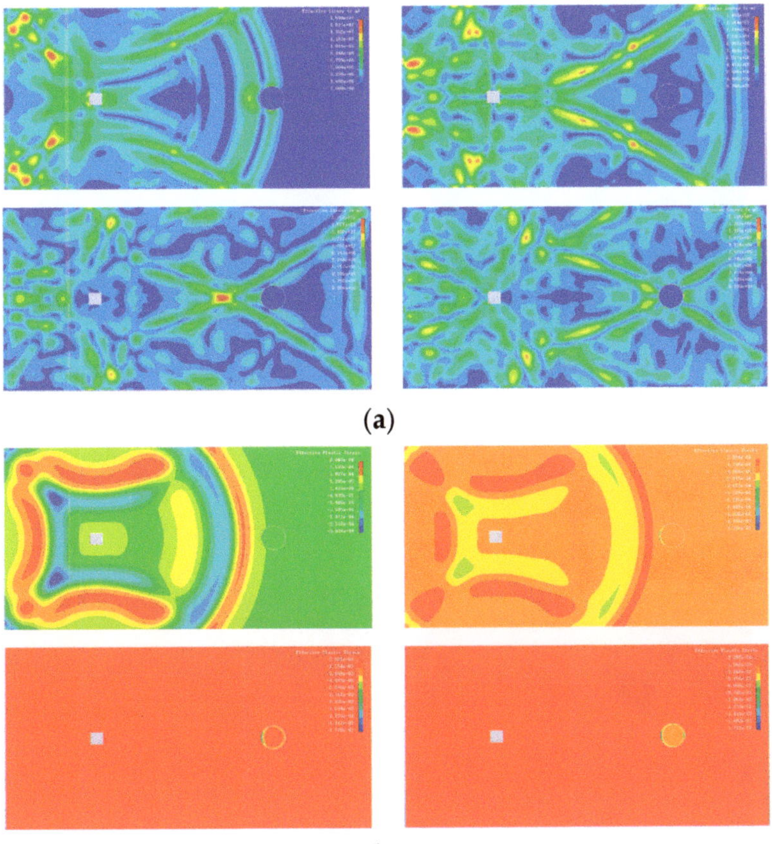

(a)

(b)

Figure 8. Effective stress and strain in a circular cavity under blasting load: (**a**) effective stress nephogram of circular cavities; (**b**) effective strain nephogram of circular cavities.

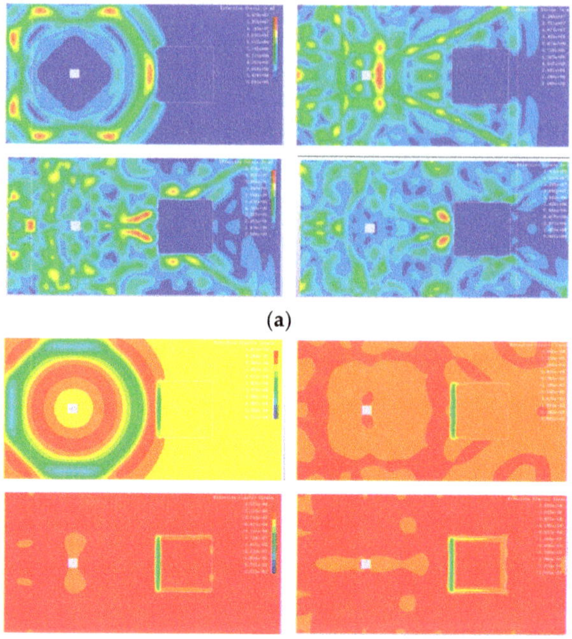

(**a**)

(**b**)

Figure 9. Effective stress and strain in a parallelepiped-shaped cavity under blasting load: (**a**) effective stress nephogram of the parallelepiped-shaped cavity; (**b**) effective strain nephogram of the parallelepiped-shaped cavity.

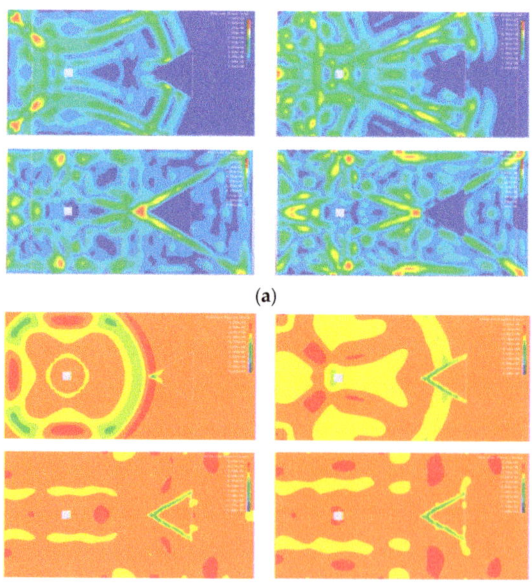

(**a**)

(**b**)

Figure 10. Effective stress and strain in a conical cavity under blasting load: (**a**) effective stress nephogram of the conical cavity; (**b**) effective strain nephogram of the conical cavity.

Figure 10 shows the process of changes in effective stress and strain within a parallelepiped cavity subjected to blast impact. Due to the large effective impact area of the parallelepiped cavity, significant stress concentration first appeared at the two corners on the left. Subsequently, the stress wave produced significant reflection and scattering between the blasthole and the cavity, resulting in a conical region of high effective stress between the blasthole and the cavity. As the stress wave propagated to the free surfaces on both sides, the incident wave and the reflected wave simultaneously acted on the parallelepiped cavity, causing sustained dynamic stress concentration on the side facing the blast, while stress concentration also appeared at the two corners on the right side of the parallelepiped cavity. From Figure 10b, it can be observed that when the blast stress wave reached the parallelepiped cavity, a significant compression strain rapidly appeared on the side facing the blast, and as the stress wave propagated, the entire cavity exhibited compression strain, although the strain on the side facing the blast was much greater than that on the upper and lower sides and the side opposite the blast, with the strain on the opposite side being the smallest. This is because the incident wave, reflected wave, and scattered waves directly acted on the side facing the blast, where the majority of the stress wave energy was consumed, while the side opposite the blast was affected the least. Finally, when the incident wave and the reflected wave produced by the right side's free surface acted on the cavity together, a significant compression strain also appeared on the opposite side, but it was still smaller than that on the side facing the blast. Additionally, the strain near the four sides of the parallelepiped was much greater than the strain within the cavity itself. In summary, during the plastic stage, the parallelepiped cavity is prone to develop cracks first at the four corners of the parallelepiped, which then gradually propagate to the four sides, thus initiating structural failure.

As shown in Figure 10a, the propagation of stress waves to the conical cavity resulted in significant scattering. The stress waves were bifurcated by the corners facing the blast, propagating along the edges of the conical cavity to the right. Between the blasthole and the conical cavity, a larger conical region appeared with high effective stress, and the corners facing the blast exhibited significant dynamic stress concentration. Subsequently, the reflected waves generated by the free surface interacted with the incident waves simultaneously on the conical cavity, causing significant stress concentration at the three corners of the conical cavity, with the effective stress at the corner facing the blast being greater than that at the other two corners. From Figure 10b, it can be observed that when the blast stress wave reached the cavity, a significant compression strain appeared at the corner facing the blast. As the stress wave propagated, both edges of the side facing the blast also showed a considerable compression strain. Finally, when both the incident wave and the reflected wave acted on the conical cavity simultaneously, a region of significant compression strain appeared at the two corners opposite the blast, but no such strain was observed near the edges on the opposite side. Therefore, during the plastic stage, the conical cavity is prone to damage and failure at the three corners and the two sides adjacent to the blast, with most of the energy being consumed by the rock mass of the cavity on the side facing the blast, thereby protecting the structure of the cavity on the opposite side.

The comprehensive analysis reveals that different cross-sectional shapes have a significant impact on the propagation of blast stress waves and the characteristics of stress and strain changes around the cavity. The circular cavity, due to its curvature, experiences only transient stress concentration at the top and bottom, and the effective strain throughout the process is far greater than that of the parallelepiped or conical cavities. This indicates that the circular cavity absorbs a large amount of blast energy for structural deformation of the cavity and subsequent cavity damage, thus reducing the energy used for the destruction of the surrounding rock mass. Therefore, the circular cavity serves as a better blast-resistant effect on the surrounding rock mass. Furthermore, both the parallelepiped and conical cavities exhibit significant stress concentration at the corners, with the effective strain of the conical cavity being less than that of the parallelepiped cavity due to the larger effective impact area of the parallelepiped cavity. However, because of the tip effect of the conical, the effective stress at the corner facing the wave is much greater than that of the parallelepiped.

4.2. Three-Dimensional Numerical Modeling and Damage Characteristics of Cavities

Based on the stress distribution characteristics of cavities under blasting, a three-dimensional elastic-plastic model was constructed to further study the destructive characteristics of cavities under blasting. Some scholars have employed the MAT_PLASTIC_KINEMATIC model for simulating the dynamic damage of rock due to its computational efficiency and alignment with the dynamic rock damage criteria [29,30]. The rock material is primarily defined using the material keyword MAT_PLASTIC_KINEMATIC (kinematic plasticity model material). This material's constitutive model can reflect the elastic and plastic failure processes of soils, rocks, concrete, and other media under the action of explosive blasting. According to the mechanical testing results of cement mortar specimens from similar model tests, the specific parameters used are listed in Table 3.

Table 3. Material parameter settings.

Density kg/m^3	Modulus of Elasticity/GPa	Poisson's Ratio	Yield Strength/MPa	Tensile Strength/MPa	Failure Strain
2000	2.60	0.23	4.65	0.47	0.06

The keyword MAT_ADD_EROSION can define various parameters of material failure, such as compression stress, tensile stress, equivalent stress, strain, and failure time. For the MAT_PLASTIC_KINEMATIC material, therefore, it is only necessary to use this keyword to define the material's tensile strength [31]. In this simulation, any model that defines the tensile strength parameter with MAT_ADD_EROSION is set to −0.47 MPa (the tensile stress value for MAT_ADD_EROSION is labeled as a negative value). Additionally, the MAT_HIGH_EXPLOSIVE_BURN material is used to define high-energy explosives and void material. This paper uses No. 2 rock emulsion explosives with the same explosive performance as the similar model test, with a density of 1050 kg/m^3 and a detonation velocity of 4200 m/s. Besides material parameters, in ANSYS/LS-DYNA, it is also necessary to set the *EOS_JWL equation of state parameters [32].

The EOS_JWL equation of state is as follows:

$$P = A[1 - \frac{\omega}{R_1 V}]e^{-R_1 V} + B[1 - \frac{\omega}{R_2 V}]e^{-R_2 V} + \frac{\omega E}{V} \tag{2}$$

where A, B, R_1, R_2, ω are parameters related to explosives determined by experiments, V_0 is relative specific volume, E_0 is initial specific energy. The specific parameters of the equation of state of the explosive are shown in Table 4.

Table 4. Parameters of the explosive equation of state.

A	B	R_1	R_2	OMEG	E_0	V_0
2.144×10^{11}	0.182×10^{9}	4.2	0.9	0.15	4.192×10^{9}	1.0

The three-dimensional numerical modeling employs the Lagrange (Lagrange) method with shared nodes, which requires not only the definition of explosive and rock materials but also the additional definition of void material. The ALE12 algorithm for *SECTION_SOLID is used to calculate the void material, which provides the necessary volume for the expansion of the explosive, thereby preventing negative volume errors during the simulation. The void material is co-nodally coupled with the rock mass without requiring additional special boundaries, ensuring that the transmission of the blasting stress wave remains unaffected. The void material (void) is defined using the same material parameters as the explosive material. The three-dimensional numerical model is shown in Figure 11.

Consistent with similar model tests, the model is rectangular in structure with dimensions of 60 cm × 40 cm × 30 cm, and the total number of elements is approximately 640,000. The blue part in the figure represents the explosive elements, with a size of 4 cm × 2 cm × 2 cm, which is close to the 20 g charge mass in the similar model test. The red part represents the void material elements, which are shaped as cubes with an edge length of 20 cm. The yellow part represents the filled area; in this figure, an individual fracture model is used as an example, with a size consistent with the corresponding similar model test. The green part represents the rock elements, with a size of 60 cm × 40 cm × 30 cm, and the void material elements partially overlap with the rock elements in the modeling. Generally, when simulating the blasting of rock mass, a non-reflecting boundary is applied at the model's boundary using the keyword BOUNDARY_NON_REFLECTING to simulate different numbers of free surfaces. The 3D model is not set with a reflection-free boundary to consistency with the experimental specimen.

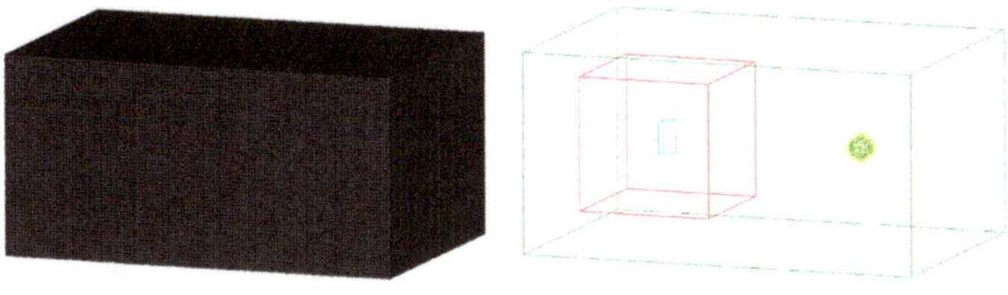

Figure 11. Three-dimensional numerical model.

Due to the presence of free surfaces, there is a limiting and nature-altering effect on the energy of the blast stress waves. This causes the stress wave energy to be confined within the expected range of blasting action through reflection, and it can alter the nature of the stress wave energy, converting the original compressive wave energy into tensile and shear wave energy. As a result, compared to the case without free surfaces, the rock mass is subjected to a longer duration of stress wave destructive action, leading to tensile and shear damage to the rock mass.

The damage development of the rock mass during a simulated typical rock blasting process is shown in Figure 12 and is specifically manifested as follows:

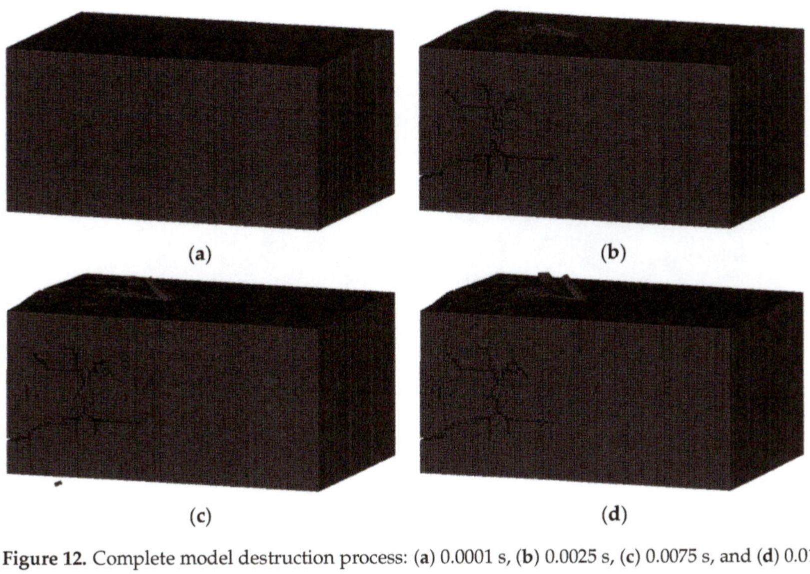

(a) (b) (c) (d)

Figure 12. Complete model destruction process: (**a**) 0.0001 s, (**b**) 0.0025 s, (**c**) 0.0075 s, and (**d**) 0.0100 s.

(1) At 0.1 ms, under the impact load of the explosive, the rock mass elements surrounding the explosive reach their failure strength and undergo massive destruction, forming a blast cavity. At 0.3 ms, the blast cavity has basically expanded to its limit, indicating that the action of the explosive impact load has essentially ended. The rock mass outside the cavity is subjected to hoop stress from the direct blast stress waves, forming radial cracks that radiate outward. At 0.7 ms, the radially spreading cracks have further developed. As the blast stress wave reaches the edge of the rock mass, due to the reflection of the stress wave at the free surface, the rock mass undergoes multi-layer fracturing under the action of the reflected wave, forming multiple cracks parallel to the free surface.

(2) At 0.9 ms, the radiating radial cracks and the parallel free surface layer cracks further expand. The reflected waves from adjacent free surfaces begin to converge, and under the superimposed stress field, the rock mass exhibits angular cracks, with the angular crack cracks beginning to develop. At 4.5 ms, in addition to the already fully developed radiating radial cracks, parallel free surface layer cracks, and angular cracks, the rock mass shows a wider variety of cracks under the action of stress waves, including edge cracks at 45° to the free surface. These various cracks are mostly interconnected, and the rock mass is overall in a state of destruction.

(3) Compared to the 4.5 ms moment, at 10 ms, the cracks have only slightly developed, and their expansion has essentially halted, indicating that the destructive effect of the rock mass blast stress waves has ceased. At this point, the cracks are primarily expanding under the quasi-static pressure of the explosive gases, and at the same time, the broken rock mass is being propelled outward by the action of the explosive gases, with the rock mass edges showing noticeable bulging. When comparing the 20 ms with the 10 ms moment, there is almost no further development of rock mass cracks, and in some cases, the edges of certain cracks have even exhibited a closure phenomenon. The main cracks have further extended, and the bulging of the rock mass at the free surface is more pronounced.

The damage situations of different cross-sectional cavity numerical models under blasting are shown in Figure 13. From Figure 13, it can be seen that the shape of the blast cavity within the rock mass model has not changed significantly with the variation in the shape of the cavity cross-section. This indicates that these conditions are set at a considerable distance, and they have little effect on the compressive and shearing destructive actions in the blasting crushing zone. At the five free surfaces near the explosive, under the action of blasting, there have been obvious tensile failure cracks, and these cracks exhibit variations in shape under different conditions.

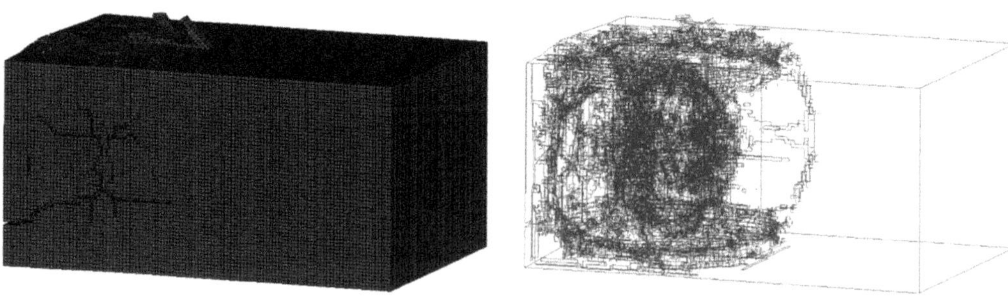

(**a**) Complete model.

Figure 13. *Cont.*

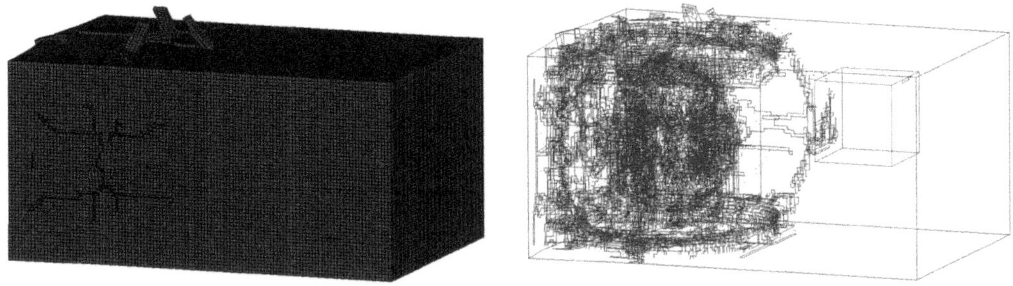

(**b**) Parallelepiped cavity model.

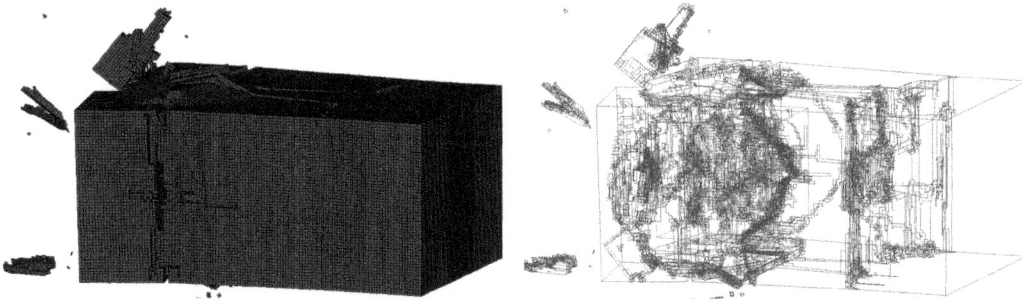

(**c**) Conical cavity model.

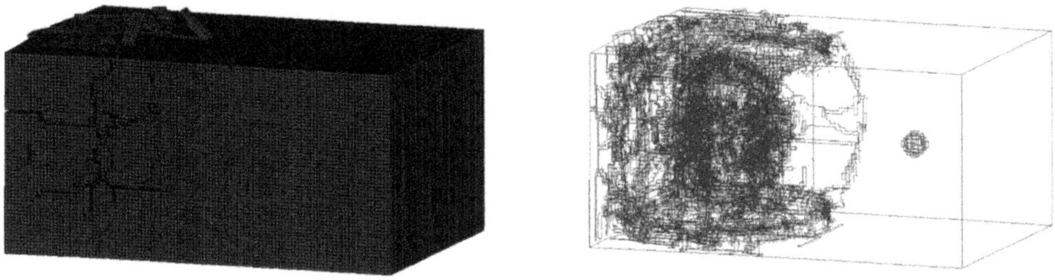

(**d**) Spherical cavity model.

Figure 13. Partial model damage.

To analyze the effects of blasting further statistically under different conditions, the information files from the simulations can be used to count the unit failure in the rock mass model. This allows us to understand the development of rock mass damage and the final damage effects during the rock blasting simulation process. Table 5 shows the number of failed units and the ratio of failed units to the total number of units for all models up to 20 ms, reflecting the impact of different conditions on the rock mass damage effects. The proportion of failed units in the parallelepiped cavity is slightly greater than that of the conical cavity and much greater than that of the circular cavity. Additionally, the proportion of failed units in the intact model is less than that of the parallelepiped and conical cavities but greater than that of the circular cavity. This means that the circular cavity absorbs more blast energy, hinders rock mass damage, and reduces the degree of rock mass destruction, while the parallelepiped and conical cavities promote rock mass damage. This is because parallelepiped and conical shapes have significant dynamic stress concentration at their corners, which induces damage and fractures in the cavities. From Figure 13, it can be observed that the cracks produced by the parallelepiped and conical cavities connect with the blasting crushing zone area, and the main cracks in the parallelepiped and conical cavities occur on the blast-facing side, which is consistent with the results of the two-dimensional numerical model. However, due to the larger effective incoming area of the parallelepiped compared to the conical, the area between the blastholes and the cavities experiences significant reflection and scattering of stress waves, thereby intensifying the damage to the rock mass in that area. Moreover, the average block sizes of the spherical, conical, and parallelepiped cavities in the experimental results are consistent with the numerical simulation results. This indicates that the spherical cavity hinders the blast stress wave from damaging the rock mass, while the conical and parallelepiped cavities promote the blasting damage to the rock mass.

Table 5. Number and proportion of failed units for different condition models.

Numerical Model Type	Complete Model	Parallelepiped Cavity Model	Conical Cavity Model	Spherical Cavity Model
Number of failed units	70,800	70,771	70,644	62,857
Number of rock units	639,566	625,742	633,580	638,661
Percentage of failed units	11.07%	11.31%	11.15%	9.84%

5. Conclusions

This study has conducted an in-depth exploration of the dynamic response and rock mass damage issues under blasting loads for different shapes of cavities. Through field tests and numerical simulations, the mechanisms of rock mass damage and destruction under blasting loads were revealed, and the following conclusions were drawn:

Firstly, the dynamic response laws of different shaped cavities under blasting loads were analyzed, and the damage mechanisms and blast block size distributions under blasting loads were studied. The influence of cavity geological bodies on the size of blasting fragments and the characteristics of rock mass damage were revealed. A combination of field tests and numerical simulations was used to verify the experimental results and further reveal the influence of different geological structures on the blasting effect. The results show that the presence of spherical, conical, and parallelepiped cavities has a significant impact on the blasting effect and rock mass damage under blasting loads. The spherical cavity, with its good anti-blast performance, can absorb a large amount of blast energy and reduce the degree of rock mass destruction, while parallelepiped and conical cavities are prone to stress concentration at the corners, promoting rock mass damage and destruction. Additionally, the study of blast block size distribution indicates that the blasting effect of the parallelepiped and conical cavities is better than that of spherical cavities. The research results of this study contribute to the enrichment of the theory of rock mass dynamic response and damage under blasting loads, providing new ideas and methods for subsequent research. These research findings serve as an important theoretical basis and practical guidance for the design and construction of underground engineering blasting, helping to improve engineering safety and promote the sustainable development of the underground engineering industry.

Author Contributions: X.C.: conceptualization, data curation, funding acquisition, methodology, writing—review and editing. M.Z.: writing—original draft, formal analysis, investigation, software. Q.L.: validation, investigation, writing—review and editing. All authors have read and agreed to the published version of the manuscript.

Funding: This research was funded by the National Natural Science Foundation of China (NO. 520640031008654), the State Key Laboratory of Precision Blasting and Hubei Key Laboratory of Blasting Engineering, Jianghan University (No.PBSKL2022C02), and the National Natural Science Foundation of China (No.52064003).

Institutional Review Board Statement: Not applicable.

Informed Consent Statement: Not applicable.

Data Availability Statement: Data will be made available on request.

Conflicts of Interest: The authors declare that they have no known competing financial interests or personal relationships that could have appeared to influence the work reported in this paper.

References

1. Liu, K.W.; Hao, H.; Li, X.B. Numerical Analysis of the Stability of Abandoned Cavities in Bench Blasting. *Int. J. Rock. Mech. Min. Sci.* **2017**, *92*, 30–39. [CrossRef]
2. Chen, J.; Chen, J.; Rui, Y.; Pu, Y. Joint Inversion of AE/MS Sources and Velocity with Full Measurements and Residual Estimation. *Rock. Mech. Rock. Eng.* **2024**, 1–16. [CrossRef]
3. Zhang, Q.; He, M.C.; Wang, J.; Guo, S.; Zhu, C.; Tao, Z.G.; Wang, C. Investigation of a Non-Explosive Directional Roof Cutting Technology for Self-Formed Roadway. *Int. J. Min. Sci. Technol.* **2022**, *32*, 997–1008. [CrossRef]
4. Zhu, X.Y.; Zhai, C.; Yu, X.; Xu, J.Z.; Sun, Y.; Cong, Y.Z.; Zheng, Y.F.; Tang, W.; Chen, A.K. Fracture Damage Characteristics of Hard Roof with Different Bedding Angles Induced by Modified Soundless Cracking Agents. *Eng. Fract. Mech.* **2023**, *289*, 109387. [CrossRef]
5. Firouzabadi, M.; Esmaeili, K.; Rashkolia, G.S.; Asadi, M. A Discrete Element Modelling of Gravity Flow in Sublevel Caving Considering the Shape and Size Distribution of Particles. *Int. J. Min. Reclam. Env.* **2023**, *37*, 255–276. [CrossRef]
6. Bottelin, P.; Baillet, L.; Mathy, A.; Garnier, L.; Cadet, H.; Brenguier, O. Seismic Study of Soda Straws Exposed to Nearby Blasting Vibrations. *J. Seism.* **2020**, *24*, 573–593. [CrossRef]
7. Rutqvist, J.; Lennart, B.; Masakazu, C.; Jan, H.; Lanru, J.; Akira, K.; Son, N. Modeling of Damage, Permeability Changes and Pressure Responses during Excavation of the TSX Tunnel in Granitic Rock at URL, Canada. *Environ. Geol.* **2009**, *57*, 1263–1274. [CrossRef]
8. Liu, R.T.; Liu, Y.K.; Xin, D.D.; Li, S.C.; Zheng, Z.; Ma, C.Y.; Zhang, C.Y. Prediction of Water Inflow in Subsea Tunnels under Blasting Vibration. *Water* **2018**, *10*, 1336. [CrossRef]
9. Chen, J.; Tong, J.; Rui, Y.C.; Cui, Y.; Pu, Y.Y.; Du, J.S.; Derek, B.A. Step-path Failure Mechanism and Stability Analysis of Water-Bearing Rock Slopes based on Particle Flow Simulation. *Theor. Appl. Fract. Mech.* **2024**, *131*, 104370. [CrossRef]
10. Yi, C.P.; Daniel, J.; Jenny, G. Effects of In-Situ Stresses on the Fracturing of Rock by Blasting. *Comput. Geotech.* **2018**, *104*, 321–330. [CrossRef]
11. Gu, H.L.; Lai, X.P.; Tao, M.; Cao, W.Z.; Yang, Z.K. The Role of Porosity in the Dynamic Disturbance Resistance of Water-Saturated Coal. *Int. J. Rock. Mech. Min. Sci.* **2023**, *166*, 105388. [CrossRef]
12. Gu, H.L.; Lai, X.P.; Tao, M.; Aliakbar, M.; Zhang, Q.L. Dynamic Mechanical Mechanism and Optimization Approach of Roadway Surrounding Coal Water Infusion for Dynamic Disaster Prevention. *Measurement* **2023**, *223*, 113639. [CrossRef]
13. Luo, H.; Tao, M.; Wu, C.Q.; Cao, W.Z. Dynamic Response of an Elliptic Cylinder Inclusion with Imperfect Interfaces Subjected to Plane SH Wave. *Geomech. Geophys. Geo-Energy Geo-Resour.* **2023**, *9*, 24. [CrossRef]
14. Zhu, W.C.; Li, Z.H.; Zhu, L.; Tang, C.A. Numerical Simulation on Rockburst of Underground Opening Triggered by Dynamic Disturbance. *Tunn. Undergr. Space Technol.* **2010**, *25*, 587–599. [CrossRef]
15. Zhu, Z.M.; Wang, C.; Kang, J.M.; Li, Y.X.; Wang, M. Study on the Mechanism of Zonal Disintegration around an Excavation. *Int. J. Rock. Mech. Min. Sci.* **2014**, *67*, 88–95. [CrossRef]
16. Li, X.B.; Weng, L. Numerical Investigation on Fracturing Behaviors of Deep-Buried Opening under Dynamic Disturbance. *Tunn. Undergr. Space Technol.* **2016**, *54*, 61–72. [CrossRef]
17. Li, C.J.; Li, X.B. Influence of Wavelength-to-Tunnel-Diameter Ratio on Dynamic Response of Underground Tunnels Subjected to Blasting Loads. *Int. J. Rock. Mech. Min. Sci.* **2018**, *2*, 23–38. [CrossRef]
18. Rui, Y.X.; Chen, J.; Chen, J.K.; Qiu, J.D.; Zhou, Z.L.; Wang, W.Z.; Fan, J.Y. A Robust Triaxial Localization Method of AE Source using Refraction Path. *Int. J. Min. Sci. Technol.* **2024**, *34*, 521–530. [CrossRef]
19. Tao, M.; Ma, A.; Cao, W.Z.; Li, X.B.; Gong, F.Q. Dynamic Response of Pre-Stressed Rock with a Circular Cavity Subject to Transient Loading. *Int. J. Rock. Mech. Min. Sci.* **2017**, *99*, 1–8. [CrossRef]
20. Li, X.B.; Li, C.J.; Cao, W.Z.; Tao, M. Dynamic Stress Concentration and Energy Evolution of Deep-Buried Tunnels under Blasting Loads. *Int. J. Rock. Mech. Min. Sci.* **2018**, *104*, 131–146. [CrossRef]
21. Li, X.H.; Zhu, Z.M.; Wang, M.; Shu, Y.; Deng, S.; Xiao, D.J. Influence of Blasting Load Directions on Tunnel Stability in Fractured Rock Mass. *J. Rock. Mech. Geotech. Eng.* **2022**, *14*, 346–365. [CrossRef]
22. Lin, M.H.; Lin, B.Q.; Yang, W.; Liu, T.; Shen, Y.; Liu, T.; Zhang, X.L.; Lin, F.; Xia, G. Study on Influence of Preset Circular Holes on Propagation of Blast-Induced Cracks Based on Experiments and Discrete Element Simulations. *Rock. Mech. Rock. Eng.* **2024**, 1–20. [CrossRef]
23. Ma, C.H.; Gao, Z.Y.; Yang, J.; Cheng, L.; Chen, L. Operation Performance and Seepage Flow of Impervious Body in Blast-Fill Dams Using Discrete Element Method and Measured Data. *Water* **2022**, *14*, 1443. [CrossRef]
24. Liu, K.W.; Zhao, X.R.; Li, X.D.; Sha, Y.Y.; Hong, Z.X.; Yang, J.C.; Ma, S.Z. Effects of Delay Time on Crack Coalescence between Two Boreholes. *Theor. Appl. Fract. Mech.* **2024**, *129*, 104210. [CrossRef]
25. Li, X.D.; Liu, K.W.; Sha, Y.Y.; Yang, J.C.; Hong, Z.X. Experimental and Numerical Investigation on Rock Fracturing in Tunnel Contour Blasting under Initial Stress. *Int. J. Impact Eng.* **2024**, *185*, 104844. [CrossRef]

26. Rial, J.A.; Moran, B. Radiation Patterns for Explosively-Loaded Axisymmetric Cavities in an Elastic Medium, Analytic Approximations and Numerical Results. *Geophys. J. Int.* **1986**, *86*, 855–862. [CrossRef]
27. Tao, M.; Zhao, R.; Du, K.; Cao, W.Z.; Li, Z.W. Dynamic Stress Concentration and Failure Characteristics around Elliptical Cavity Subjected to Impact Loading. *Int. J. Solids Struct* **2020**, *191*, 401–417. [CrossRef]
28. Wang, W.J.; Fan, L.; Ma, Y.J.; Yuan, C. Research on surrounding rock control technology of deep roadway based on butterfly failure theory. *Coal Sci. Technol.* **2023**, *51*, 157–167.
29. Tang, R.; Bawden, W.F.; Katsabanis, P.D. A New Constitutive Model for Blast Damage. *Int. J. Rock Mech. Min. Sci. Geomech. Abstr.* **1996**, *33*, 245–254.
30. Li, X.H.; Liu, K.W.; Yang, J.C.; Li, X.D. Analysis of Blasting Vibration Effects Under Different Ground Stress. *Gold. Sci. Technol.* **2019**, *27*, 241–248. (In Chinese)
31. Filippi, C.D.L.; Grandjean, A.B.; Côte, P. New Robust Observables on Rayleigh Waves Affected by an Underground Cavity, from Numerical to Experimental Modelling. *Geophys. J. Int.* **2019**, *218*, 1903–1918. [CrossRef]
32. Yi, C.; Sjöberg, J.; Johansson, D.; Petropoulos, N. A numerical study of the impact of short delays on rock fragmentation. *Int. J. Rock Mech. Min. Sci.* **2017**, *100*, 250–254. [CrossRef]

applied sciences

MDPI

Article

The Vibration Response to the High-Pressure Gas Expansion Method: A Case Study of a Hard Rock Tunnel in China

Huaide Peng [1,2], Jia Sheng [2], Qi Da [1,*], Bing Dai [1], Lei Zhang [1] and Lihai Tan [2,3]

[1] School of Resource Environment and Safety Engineering, University of South China, Hengyang 421001, China; 2018000052@usc.edu.cn (H.P.); daibingusc@usc.edu.cn (B.D.); zhagleicpu1@usc.edu.cn (L.Z.)

[2] State Key Laboratory of Safety Technology of Metal Mines, Changsha Institute of Mining Research Co., Ltd., Changsha 410012, China; shengjia001@163.com (J.S.); lt716@uowmail.edu.au (L.T.)

[3] School of Civil, Mining and Environmental Engineering, University of Wollongong, Wollongong, NSW 2522, Australia

* Correspondence: 20222002110041@stu.usc.edu.cn; Tel.: +86+17673806014

Abstract: The vibration of rock breaking in tunnel excavation may cause serious damage to nearby buildings if it is not controlled properly. With reference to a hard rock tunnel in China, the vibration response to the high-pressure gas expansion method (HPGEM), an emerging rock-breaking approach, was investigated with field tests, theoretical derivations, and numerical simulations, then comparisons with the traditional dynamite blast were performed. Firstly, the vibration velocity prediction formulas of the two methods were fitted based on the field tests. Subsequently, the accuracy of the formula was verified by numerical simulation, and the vibration attenuation law of the HPGEM was explored. Comparisons were made between the blast and HPGEM, particularly the differences in peak particle velocity (PPV) for different agent qualities, distance from the blasting center, and engineering conditions. Furthermore, this study also analyzed the relationship between the agent qualities and the rock-breaking volume under different cases, finding that the HPGEM has slight vibration and good rock-breaking effect. The HPGEM is thus fully capable of replacing dynamite blasting to carry out rock-breaking operations in certain special areas.

Keywords: high-pressure gas expansion; dynamite blasting; peak particle velocity; rock breaking; tunnel excavation

1. Introduction

The vibration generated by the rock-breaking process in tunnel excavation is one of the key factors that jeopardize the safety of adjacent important buildings (structures). Currently, dynamite blasting is still one of the most commonly used rock-breaking methods [1,2]. However, it inevitably produces high vibration velocity due to its own principle of action; thus, it is no longer applicable in some special construction environments. To solve this problem, some new methods such as the tunnel boring machine (TBM) [3], mechanical excavation [4,5], and gas blasting [6,7], etc., have emerged (Figure 1). Nevertheless, these methods still have some shortcomings. TBM equipment is costly and requires a large construction area [5]. Mechanical excavation is inefficient and has a limited scope of application [6]. Gas blasting has high requirements for equipment and operation, and high requirements for the environment [7]. The above problems bring challenges to the safe excavation of hard rock tunnels near important buildings or facilities in urban development.

The "High-Pressure Gas Expansion Method" (HPGEM) [8,9] was proposed to address the above-mentioned shortcomings. This method involves the simultaneous chemical and physical reactions of the gas generator in the expansion pipe as the active ingredients under sealed conditions. In an extremely short period of time, a large number of high-temperature and high-pressure gases are released to expand and affect the surrounding media, thus achieving rock breaking (Figures 2 and 3). Existing studies have shown that the rock fragmentation morphologies found after HPGEM and traditional explosive blasting are quite different. The fragmentation of rock slag after HPGEM is obviously larger than that from the drilling and blasting method, and the failure surface of the rock slag basically retains the original structural plane, while the failure surface of rock slag produced by the drilling and blasting method is mostly a fresh fracture surface. More specific HPGEM principles can be found in the authors' previously published literature [10].

Figure 1. Commonly used rock-breaking techniques.

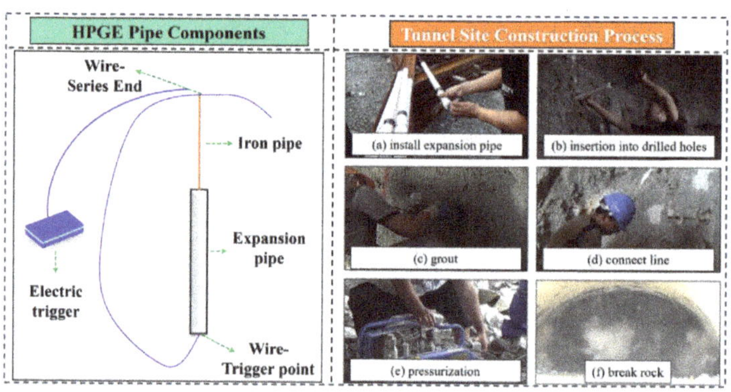

Figure 2. High-pressure gas expansion tube and on-site construction process.

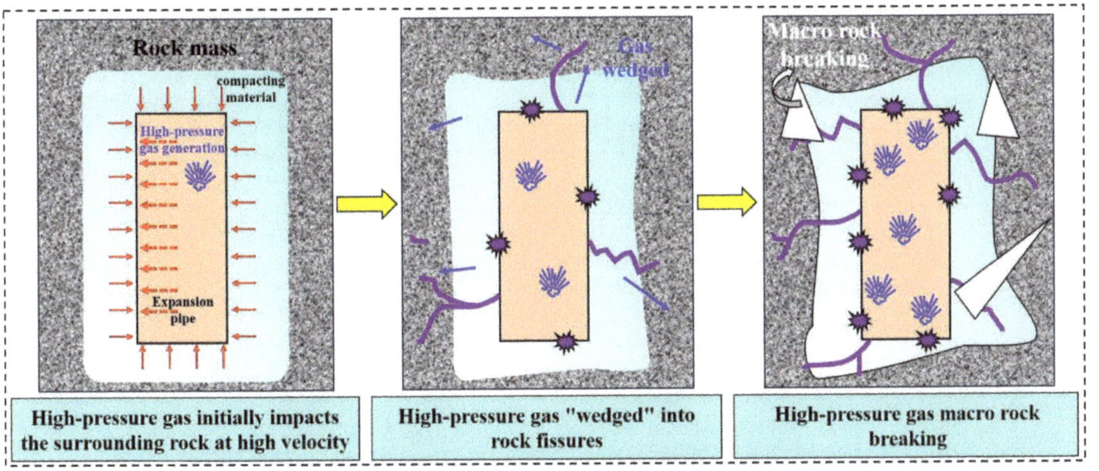

Figure 3. Rock-breaking mechanism of HPGEM (modification based on [10]).

As an emerging rock-breaking method, there are few studies on HPGEM. Peng et al. [9] applied HPGEM to a tunnel excavation to find the optimal blasting design parameters. In addition, they constructed a set of rock mass classification methods applicable to HPGEM [10]. Based on a tunnel in Hangzhou, Liu et al. [11] proposed a set of safe and efficient solutions for urban hard rock excavation by analyzing the rock-breaking effect and vibration effects of the HPGEM on the surrounding buildings. Subsequently, to investigate the applicability of HPGEM in hard rock tunnels near historical sites, Liu et al. [12] obtained suitable rock-breaking parameters for tunnel excavation by means of theoretical analyses, on-site experiments, and vibration monitoring. They found that if the plugging holes were of high quality and not drilled, HPGEM could have the advantages of low vibration, low noise, and fewer flying rocks, which provided a solution for hard rock tunnel excavation near ancient buildings and historical sites.

There have been many studies on the hazards caused by blasting vibration. As early as the 1970s, standards were issued for the control of blasting vibration damage to buildings [13]. Singh [14] carried out dynamic monitoring of the stratum before and after the explosion to obtain the maximum peak particle velocity, and then investigated the influence of blasting in open-cast coal mines on the stability of an adjacent underground coal mine roadway. Zhou et al. [15] assessed the damage to residential structures induced by blasting vibration in open pit mines by developing a gradient hoist model, and damage categories were more accurately investigated. Noren-Cosgriff et al. [16] evaluated the damage to residential structures caused by blasting vibration in a quarry. The test results confirmed that the limits of the standards in most countries were too conservative. Yue et al. [17] used ANSYS/LS-DYNA software to study the effects of borehole arrangement, delay time and decoupling charge on the rock damage and vibration attenuation of multi-hole blasting, and then optimized the blasting design to improve the blasting effect. Wang et al. [18] discussed the propagation characteristics and attenuation prediction equations for blasting vibration in rock tunnels by on-site monitoring and numerical simulation. They also improved the empirical equations of scalar distance to predict the peak mass vibration in the whole space of the rock tunnel, and finally proposed a new equation for predicting the peak mass vibration of the adjacent tubes in the plane of the center of blasting. Lv et al. [19] obtained the propagation mechanism and the damage distribution characteristics of blasting vibration along the elevation direction of the high sidewall of a deep underground cavern. The propagation mechanism of blasting vibration in the direction of blasting vibration was analyzed, and the prediction formula of particle vibration velocity along the elevation direction under the blasting vibration condition was derived. Aiming at problems such as the lack of comprehensive stage analysis in the field of blasting vibration, Zhang et al. [20] elucidated the research progress of blasting vibration, clarified the research frontiers, and discussed in depth the social significance of the intelligent blasting system in the context of artificial intelligence. Based on a tunnel project in Shenzhen, Wu et al. [21] proposed a four-part excavation method with a vibration buffer rock layer. Numerical simulation was utilized to model the damage prevention mechanism of the vibration buffer rock layer to understand the propagation of cracks in the interbedded rock. Finally, combined micro-vibration control blasting technology was realized by combining different loading structures and blasting equipment designed according to the change in the thickness of the interbedded rock.

In summary, as an emerging rock-breaking method, the application effect and vibration response law of HPGEM in practical engineering still need to be studied. In view of this, based on a hard rock tunnel in China, this paper analyzed the vibration velocity and rock-breaking capacity of the dynamite blasting method and HPGEM. The forces and vibrations of the buildings near the excavated tunnel under different scenarios were identified by means of numerical simulation and on-site monitoring. The vibration attenuation law during rock-breaking by the HPGEM was explored, and the relevant fitting equations were proposed. Finally, engineering recommendations were made for the adoption of these two methods.

2. Project Overview

This study is based on the background of a double-hole tunnel of a highway in China, which is close to underground pipelines and a high-speed rail line; therefore, there are strict restrictions on construction vibration in this project. An overview of the study area is presented in Figure 4. The Quaternary overburden in the study area is mainly Pleistocene diluvial silty clay ($Q_P{}^{d1}$), and the underlying is the upper Jurassic C-1 member ($J_3{}^{c\text{-}1}$) tuff. The surrounding rock is hard, the development degree of joints and fissures is different, there is a dripping phenomenon, and the weathering degree is from never weathered to slightly weathered. There is no obvious surface water in the tunnel site. The groundwater is mainly bedrock fissure water. The connectivity of joint fissures is general. It is the main storage space of groundwater and mainly depends on atmospheric precipitation. The low-lying parts are discharged in the form of underground runoff, while the slope parts are discharged from the surface through seepage along fissures or terrain cutting. Therefore, the project has a strict restriction on construction vibration. According to the previous design and construction experience and the seismic intensity table in the blasting safety regulations, as well as the requirements of the project specification documents, the safety seismic standards for the underground pipelines and high-speed rail line were formulated to require that the vibration velocity should be controlled within 7.0 cm/s. The vibration velocity of the high-speed rail line should be controlled within 1.0 cm/s. This brings great challenges to tunnel boring work. Meanwhile, the geological survey demonstrates a mean compressive strength of 69.8 MPa for the rock mass, but the actual strength of the rock mass is more than 140.0 MPa. Overall, the tunnel is a hard rock tunnel, and blasting operations with explosives are extremely difficult in these circumstances. Meanwhile, the use of other blasting methods is either costly or inefficient. Therefore, there is an urgent need for a new rock-breaking technique with good rock-breaking effect and tiny vibration velocity to ensure the safety and efficiency of the construction project.

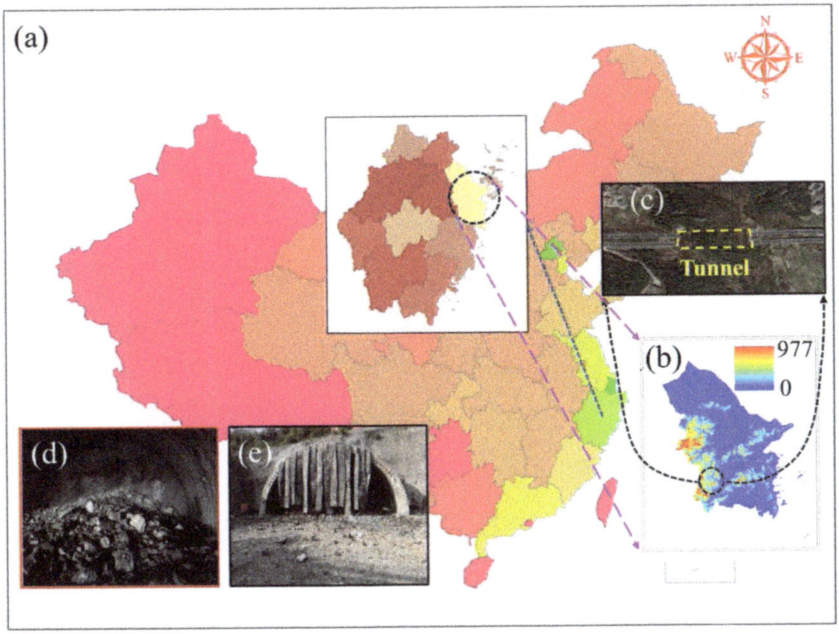

Figure 4. Overview of the study area. (**a**) Geographic location of the study area; (**b**) elevation map of the study area; (**c**) satellite view of the study tunnel area; (**d**) inside the tunnel; (**e**) outside the tunnel.

3. Rock Breaking Design Program

3.1. Dynamite Blasting

3.1.1. Schematic Design

Since HPGEM is an emerging rock-breaking method with little engineering experience to draw on, this study first carried out the design of the rock-breaking scheme by explosive blasting (No. 2 rock emulsion explosive), to provide a reference point for the subsequent design of the HPGEM. Due to the constraints of the construction project's environment, dynamite blasting cannot be performed in the case study tunnel. Therefore, all blasting tests in this study were conducted in another tunnel about 1 km away from the case study tunnel (with almost the same geological conditions as in the study area). The blasting vibration velocities in three directions are monitored during the blasting operation. The peak particle velocity [22,23] is usually taken as the key factor for determining the impact of blasting vibration. Determination of the PPV is the basis for measuring whether the blasting operation causes damage to adjacent buildings (structures), which the Sadowski formula is commonly used to predict [24,25] (Equation (1)).

$$PPV = K\left(\frac{\sqrt[3]{Q}}{R}\right)^{\alpha} \tag{1}$$

where the characteristic coefficient K and α are the coefficients related to the site topography (geological conditions and other factors are shown in Table 1). PPV is the peak particle velocity (cm/s), R is the distance from the blasting center (m), Q is the maximum explosive charge in all sections (kg).

Table 1. Relationship between the values of K, α, and lithology.

Lithology	K	α
Hard rock	50~150	1.3~1.5
Medium hard rock	150~250	1.5~1.8
Soft rock	250~350	1.8~2.0

Taking the logarithm of both sides of Equation (1) gives the following:

$$lgPPV = lgK + \alpha lg\left(\frac{\sqrt[3]{Q}}{R}\right) \quad (2)$$

$$\text{Let } y = lgPPV; x = lg\left(\frac{\sqrt[3]{Q}}{R}\right); b = lgK; k = \alpha$$

Then:

$$y = kx + b \quad (3)$$

The regression calculation yields the following:

$$k = \frac{n\sum_{i=1}^{n} x_i y_i - \sum_{i=1}^{n} x_i \sum_{i=1}^{n} y_i}{n\sum_{i=1}^{n} x_i^2 - \left(\sum_{i=1}^{n} x_i\right)^2} \quad (4)$$

$$b = \frac{1}{n}\left(\sum_{i=1}^{n} y_i - \alpha \sum_{i=1}^{n} x_i\right) \quad (5)$$

The correlation coefficient can be expressed as:

$$\gamma = \frac{\sum_{i=1}^{n} (x - \overline{x}) \cdot (y - \overline{y})}{\sqrt{\sum_{i=1}^{n} (x - \overline{x})^2 \cdot \sum_{i=1}^{n} (y - \overline{y})^2}} \quad (6)$$

where n is the total number of monitoring; and, each x_i and y_i represents the size of a particular experiment x and y.

If the full section excavation is used, the PPV will exceed the specified limit. Therefore, the proposed tunnel is divided into a drilling and blasting excavation area and a mechanical and drilling and blasting excavation area, according to the distance from the highway and underground pipelines. Excavation operations are carried out in the areas within 34 m from the left and right sides of the highway centerline, and drilling and blasting excavation is performed outside the area. Since the PPV is the direct cause of damage to buildings (structures), the analysis is conducted on vibration monitoring and the blasting effect of the drilling and blasting method. In this blasting design, the wedge cutting technique was applied to blast excavate a section of the upper and lower benches. The spacing of the cut holes was 0.9 m, with a plum-shaped arrangement and a row spacing of 0.7 m. Each hole was charged with 0.4 kg of explosives (calculated based on the design of the drilling and blasting excavation). The first section of the charge was slightly larger, with a charge of 0.6 kg per hole, and an advance of 1.0 m per cycle. Refuge blasting is used at the center of the empty hole by spiral hollowing, with a maximum charge of 0.6 kg per hollowing hole. Each cut and peripheral hole is charged with 0.2 kg of explosives, and the guide mine for each advance cycle is 0.8 m long. Sections are divided according to the distribution of drilling and blasting holes, and the detonator section is shown in Figure 5.

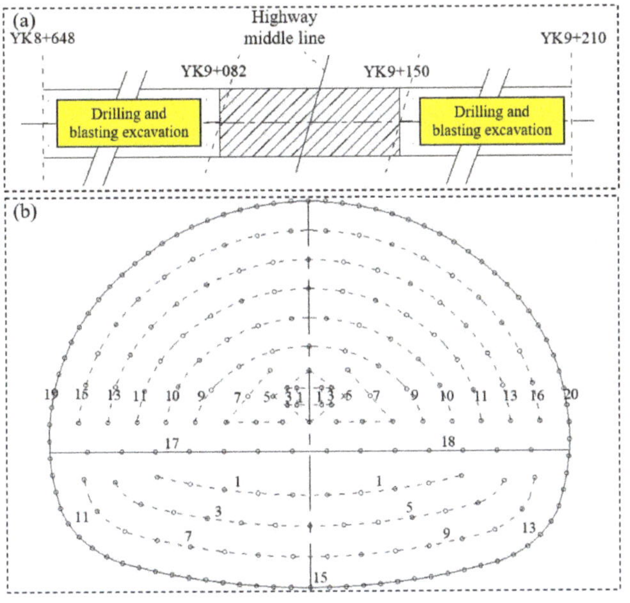

Figure 5. Schematic diagram of tunnel excavation section, borehole distribution and detonator segmentation of drilling and blasting method. (**a**) Tunnel excavation sections; (**b**) drilling and blasting hole distribution and detonator segmentation.

3.1.2. Monitoring and Fitting

An attempt is first made to obtain the blasting vibration fitting equations for the study area. The TC-4850 blast vibrometer (Zhongke Electric, Yueyang, China), TC-3850 blast vibrometer (Zhongke Electric, Yueyang, China), CDJ-1 type velocity sensor (China Sichuan Chengdu Zhongke measurement and Control Co., LTD, Chengdu, China), and three-direction velocity sensor (China Sichuan Chengdu Zhongke measurement and Control Co., LTD, Chengdu, China) were utilized to perform on-site PPV monitoring. The obtained PPV monitoring results were divided into three velocity directions: vertical, radial, and tangential. Since the PPV mainly occurs in the vertical direction [26,27], only the vibration velocity in the vertical direction is regressed and analyzed in this study. The vibration test results and vertical vibration velocity are shown in Table 2.

Table 2. Vibration test results.

Monitoring Point No.	R (m)	Q (kg)	PPV (cm/s)	Main Vibration Frequency (Hz)
1	26	20	4.52	108.108
2	28	22	4.25	98.199
3	35	18	2.93	116.280
4	42	24	2.64	166.166
5	31	15	2.98	85.107
6	39	18	2.69	109.863
7	28	17	3.67	94.933
8	13	17	11.78	88.889
9	53	14	1.56	200.911
10	50	12	1.39	192.667
11	40	11	1.96	145.752
12	24	12	3.79	90.909
13	35	12	2.47	100.122

Based on Table 2 and Equations (1)–(6), the relevant parameters can be determined as: $K = 103.02$, $\alpha = 1.377$. It can be found from Table 1 that the K is in the range of 50–150 and the α is in the range of 1.3–1.5 for hard rock, which is in line with the actual rock mass conditions. The fitting equation for the vibration velocity of the explosive blasting currently is as follows:

$$PPV = 103.02 \times \left(\frac{\sqrt[3]{Q}}{R}\right)^{1.377} \tag{7}$$

Then based on Equation (7), for the inverse highway roadbed (HR), underground pipeline (UP), and high-speed rail line (HSRT) to reach the PPV control value at the nearest distance to the maximum section of the explosive charge, that is, the limit cases of the maximum section of the explosive charge without considering the possibility of practical operation, the values are, specifically, as shown in Table 3.

Table 3. Maximum section explosive charge in limit cases.

Location	PPV Control Value (cm/s)	R at Minimum PPV (m)	Q (kg)
HR	7	5.70	0.53
UP	7	4.64	0.29
HSRT	1	50.00	5.15

(1) Drilling and blasting

Based on the tunnel segmentation schematic, the PPV in the upper bench blasting will occur in the 13th segment at the tunnel face of section YK9+082. The Q of the 13th section, the explosive charge per hole, and the charge of other segments is calculated according to Table 3 and Equation (7). Finally, the vibration velocity of the highway roadbed, underground pipelines and high-speed rail line caused by blasting with explosives in each section is back-calculated to verify whether the PPV control value is exceeded, and the calculated seismic wave propagation distance is taken as the average value of the distances of each section of the gun holes. The results show that, for the upper bench blasting at the tunnel face at section YK9+082 in the 13th section (Q is 6.8 kg), the PPV for the highway roadbed, underground pipelines and high-speed rail line are calculated to be 6.95 cm/s, 6.36 cm/s and 0.97 cm/s, respectively, which are close to but less than the limit of the specified PPV. Similarly, the PPV of each section at the lower bench of the tunnel face at section YK9+082 during blasting are calculated. Since the underground pipeline is closer to the tunnel lower bench, it is not necessary to calculate the PPV of each section of Q for the highway roadbed during blasting of the lower bench. For the upper bench blasting of the tunnel face at section YK9+082, the calculated PPVs for the underground pipelines and high-speed rail line are found in the first segment (with a Q of 4 kg), which are 6.33 cm/s and 0.76 cm/s, respectively, which are less than the specified PPV control value. Therefore, the settings of the blasting parameters for the upper and lower benches are reasonable. Their specific calculation results are shown in Figure 6.

(2) Collaborative dredging

The project overview confirms that the minimum clear distance of the underground pipeline from the tunnel bottom at the highway central divider is only 4.64 m. Therefore, in this blasting design, it is assumed in the limit case that the highway central divider, the blasting tunnel face and the underground pipeline are in the same plane. Theoretically the maximum PPV will occur in this section. If the PPV at the tunnel face satisfies the vibration velocity provisions, the PPV at the section between the YK9+082 and YK9+150 should be within the safety limit. The calculation method is the same as the previous drilling and blasting method, and the calculation results are shown in Figure 7. It can be seen that when the underground pipeline is located directly above the guided blasting, the PPV of the highway roadbed, underground pipeline and high-speed rail line are in the 15th segment (with Q of 1 kg), the 13th segment (with Q of 0.6 kg), as well as in the 13th and 15th segments (with Q of 1 kg), respectively, which are 4.83 cm/s, 6.10 cm/s and

0.47 cm/s, respectively, which are less than the specified PPV limit, indicating that the blasting parameters of the lower bench meet the safety requirements.

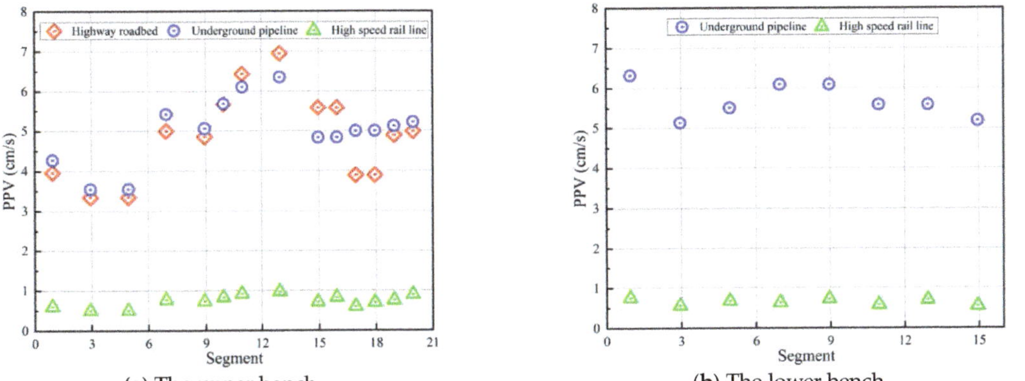

(**a**) The upper bench (**b**) The lower bench

Figure 6. Calculation results of PPV during the construction of upper and lower step drilling and blasting method (YK9+082).

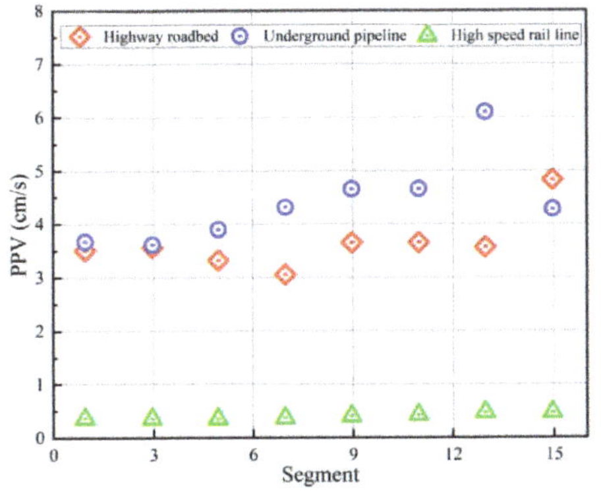

Figure 7. Calculation results of PPV with the guide hole cooperative excavation method (YK9+082).

(3) Numerical simulation of explosive blasting under three operating conditions

Conventional blasting for the proposed tunnel is a design based on the protection of adjacent projects. The highway in operation, the fully enclosed high-speed railway line, and the inaccessible underground pipeline make it impossible to carry out large-scale on-site monitoring. Therefore, it is necessary to verify the design through numerical simulation and analyze the stresses of the structure under the limit conditions, to provide a basis for a comparative analysis between the HPGEM and dynamite blasting. Considering that the excavation of the left tunnel has been completed, and the two tunnels are not adjacent tunnels, only the right tunnel of the proposed tunnel is considered in the numerical model. Numerical simulations were performed to investigate the vibration velocity and force of the highway roadbed, underground pipeline, and high-speed rail line during dynamite blasting under three working conditions.

According to the relative position relationship of each structure and the geological conditions, the model size is 160 m × 80 m × 100 m (long, high, wide). The element of the model is solid-164. The three main materials used in the model are explosives, rock, and concrete. C30 concrete is considered for the highway roadbed, and the explosives are the No. 2 rock emulsified explosives. The equation of state is defined as follows:

$$P = A\left(1 - \frac{\omega}{R_1 V}\right)e^{-R_1 V} + B\left(1 - \frac{\omega}{R_2 V}\right)e^{-R_2 V} + \frac{\omega E}{V} \tag{8}$$

where P is the blasting pressure (MPa); V is the relative volume; E is the internal energy per unit volume (GPa); and ω, A, B, R_1, R_2 are the material constants.

Actual geotechnical media are discontinuous and inhomogeneous, and it is currently almost impossible to describe these envelopes mathematically directly and objectively. The rock mass is usually regarded as a homogeneous and continuous elastic medium. In this paper, ANSYS 16.0 is used to simplify the model, that is, rock and soil are regarded as elastic–plastic materials. The material of the model is chosen to simulate the geotechnical medium, and the yield condition is shown in Equation (9). The constitutive relationship and numerical model are presented in Figure 8. The concrete of roadbed is simulated by the material of the model. According to the data provided by the geological report and indoor test, the main parameters of the selected material of the model are illustrated in Tables 4 and 5.

$$\varphi = \frac{1}{2}\xi_{ij}^2 - \frac{\sigma_y^2}{3} = 0 \tag{9}$$

$$\xi_{ij} = s_{ij} - a_{ij},\ \sigma_y = \left[1 + \left(\frac{\dot{\varepsilon}}{C}\right)^{\frac{1}{p}}\right](\sigma_0 + \beta E_p \varepsilon_{eff}^p).$$

where s_{ij} is the Cauchy stress tensor; p and C are input constants; σ_0, β are the initial yield stress and hardening parameter, respectively; $\dot{\varepsilon}$ is the strain rate, $\dot{\varepsilon} = \sqrt{\dot{\varepsilon}_{ij}\dot{\varepsilon}_{ij}}$; E_p is the plastic hardening modulus, $E_p = \frac{E_t E}{E - E_t}$, with E being the elastic modulus, E_t being the tangential modulus; and ε_{eff}^p is the effective plastic strain.

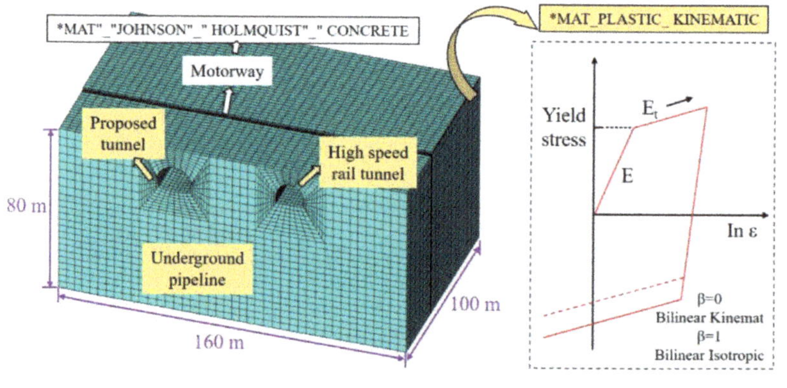

Figure 8. Numerical model and constitutive relationship.

In general, the ramp-up time of the blast load is 8–12 ms, and the unloading time is 80–100 ms. To better analyze the propagation law of the blasting vibration seismic wave in the modeled medium, the computation time of the dynamite explosion simulation in this study is set to 0.3 s and the blast vibration seismic wave propagates in the modeled medium.

(1) Analysis of Blasting Vibration Velocity

The maximum vibration velocity nephograms of the highway roadbed in each direction during blasting under three working conditions are extracted, as presented in Figure 9. Figure 10 presents the maximum vibration velocities of the highway roadbed, underground pipeline, and high-speed rail line under each condition, where they are compared with the fitted values. It can be seen from Figure 9 that the maximum vibration velocities of the highway roadbed under the three conditions are all located at the nearest point on the side of the head-on blast, and their values are within the specified range. It can be seen from Figure 10 that the simulated and the calculated values of vibration velocity under the three conditions are similar, indicating that the numerical simulation results are credible, and confirming the feasibility of applying the parameters in the empirical formula of Sadowski to the proposed tunnel project.

Table 4. Main parameters of numerically simulated explosives.

Parameters	β (g/cm^3)	D (cm/μs)	P (MPa)	A (MPa)	B (MPa)
Value	1.2	0.48	0.097	2.144	1.82×10^{-3}
Parameters	**R_1**	**R_2**	**ω**	**E_0 (J/cm^3)**	**V_0**
Value	4.2	0.9	0.15	3.60×10^3	0

Table 5. Main parameters of rock and concrete used in numerical simulation.

Parameters	ρ (g/cm^3)	E (MPa)	γ	ET (MPa)	σs (MPa)
Rock	2.6	6.00×10^{10}	0.27	1.30×10^3	4.00×10^3
Concrete	2.25	3.00×10^{10}	0.2	2.50×10^2	2.40×10^3

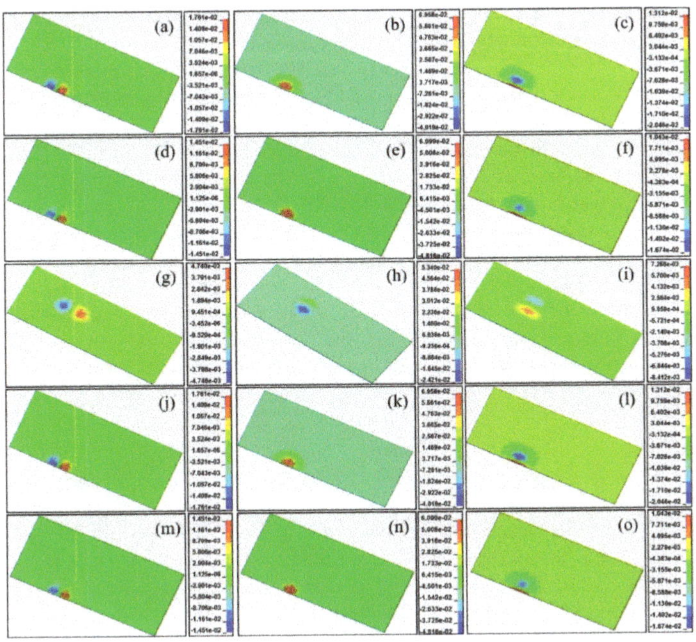

Figure 9. Numerical simulation results of highway roadbed under different working conditions: (**a**) Vibration velocity–upper bench–X direction; (**b**) vibration velocity–upper bench–Y direction; (**c**) vibration velocity–upper bench–Z direction; (**d**) vibration velocity–lower bench–X direction; (**e**) vibration velocity–lower bench–Y direction; (**f**) vibration velocity–lower bench–Z direction; (**g**) vibration velocity–guide hole–X direction; (**h**) vibration velocity–guide hole–Y direction; (**i**) vibration velocity–guide hole–Z direction; (**j**) first principal stress–upper bench; (**k**) first principal stress–upper bench; (**l**) first principal stress–guide hole; (**m**) third principal stress–upper bench; (**n**) third principal stress–upper bench; (**o**) third principal stress–guide hole.

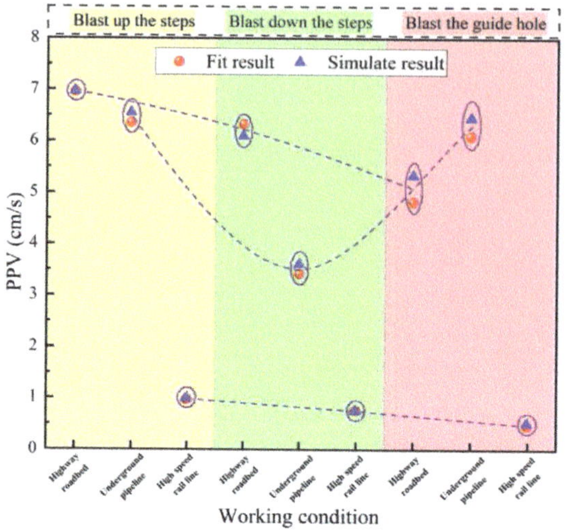

Figure 10. Simulated and fitted values of vibration velocity for different blasting conditions.

Since structural stress affects structural stability, it is necessary to analyze the first and third principal stresses of the structure under various working conditions. Due to limited space, only the stress state of the highway roadbed is analyzed in this study, where the first principal stress is tensile stress, with positive values; the third principal stress is compressive stress, with negative values, also as shown in Figure 9. The maximum stresses of the highway roadbed in the three conditions are also located at the nearest point on the blasting side, with maximum tensile stresses of 0.87 MPa, 0.793 MPa and 0.989 MPa, respectively, and maximum compressive stresses of 1.83 MPa, 1.081 MPa and 1.39 MPa, respectively. According to the specification, the design tensile strength of C30 concrete is 1.39 MPa, and the design compressive strength is 13.8 MPa. Under the action of blasting dynamic load, its tensile and compressive strengths will be much higher than these two design values. For the highway roadbed under the action of blasting stress, the value is lower than its own ultimate destructive strength, so the highway roadbed is safe and stable.

In summary, the analysis of explosive blasting excavation based on theoretical calculations and numerical simulations confirms that the maximum vibration velocity and the maximum stress subjected to this scheme are within safe limits, and the scheme is feasible.

3.2. HPGEM

3.2.1. Schematic Design

The HPGEM allows rock-breaking tests to be conducted directly on the tunnel face of the project tunnel. Firstly, under the HPGEM rock-breaking scheme, wedge-shaped hollowing rock-breaking testing is carried out to determine the optimal type of hollowing. After the rock-breaking test, the amount of rock breakage was measured on site, and the statistical results are summarized in Table 6. It was found that the wedge-shaped hollowing had formed a cavity at the location of the broken rock, and there was no obvious "undercutting" in the cavity (Figure 11). After the large-diameter hollow straight-hole hollowing test, the same cavity was formed at the rock-breaking location, and there was basically no "undercutting" at the bottom of the cavity. Therefore, the hollowing results are as expected, which indicates that both types of hollowing tests are successful. Although the effect of both types of hollowing can meet construction requirements, it can be seen from Table 6 that, in this hollowing test, the unit consumption of wedge-shaped hollowing was 1.25 kg/m^3, which is lower than that of the large-diameter hollow hole hollowing of 1.39 kg/m^3. Hence, wedge-shaped hollowing is superior to the large-diameter hollow hole hollowing, and wedge-shaped hollowing was adopted in the subsequent test.

Table 6. Comparison of results of rock removal tests.

Hollowing Type	Expansion Pipe Number	Gas Generated Agent Quality (kg)	Rock Breakage Amount (m^3)	Unit Consumption (kg/m^3)
Wedge-shaped hollowing	6	3	2.40	1.25
Large-diameter hollow Straight-hole hollowing	8	4	2.87	1.39

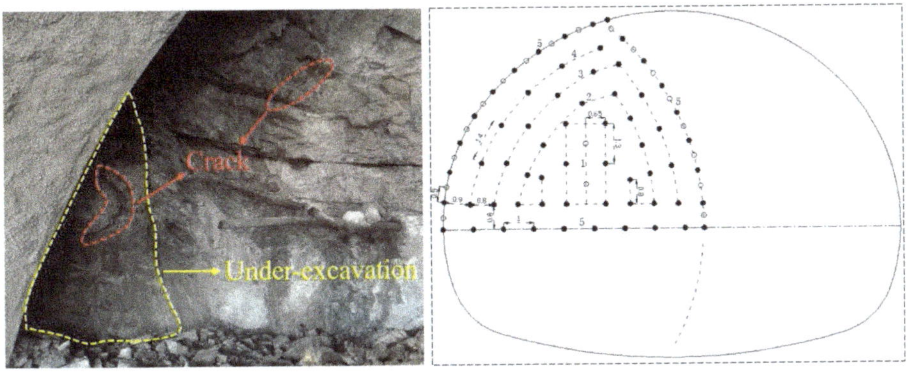

Figure 11. Rock-breaking effect of the HPGEM.

Subsequently, different expansion tube burial depths and hole spacings were set to optimize the matching of the expansion pipe hole network parameters with the lithology of the rock mass at the tunnel face. First, three groups of high-pressure gas expansion rock-breaking tests were designed, with three top-down arranged boreholes in each group. The spacing of each group of boreholes was more than 2 m, without interfering with each other. The hole spacings of group 1, group 2 and group 3 were 1.0, 1.2 and 1.4 m, respectively. The depth of the holes was 1.5 m, and the buried depth of the expansion pipe was 1.4 m. Secondly, according to the effect of the first three groups of tests, group 4 and group 5 tests were designed, where the depths of holes in these two groups were 1.4 and 1.6 m, respectively; and the buried depths of the expansion pipe were 1.3 m and 1.5 m, respectively. The rest of the parameters are the same as those of the group 2 tests. The results of the first three groups of tests show that the optimal hole spacing for HPGEM in this tunnel is 1.2 m, and the optimal hole depth is 1.4 m. For monitoring the vibration data during the test, the monitoring program adopted the same monitoring basis and monitoring instrument. The monitoring method and monitoring point arrangement are as described in the previous section. The vibration monitoring points are mainly arranged on the ground in front of the tunnel working face. The monitoring points for rock breakage are arranged at intervals of a few meters.

3.2.2. Monitoring and Fitting

The vibration monitoring data are shown in Table 7. It was found that there is an obvious exponential function relationship between the two sets of data. With reference to the process of fitting the Sadowski formula for the vibration velocity of dynamite blasting, the Statistics Analysis System (SAS) software (v.9.4) was used to statistically analyze the above data to fit the formula for calculating the vibration velocity of the rock breakage of the HPGEM of the tunnel in this Equation (10):

$$PPV = e^{3.3205} \cdot \frac{{Q_1}^{0.7012}}{R^{1.5695}} \tag{10}$$

where Q_1 is the mass of the explosive (kg), and R is the distance from the vibration monitoring point to the rock-breaking point (m).

Table 7. Measured and calculated PPV of the tunnel with HPGEM.

Number	Q_1 (kg)	R (m)	Monitor PPV (cm/s)	Calculate PPV′ (cm/s)
1	0.5	3	3.389	3.034
2	0.5	4.3	1.713	1.724
3	0.5	43.6	0.057	0.045
4	1	7	1.600	1.305
5	1	5.9	1.810	1.706
6	1	43.6	0.084	0.074
7	1.5	45	0.118	0.095
8	1.5	65	0.039	0.051
9	1.5	6.2	2.237	2.098
10	1.5	11	0.667	0.850
11	1.5	18	0.225	0.391
12	5	18	0.740	0.910
13	7	15	1.659	1.544
14	15	13.6	2.438	3.070
15	20	27.5	1.480	1.245

The calculated vibration velocity results were compared with the actual measured data to verify the reliability of the fitting formula. It can be seen from Table 7 and Figure 12 that the fitted calculation values match well with the actual measured values, with a correlation coefficient R_2 of 0.946, indicating that the Equation (10) is reasonable. As shown in Figure 12, the fitting formula demonstrates a better fitting degree at lower PPV values. However, as the PPV values increase, the data exhibits greater dispersion. Therefore, the fitting formula may perform better in the context of smaller PPV values.

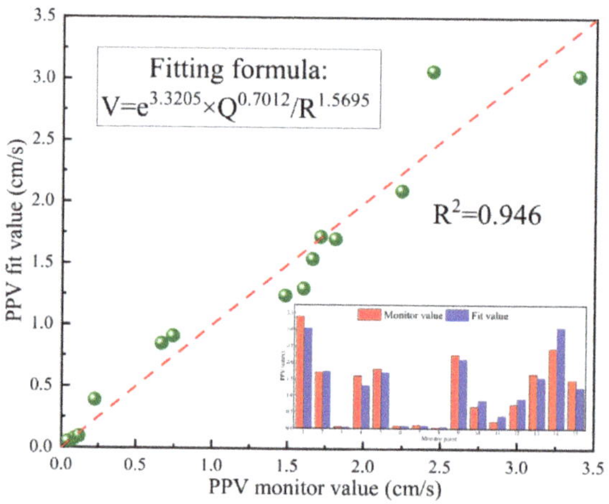

Figure 12. Comparison of measured and calculated PPV by HPGEM.

4. Comparison of the Dynamite Blasting and HPGEM

The vibration velocity under the same geological conditions is mainly affected by the pharmaceutical quality and the R. Therefore, under the same geological conditions, the PPV of the two rock-breaking methods with different R of the same Q or different Q of the same R is compared and analyzed.

Due to the limitations of the site conditions and the differences between the two rock-breaking methods, the field test R and Q cannot be fully consistent. Considering that the reasonableness of the fitting equations for the vibration velocity of the two rock-breaking methods has been verified; therefore, the two fitting equations can be used for the calculation of vibration velocity. To facilitate the comparison, the Q is set to 5, 10, 15, 20 and 25 kg, while the R is set to 5, 10, 15, 20 and 25 m. The specific data are as displayed in Table 8. The vibration velocities of the explosives and aerosols under different Q and R are plotted in Figure 13, in which PPV_1 and PPV_2 are the vibration velocities of the explosives and HPGEM, respectively.

Table 8. Comparison of PPV of dynamite blasting and HPGEM with different Q and R.

Number	Q (kg)	R (m)	PPV_1 (cm/s)	PPV_2 (cm/s)
1	5	5	23.509	6.839
2	5	10	9.051	2.304
3	5	15	5.179	1.219
4	5	20	3.485	0.776
5	5	25	2.563	0.547
6	10	5	32.314	11.120
7	10	10	12.442	3.746
8	10	15	7.119	1.983
9	10	20	4.790	1.262
10	10	25	3.523	0.889
11	15	5	38.923	14.776
12	15	10	14.986	4.979
13	15	15	8.575	2.635
14	15	20	5.770	1.677
15	15	25	4.244	1.182
16	20	5	44.417	18.079
17	20	10	17.102	6.091
18	20	15	9.785	3.224
19	20	20	6.584	2.052
20	20	25	4.843	1.446
21	25	5	49.207	21.141
22	25	10	18.946	7.123
23	25	15	10.840	3.770
24	25	20	7.294	2.400
25	25	25	5.365	1.691

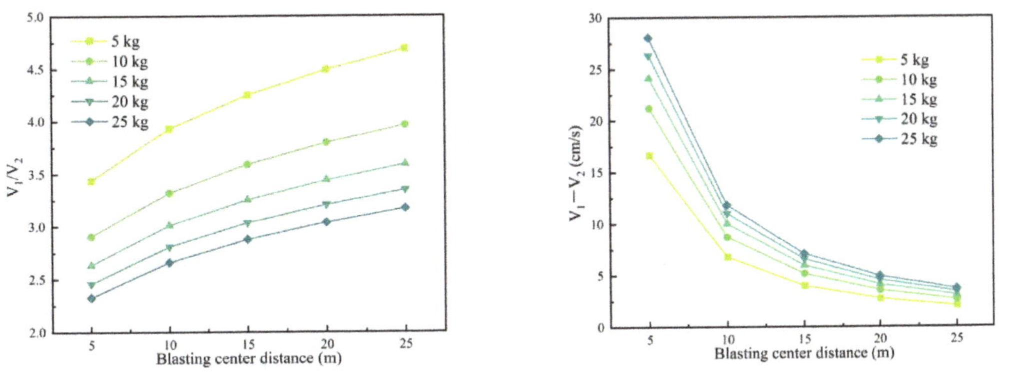

(**a**) Specific value (**b**) Difference in value

Figure 13. Difference and ratio of vibration velocities with different Q and R.

As can be seen from Figure 13, the vibration velocity of the dynamite blasting method and HPGEM exhibits a similar pattern of change. In the case of the same agent quality and blasting center distance, the vibration velocity formed by explosives and gas generator explosion is much higher than the HPGEM. In the case of the same R, the greater the Q, the larger the vibration velocity, the larger the difference in vibration velocity between the dynamite blasting and HPGEM. For the same Q, a larger R results in a smaller difference in vibration velocity between the explosive blasting and HPGEM. The smaller the R, the more obvious the difference brought about by the difference in the Q. The opposite is less obvious.

For the same Q and R, the vibration velocity of the explosive blasting method is 2.33–4.7 times that of HPGEM. For the same R, the greater the Q, the smaller this multiplier relationship. But for the same Q, with the increase in the R, the multiplier becomes greater. The smaller the maximum section of the charge (for HPGEM, it is a detonation of the mass of the gas generating agent) and the further away the protected structure is from the point of explosion, the better the vibration rate can be controlled and reduced by the HPGEM. For scenarios where the maximum cross-section of the charge is very large and is close to the protection limits, the use of HPGEM can also significantly reduce the vibration rate, but the effect is relatively weak. The reason for this phenomenon is that the explosives' blasting vibration mainly generates from explosive shock waves (of course, there are also high-temperature and high-pressure gas effects). Shock waves propagate quickly and cause damage to the nearby zone of the explosion, and then decay into stress and seismic waves, while stress waves propagate relatively slowly in the rock mass. The rock mass was moved by the superposition of explosive gases resulting from the shock waves and explosives' explosion, while the disturbance of the rock mass by shock waves precedes the uptake of explosive gases in the rock mass. The vibration generated by the HPGEM is basically from the impact of the high-temperature and high-pressure gas on the expansion of the rock wall, but the impact of the shock wave is relatively weak. As the expansion of the fissures in the rock and the extension of the pressure decrease rapidly, the disturbance of the rock mass will decrease. In general, the further the distance, the faster the disturbance decreases.

After understanding the change rule of vibration velocity of the two rock-breaking methods, it is beneficial to compare and analyze the vibration velocities of the two rock-breaking methods when they are employed in difficult engineering sites. Based on the above conclusions and the safety requirements mentioned, the vibration velocities of the HPGEM with the same Q in the limit case were obtained for comparative analysis, as shown in Table 9. Similarly, according to the explosive blasting program in the field, the vibration velocity of the HPGEM with the same Q was also acquired, as displayed in Table 10.

Table 9. Comparison of PPV_1 and PPV_2 in the limit case.

Location	Q (kg)	R (m)	PPV_1 (cm/s)	PPV_2 (cm/s)	V/V_1
HR	0.53	5.70	7	1.152	0.165
UP	0.29	4.64	7	1.032	0.147
HSRL	5.15	50.00	1	0.188	0.188

Table 10. Comparison results of PPV_1 and PPV_2 in the practical case.

Location		Q (kg)	R (m)	PPV_1 (cm/s)	PPV_2 (cm/s)	PPV_2/PPV_1
Upper bench	HR	6.80	13.45	6.950	1.795	0.258
	UP	6.80	14.324.00	6.355	1.626	0.256
	HSRL	6.80	56.00	0.972	0.191	0.197
Lower bench	HR	4.00	12.037.00	6.330	1.473	0.233
	UP	4.00	56.00	0.762	0.132	0.173
Pilot	HR	1.00	9.23	4.823	0.845	0.175
	UP	1.00	7.79	6.100	1.103	0.181
	HSRL	1.00	50.00	0.471	0.060	0.127

As can be seen from Tables 9 and 10, in the case of the same Q of explosives and the gas generating agent explosion, the vibration velocity of the HPGEM with various engineering difficulties is much lower than that of the explosives' blasting method. Moreover, it can be seen from the ratio of vibration velocity that the vibration velocity control efficiency of the HPGEM in the actual operable procedure is basically lower than in the limit case. This is due to the different locations of the three major engineering difficulties. Considering that the excavation design program must be practical to operate, the Q and the R will be adjusted accordingly. However, the smaller vibrations produced by the HPGEM demonstrated its advantage in controlling vibration hazards.

According to the maximum allowable vibration velocity set at each engineering difficulty in the limit case, the maximum section charges of the two rock-breaking methods at each engineering difficulty were back-calculated using the fitting formula obtained from dynamite blasting and HPGEM in the test tunnel (for the HPGEM, it is the quality of single consumption). The results of the comparative analysis are shown in Table 11. The minimum value of 0.6 kg/m^3 is taken for the single consumption of dynamite blasting for rocks with hardness coefficients f = 9–15, which is less than the actual single consumption of dynamite blasting. In this case, the maximum segmented Q can produce the maximum amount of rock breakage. Comparison is made with the calculation results of the HPGEM tests to demonstrate the advantages of the HPGEM in terms of the amount of rock breaking and the control of vibration hazards.

Table 11. Comparison of maximum Q at one time and rock-breaking volume at the PPV in the limit case.

Location	PPV_1 (cm/s)	R (m)	Dynamite Blasting			HPGEM			m_2/m_1	A_2/A_1
			Q (kg)	Unit Consumption (kg/m^3)	Rock Breakage Amount A_1 (m^3)	Q_1 (kg)	Unit Consumption (kg/m^3)	Rock Breakage Amount A_2 (m^3)		
HR	7	5.7	0.53	0.6	0.881	6.930	0.87	7.965	13.10	9.04
UP	7	4.64	0.29		0.475	4.372		5.026	15.33	10.57
HSRL	1	50	5.15		8.576	55.776		64.110	10.85	7.48

As can be seen from Table 11, at the maximum vibration velocity in the limit case, to protect the highway roadbed, underground pipelines and high-speed rail line, the maximum masses of gas agent required for a single initiation of blasting in the high-pressure gas expansion method are 13.103, 15.327 and 10.839 times the maximum sectional charge in dynamite blasting, respectively; while the volumes of the rock breakage in the high-pressure gas expansion method are 9.037, 10.570 and 7.475 times those in dynamite blasting, respectively. In other words, the high-pressure gas expansion method can use more mass of agent than dynamite blasting at one time, without having to be overly limited to segmentation, and can obtain far greater volumes of the rock breakage than dynamite blasting. At the same vibration rate, high pressure gas expansion is also more conducive to increasing the volume of rock breakage.

Considering the feasibility of the actual tunnel excavation program, comparisons of the maximum vibration velocity and the amount of rock breakage under the single maximum charge quantity in the limit case are shown in Table 12, where the unit explosive is the corresponding to actual unit consumption. To protect the highway roadbed, underground pipelines and high-speed rail line, the upper bench, the lower bench, and guide blasting are used for three detonations. In the upper-bench rock breaking, the maximum mass of gas generating agent used in the high-pressure gas expansion method is 6.89, 6.98 and 10.15 times the maximum charge in the dynamite blasting, while the rock-breaking volume is 7.23, 7.33 and 10.65 times, respectively. In the lower-bench rock breaking, the maximum mass of gas generating agent used in the high-pressure gas expansion method is 8.0 and 12.2 times of the maximum charge in the dynamite blasting, while the rock-breaking volume is 6.67 and 10.16 times, respectively. In the guide blasting rock breaking, the maximum mass of gas generating agent used in the high-pressure gas expansion method is 11.99, 11.46, and 19.06 times the maximum charge in the dynamite blasting, while the rock-breaking volume is 12.34, 11.80, and 19.63 times, respectively. In the limit case, the high-pressure gas expansion method can break rock in a single blasting using a much larger charge than dynamite blasting, and can obtain a much larger rock-breakage volume than dynamite blasting. However, the variation in the ratio between high-pressure gas expansion and dynamite blasting in the practical scenario is much greater than that in the limit case, especially in the case of guide rock breaking. The further away from the breaking point, the more advantageous the high-pressure gas expansion method is over dynamite blasting in controlling the vibrations of the fragmented rock and in increasing the volume of fragmented rock. This is mainly due to the rock-breaking mechanism of the high-pressure gas expansion method and the attenuation law of the vibration rate.

Table 12. Comparison of the maximum Q and rock-breaking volume at one time under the PPV under the practical program.

Location		PPV_1 (cm/s)	R (m)	Dynamite Blasting			HPGEM			m_2/m_1	A_2/A_1
				Q (kg)	Unit Consumption (kg/m^3)	Rock-Breakage Amount A_1 (m^3)	Q (kg)	Unit Consumption (kg/m^3)	Rock-Breakage Amount A_2 (m^3)		
Upper bench	HR	6.950	13.45	6.80	0.91	7.45	46.86	0.87	53.86	6.89	7.23
	UP	6.355	14.32	6.80		7.45	47.49		54.58	6.98	7.33
	HSRL	0.972	56.00	6.80		7.45	69.03		79.34	10.15	10.65
Lower bench	HR	6.330	12.04	4.00	0.73	5.52	31.99		36.77	8.00	6.67
	UP	0.762	56.00	4.00		5.52	48.78		56.07	12.20	10.16
Guide hole	HR	4.823	9.23	1.00	0.9	1.12	11.98		13.77	11.99	12.34
	UP	6.100	7.79	1.00		1.12	11.45		13.17	11.46	11.80
	HSRL	0.471	50.00	1.00		1.12	19.06		21.90	19.06	19.63

During the tunnel blasting excavation, blast vibration monitoring of the adjacent ground or underground buildings (structures), and comparative analysis of the monitoring results with the standard specifications can ensure the safety of adjacent buildings (structures), and can also optimize the blasting parameters. Therefore, the use of the above blasting design for blasting excavation of the test tunnel can meet the requirements of all parties to the project on vibration velocity, and safe construction of the tunnel can also be achieved. Due to the complexity of the site construction environment, the density of the surrounding personnel and vehicles, as well as the difficulty in unifying the minds of all parties, it is difficult to obtain approval for the excavation plan using traditional blasting with explosives. Therefore, combined with the comparative analysis of the above two rock-breaking methods, it is recommended that the high-pressure gas expansion method be used for tunnel excavation. The upper bench, lower bench and guide hole can be blasted according to the above blasting design.

Since the high-pressure gas expansion method can achieve the excavation of larger rock-breakage volume within the maximum allowable vibration velocity, the tunnel can be divided into upper and lower bench blasting areas. Outside the range of section YK9+082–YK9+150, the high-pressure gas expansion method can be adopted for tunnel excavation, and in the range of section YK9+08–YK9+150, guide hole excavation can be expanded, or the bench tunneling method can be used, In this way, the efficiency of rock breaking can be improved, the construction period can be shortened, and the cost saving can be maximized.

5. Conclusions

To investigate the safety performance of the high-pressure gas expansion method in the process of rock breaking, a hard rock tunnel in China was taken as the research object, and a comparative analysis of the vibration caused by this new rock-breaking method and traditional rock breaking using explosives was carried out by means of on-site test, theoretical derivation, and numerical simulation. The main conclusions are as follows:

(1) Firstly, field tests were conducted on traditional blasting with explosives. The peak vibration velocity prediction formula was fitted based on Sadowski's formula. Combined with the maximum allowable vibration velocity given by the design, the maximum segmental loading of the project was calculated, and the force and vibration of the buildings near the tunnel during the maximum segmental loading of blasting under three working conditions were obtained through numerical simulation.
(2) With reference to the dynamite blasting method, field tests of the high-pressure gas expansion method of rock breaking were carried out to determine its optimal hollowing method and rock-breaking effect. Based on the vibration velocities monitored in the field, the vibration attenuation law was explored and the prediction formula for the peak vibration velocity was established.
(3) A comparative analysis of the dynamite blasting method and the high-pressure gas expansion method was carried out, with emphasis on the effects of different agent qualities, blasting center distances, and the application of engineering difficulties on the PPV. The relationship between the maximum agent quantity and the volume of rock breakage under the limit case, and a practical implementation scheme are discussed.
(4) The study demonstrates that the high-pressure gas expansion method has low vibration and excellent rock-breaking effect, and is fully capable of replacing explosives in some special areas.

Author Contributions: Conceptualization, H.P., J.S., Q.D., B.D., L.Z. and L.T.; methodology, Q.D. and H.P.; software, Q.D. and H.P.; validation, J.S.; formal analysis, Q.D., J.S., B.D. and H.P.; investigation, Q.D., L.T. and B.D.; resources, Q.D. and H.P.; data curation, H.P.; writing—original draft preparation, Q.D. and H.P.; writing—review and editing, H.P., J.S., Q.D., B.D., L.Z. and L.T.; visualization, B.D. and Q.D.; supervision, H.P. and L.T.; project administration, H.P. and Q.D. All authors have read and agreed to the published version of the manuscript.

Funding: This research was funded by the National Nature Science Foundation of Hunan Province (2022JJ40349); Open Project of State Key Laboratory of Safety Technology of Metal Mines, Changsha Institute of Mining Research Co., Ltd. (kfkt2023-03); the Postgraduate Scientific Research Innovation Project of Hunan Province (QL20220213); Key Project of Education Department of Hunan Province (22A0293); General Project of Education Department of Hunan Province (22C0235).

Data Availability Statement: The raw data supporting the conclusions of this article will be made available by the authors on request.

Conflicts of Interest: Authors Huaide Peng, Jia Sheng and Lihai Tan were employed by the company Changsha Institute of Mining Research Co., Ltd. The remaining authors declare that the re-search was conducted in the absence of any commercial or financial relationships that could be construed as a potential conflict of interest.

References

1. Nicholls, H.R.; Johnson, C.F.; Duvall, W.I. *Blasting Vibrations and Their Effects on Structures*; US Government Printers: Washington, DC, USA, 1971; ISBN 1-61344-013-8.
2. Xia, Y.; Jiang, N.; Zhou, C.; Meng, X.; Luo, X.; Wu, T. Theoretical Solution of the Vibration Response of the Buried Flexible HDPE Pipe under Impact Load Induced by Rock Blasting. *Soil Dyn. Earthq. Eng.* **2021**, *146*, 106743. [CrossRef]
3. Maidl, B.; Schmid, L.; Ritz, W.; Herrenknecht, M. *Hardrock Tunnel Boring Machines*; John Wiley & Sons: Hoboken, NJ, USA, 2008; ISBN 978-3-433-01676-3.
4. Zheng, Y.L.; Zhang, Q.B.; Zhao, J. Challenges and Opportunities of Using Tunnel Boring Machines in Mining. *Tunn. Undergr. Space Technol.* **2016**, *57*, 287–299. [CrossRef]
5. Bilgin, N.; Copur, H.; Balci, C. *Mechanical Excavation in Mining and Civil Industries*; CRC Press: Boca Raton, FL, USA, 2013; ISBN 1-4665-8474-2.
6. Kim, H.; Ha, S.; Rehman, H.; Yoo, H. Analysis of Mechanical Excavation Characteristics by Pre-Cutting Machine Based on Linear Cutting Tests. *Appl. Sci.* **2023**, *13*, 1205. [CrossRef]
7. Torno, S.; Toraño, J.; Ulecia, M.; Allende, C. Conventional and Numerical Models of Blasting Gas Behaviour in Auxiliary Ventilation of Mining Headings. *Tunn. Undergr. Space Technol.* **2013**, *34*, 73–81. [CrossRef]

8. Wang, B.; Li, H.; Shao, Z.; Chen, S.; Li, X. Investigating the Mechanism of Rock Fracturing Induced by High-Pressure Gas Blasting with a Hybrid Continuum-Discontinuum Method. *Comput. Geotech.* **2021**, *140*, 104445. [CrossRef]
9. Huai-de, P.; Dun-wen, L.; Fu-jiao, C.; Ying-hua, J. Test on High Pressure Gas Expansion Rock Fragmentation in Hard Rock Tunnel. *Rock Soil Mech.* **2018**, *39*, 242–248.
10. Peng, H.; Li, D.; Sheng, J.; Dai, B.; Da, Q. A Rock Mass Classification Method for Tuff Tunnel Based on the High-Pressure Gas Expansion Method. *J. Eng.* **2023**, *2023*, 1–10. [CrossRef]
11. Liu, D.; Tang, Y.; Cai, C.; Jian, Y. A Rock Fracturing Method Using High-Pressure Gas Expansion: Case Study on Its Application in Hangzhou-Lin'an Intercity Railway. *Adv. Civ. Eng.* **2021**, *2021*, e6654471. [CrossRef]
12. Liu, D.; Wang, C.; Tang, Y.; Chen, H. Application of High-Pressure Gas Expansion Rock-Cracking Technology in Hard Rock Tunnel near Historic Sites. *Appl. Sci.* **2023**, *13*, 1017. [CrossRef]
13. Duvall, W.I.; Fogelson, D.E. *Review of Criteria for Estimating Damage to Residences from Blasting Vibrations*; US Department of the Interior, Bureau of Mines: Washington, DC, USA, 1962; Volume 5968.
14. Singh, P.K. Blast Vibration Damage to Underground Coal Mines from Adjacent Open-Pit Blasting. *Int. J. Rock Mech. Min. Sci.* **2002**, *39*, 959–973. [CrossRef]
15. Zhou, J.; Shi, X.; Li, X. Utilizing Gradient Boosted Machine for the Prediction of Damage to Residential Structures Owing to Blasting Vibrations of Open Pit Mining. *J. Vib. Control* **2016**, *22*, 3986–3997. [CrossRef]
16. Norén-Cosgriff, K.M.; Ramstad, N.; Neby, A.; Madshus, C. Building Damage Due to Vibration from Rock Blasting. *Soil Dyn. Earthq. Eng.* **2020**, *138*, 106331. [CrossRef]
17. Yue, H.Z.; Yu, C.; Li, H.B.; Zhou, C.B.; Chen, S.H.; Shao, Z.S. The Effect of Blast-Hole Arrangement, Delay Time, and Decoupling Charge on Rock Damage and Vibration Attenuation in Multihole Blasting. *Shock Vib.* **2022**, *2022*, e2110160. [CrossRef]
18. Wang, X.; Li, J.; Zhao, X.; Liang, Y. Propagation Characteristics and Prediction of Blast-Induced Vibration on Closely Spaced Rock Tunnels. *Tunn. Undergr. Space Technol.* **2022**, *123*, 104416. [CrossRef]
19. Lv, J.; Liu, Z.; Huang, J. The Vibration Propagation and Damage Distribution Characteristics of Deep-Buried Underground Cavern Blasting. *Mech. Time-Depend. Mater.* **2023**, 1–25. [CrossRef]
20. Zhang, Y.; He, H.; Khandelwal, M.; Du, K.; Zhou, J. Knowledge Mapping of Research Progress in Blast-Induced Ground Vibration from 1990 to 2022 Using CiteSpace-Based Scientometric Analysis. *Environ. Sci. Pollut. Res.* **2023**, *30*, 103534–103555. [CrossRef] [PubMed]
21. Wu, X.; Gong, M.; Wu, H.; Hu, G.; Wang, S. Vibration Reduction Technology and the Mechanisms of Surrounding Rock Damage from Blasting in Neighborhood Tunnels with Small Clearance. *Int. J. Min. Sci. Technol.* **2023**, *33*, 625–637. [CrossRef]
22. Dehghani, H.; Ataee-pour, M. Development of a Model to Predict Peak Particle Velocity in a Blasting Operation. *Int. J. Rock Mech. Min. Sci.* **2011**, *48*, 51–58. [CrossRef]
23. Wei, J.; Wang, W.; Wu, J. Numerical Investigation of inside Peak Particle Velocities for Predicting the Vibration Influence Radius of Vibratory Pile Driving. *Soil Dyn. Earthq. Eng.* **2022**, *153*, 107103. [CrossRef]
24. Hu, J.-H.; Shang, J.-L.; Luo, X.-W.; Zhou, K.-P. Monitoring of Single-Hole Blasting Vibration and Detection of Its Attenuation Law by Using Multiple Linear Regression. *Zhendong Yu Chongji J. Vib. Shock* **2013**, *32*, 49–53.
25. Jiang, N.; Zhou, C.; Lu, S.; Zhang, Z. Effect of Underground Mine Blast Vibrations on Overlaying Open Pit Slopes: A Case Study for Daye Iron Mine in China. *Geotech. Geol. Eng.* **2018**, *36*, 1475–1489. [CrossRef]
26. Ghasemi, E.; Kalhori, H.; Bagherpour, R. A New Hybrid ANFIS–PSO Model for Prediction of Peak Particle Velocity Due to Bench Blasting. *Eng. Comput.* **2016**, *32*, 607–614. [CrossRef]
27. Nguyen, H.; Bui, X.-N.; Bui, H.-B.; Cuong, D.T. Developing an XGBoost Model to Predict Blast-Induced Peak Particle Velocity in an Open-Pit Mine: A Case Study. *Acta Geophys.* **2019**, *67*, 477–490. [CrossRef]

MDPI AG
Grosspeteranlage 5
4052 Basel
Switzerland
Tel.: +41 61 683 77 34

Applied Sciences Editorial Office
E-mail: applsci@mdpi.com
www.mdpi.com/journal/applsci

www.ingramcontent.com/pod-product-compliance
Lightning Source LLC
Chambersburg PA
CBHW061805210926
51850CB00006B/619
9783725860159